Nanotechnology in Advanced Electrochemical Power Sources

Nanotechnology in Advanced Electrochemical Power Sources

edited by

S. R. S. Prabaharan
M. S. Michael

Pan Stanford Publishing

Published by

Pan Stanford Publishing Pte. Ltd.
Penthouse Level, Suntec Tower 3
8 Temasek Boulevard
Singapore 038988

Email: editorial@panstanford.com
Web: www.panstanford.com

British Library Cataloguing-in-Publication Data
A catalogue record for this book is available from the British Library.

Nanotechnology in Advanced Electrochemical Power Sources

ISBN 978-981-4241-43-4 (Hardcover)
ISBN 978-981-4303-05-7 (eBook)

Printed in the USA

Contents

Senior Authors xiii
Foreword xv
Preface xvii

1 Redox Reaction in Size-Controlled Li_xFePO_4 **1**
Atsuo Yamada
1.1 Background 2
1.2 Binary Phase Diagram 3
1.3 Size Effect *versus* Surface Effect 4
1.4 Air Poisoning 6
1.5 Careful Verification of Phase Diagram 9
1.5.1 Electrochemical Approach 9
1.5.2 Structural Approach 10
1.5.3 Consistency between Electrochemical and Structural Data 16
1.5.4 Thermodynamic Approach 19
1.6 Summary 21

2 Orthosilicate-Based Cathode Materials for Lithium-Ion Batteries **23**
Robert Dominko
2.1 Background 24
2.2 Structure–Property Relationship 27
2.2.1 X-Ray Diffraction Investigation 27
2.2.2 ^{6}Li MAS NMR Investigation 29
2.3 Synthesis and Microscopy 31
2.4 Electrochemical Properties (*In situ* Structural Studies: XRD and XAS) 32

2.5 Effect of Carbon Coating 44
2.6 Conclusions 46

3 Nanoscale Conversion Materials for Electrochemical Energy Storage 51
Maximilian Fichtner
3.1 Introduction 51
3.2 Current Electrode Materials 53
3.3 Conversion Electrodes 55
3.3.1 Concept and Examples 56
3.3.2 Anode Materials 57
3.3.2.1 Metal–air batteries 57
3.3.2.2 In Li-ion batteries 64
3.3.3 As Cathode Materials 72
3.3.3.1 Sulfides 73
3.3.3.2 Fluorides 74
3.3.4 Solid-State Reactions in Conversion Electrodes 79
3.3.4.1 Thermodynamics 80
3.3.4.2 Kinetics 83
3.4 Summary 86

4 Nanoengineered Lithium–Air Secondary Batteries: Fundamental Understanding and the Current Status of Development 89
M. S. Michael, Junichi Kawamura, Naoaki Kuwata, and S. R. S. Prabaharan
4.1 Introduction 90
4.2 Why Li–Air Battery? 92
4.3 Other Metal–Air Battery Systems 93
4.4 Rechargeable Li–Air Battery System 94
4.4.1 Li–Air Battery Fundamentals 94
4.4.2 Principle of Operation of Li–Air Battery 98
4.4.2.1 Non-aqueous Li–air system 98
4.4.2.2 Aqueous Li–air system 99
4.4.2.3 Aqueous/nonaqueous hybrid electrolyte 99
4.4.2.4 Solid-state electrolyte 100
4.5 Current Problems Encountered in Li–Air Rechargeable System 100

4.5.1 Non-aqueous Solvents 100
4.5.2 Gas Diffusion Electrode 106
4.5.3 Role of Electrocatalysts in the Li–Air System 111
4.5.3.1 Nano-electrocatalysts for efficient air cathode 112
4.6 Developmental Challenges 118
4.7 Our Proprietary Cell Design 120
4.8 Summary 120

5 Nano Anode Materials for Lithium-Ion Batteries 127
Lijun Fu, Yi Shi, Shu Tian, Yuping Wu, and Teunis van Ree
5.1 The Development of Anode Materials for Lithium Ion Batteries 128
5.2 Nanocarbon Anode Materials 128
5.2.1 Carbon Nanotubes 129
5.2.2 Carbon Nanofibers 134
5.3 Nano Si-Based Anode Materials 135
5.4 Nano Sn-Based Anode Materials 139
5.4.1 Metallic Sn 139
5.4.2 SnO_2 142
5.5 Titanium Oxide-Based Nano Anode Materials 146
5.5.1 Anatase TiO_2 146
5.5.2 Rutile TiO_2 150
5.5.3 TiO_2(B) 151
5.5.4 Nano $Li_4Ti_5O_{12}$ 153
5.6 Other Oxides-Based Nano Anode Materials 156
5.6.1 Co-Based Oxides 156
5.6.2 Ni-Based Oxides 161
5.6.3 Cr_2O_3 162
5.6.4 Nano MoO_2 164
5.6.5 Fe-Based Oxides 167
5.7 Future Aspects 169

6 Interface between Transition Metal Oxides–Based Electrodes and Lithium Salts Electrolytes: A Physicochemical Approach 177
Sylvain Franger
6.1 Introduction 178
6.2 Experimental Section 179

6.3 Physicochemical Approach Proposed 180
6.3.1 Background 180
6.3.2 Kinetic Equations 182
6.4 Confrontation/Discussion 184
6.4.1 Limitation of this Model at High Voltages (>4.3 V *vs.* Li) Especially in Presence of Propylene Carbonate 188
6.4.2 Limitation of this Model with the Specific Surface of the Particles: the Case of the Nanostructured Materials 192
6.5 Summary/Perspectives 194

7 Electron Spin Resonance Studies of Lithium-Ion Battery Materials 199
Nail Suleimanov, S. R. S. Prabaharan, Dinar Abdullin, and M. S. Michael
7.1 Introduction 200
7.2 Electron Spin Resonance Phenomenon 202
7.3 Applications of ESR to Study Layered Cathode Materials 204
7.3.1 Structural Imperfections of $LiCoO_2$ 204
7.3.2 ESR Investigation of $LiM_yCo_{1-y}O_2$ (M = Al, Ga) System 205
7.4 ESR Investigation of NASICON Structured $Li_2Mn_2(MoO_4)_3$ 206
7.4.1 Experiment 206
7.4.2 ESR Analysis 208
7.4.2.1 Micro-scaled $Li_2Mn_2(MoO_4)_3$ 208
7.4.2.2 Nanoscaled $Li_xMn_2(MoO_4)_3$ (x = 0–2.0) 215
7.4.3 Comparison of Resonance Parameters of Microscaled and Nanoscaled $Li_2Mn_2(MoO_4)_3$ 216
7.5 Summary and Outlook 218

8 Graphene and Graphene-Based Nanocomposites for Electrochemical Energy Storage 221
Marappan Sathish and Itaru Honma
8.1 Introduction 222

8.2 Graphene Nanosheets and Functionalized Graphene Nanosheets Preparation 224
8.3 Graphene Nanosheets for Supercapacitors 226
8.4 Graphene Nanosheets for Li-Ion Battery 234
8.5 Graphene Nanosheets for Li–Air Battery 239
8.6 Summary 244

9 Carbon Nanotubes for Energy Storage Application 249
Elzbieta Frackowiak, Grzegorz Lota, and Krzysztof Fic
9.1 Introduction 250
9.2 Carbon Nanotubes/Conducting Polymers Composites for Capacitor 251
9.3 Carbon Nanotubes/Metal Oxides Composites for Capacitor 252
9.4 Pseudoeffects from Heteroatoms in the Carbon Network 255
9.5 Pseudocapacitance Effect Related with Hydrogen Electrosorbed in Carbon 266
9.6 Carbon Nanotubes as a Support for Catalyst in Fuel Cells 271
9.7 Summary 276

10 Manganese Oxide/Carbon Nanotube Nanocomposites for Electrochemical Energy Storage Applications 281
Sang-Bok Ma and Kwang-Bum Kim
10.1 Introduction 282
10.2 Synthesis and Characterization of MnO_2/CNT Nanocomposites 285
10.2.1 In situ Monitoring of Reduction Potential and pH 285
10.2.2 UV-VIS Analysis 287
10.2.3 Effect of pH 287
10.2.4 CNT as a Reducing Agent 290
10.2.5 Morphology of MnO_2 on CNT 291
10.2.6 Phase and Structure of MnO_2 on CNT 292
10.3 Electrochemical Properties of MnO_2/CNT Nanocomposites 294

10.3.1 Electrochemical Properties of MnO_2 Coated onto CNT 294
10.3.2 Effect of Initial Solution pH 296
10.3.3 Effect of CNT as a Conducting Agent 298
10.3.4 Effect of Heat Treatment of MnO_2/CNT Nanocomposite 299
10.4 Nanosized Lithium Manganese Oxide Dispersed on Carbon Nanotubes 303
10.4.1 Synthesis of LMO Dispersed on CNT 303
10.4.2 Phase and Structure of LMO Dispersed on CNT 306
10.4.3 Electrochemical Properties of LMO Dispersed on CNT 307
10.4.4 Structural Stability of LMO Dispersed on CNT 308
10.5 Summary 312

11 Manganese Oxides for Supercapacitors **317**
Shinichi Komaba, Naoaki Yabuuchi, and Tomoya Tsuchikawa
11.1 Manganese Oxides for Electrode Materials 317
11.2 Synthesis of Manganese Dioxides as Electrode Materials of Redox Capacitor 321
11.3 From Symmetric to Asymmetric Supercapacitor 328
11.4 Electrolyte Additives for Improvement of Cyclabiltiy 331
11.5 Summary and Outlook 335

12 High-Voltage Electrode/Electrolyte Interface in ECs and Hybrid Capacitors **339**
Katsuhiko Naoi and Kazumi Chiba
12.1 Background 340
12.1.1 High-Voltage Electrode/Electrolyte Interface 340
12.1.1.1 Cell voltage range of 2.5–3.0 V (Region I) 344
12.1.1.2 Cell voltage range of 3.0–3.7 V (Region II) 345

12.1.1.3 Cell voltage range of 3.7–4.0 V (Region III) 346
12.2 High-Voltage Electrolyte Structure at Interface 346
12.3 Summary 349

Index 353

Senior Authors

Atsuo Yamada
Department of Engineering, Building 5-607, The University of Tokyo, 7-3-1 Hongo, Bunkyo-ku, Tokyo 113-8656, Japan

Elzbieta Frackowiak
Institute of Chemistry and Technical Electrochemistry, Poznan University of Technology, 60-965 Poznan, Piotrowo 3, Poland

Itaru Honma
Institute of Multidisciplinary Research for Advanced, Tohoku University, 2-1-1, Katahira, Sendai, Miyagi 980-8577, Japan

Junichi Kawamura
Institute of Multidisciplinary Research for Advanced Materials (IMRAM), Tohoku University, Katahira 2-1-1, Aobaku, Sendai, Japan

Katsuhiko Naoi
Institute of Symbiotic Science and Technology, Tokyo University of Agriculture & Technology, Tokyo, Japan

Kwang-Bum Kim
Department of Materials Science and Engineering, Yonsei University, 134 Shinchon-dong, Seodaemun-gu, Seoul 120-749, Korea

Maximilian Fichtner
Institute of Nanotechnology, Karlsruhe Institute of Technology (KIT), P.O. Box 3640, D-76021 Karlsruhe, Germany

Nail Suleimanov
Zavoisky Physical-Technical Institute of Russian Academy of Sciences, 420029 Kazan, Russian Federation

Robert Dominko
National Institute of Chemistry, Hajdrihova 19, SI-1000 Ljubljana, Slovenia, and ALISTORE-ERI, 80039 Amiens Cedex, France

Shinichi Komaba
Department of Applied Chemistry, Tokyo University of Science, Kagurazaka 1-3, Shinjuku, Tokyo 162-8601, Japan

S R S Prabaharan
School of Electronics Engineering, VIT University, Chennai Campus, Vandalur-Kelambakkam Road, Chennai, 600 127, India

Sylvain Franger
Physico-Chimie de l'Etat Solide, Institut de Chimie Mol'eculaire et des Mat'eriaux, d'Orsay, UMR CNRS 8182, Universit'e Paris XI, 91405 Orsay cedex, France

Teunis van Ree
Department of Chemistry, University of Venda, Thohoyandou, 0950 South Africa

Yuping Wu
New Energy and Materials Laboratory (NEML), Department of Chemistry, Fudan University, Shanghai 200433, China

Foreword

Electrochemical energy storage is regarded as a vital means of providing portable power to the mankind in variety of ways. The current trend in developing advanced power/energy sources has rendered impetus to advancement as anticipated by scientists and engineers around the world. Thus, tremendous research opportunities exist in both government and private funded activities all over the world. To facilitate that nanotechnology has complemented laurels to enhance such opportunities to meet the anticipated demand by improving the existing storage limits of battery system especially lithium-containing batteries of present and the future technologies and supercapacitors.

In this context, the book entitled *Nanotechnology in Advanced Electrochemical Power Sources* is timely which encompasses chapters contributed by various world renowned experts in the field of lithium-containing battery electrodes, electrolytes and supercapacitors. State-of-the-art understanding pertaining to nanoscale aspects of the above mentioned advanced energy storage devices has been the focus of this volume. While the fundamental issues relating to the device performance of various positive and negative electrode materials for instance, olivine and silicate cathodes, high-capacity anodic materials such as Sn alloys, silicon and conversion type electrode, modified electrode with CNTs, graphene-based supercapacitors, hierarchically designed porous air cathodes and their nanoscale advantages in terms of device performances are also covered. There are 12 full length chapters including a dedicated chapter on Li-air secondary batteries, a futuristic endeavor in developing a high energy density battery having theoretical specific capacity closest to gasoline. The intended audience of this book includes students, materials scientists, electrical engineers, new comers who would like to learn the nuances of the exhilarating branch having enormous societal advantages for the mankind.

B.V.R. Chowdari
National University of Singapore,
Former president of IUMRS, President MRS (Singapore)
Executive Director NUS-India Research initiatives and Chairman of ICMAT conferences

Preface

The challenge to provide adequate power supplies to the human race on an indefinite basis without causing long-term damage to the environment requires versatile means of energy conversion and storage. The advances in large-scale original equipment manufacturer (OEM) gadgets, smart cellular phones, tablet computers (tabs), power-hungy long-range communication devices, and modern-day electric mobility systems (plug-in electric cars, for instance) demand for huge power and energy in the form of rechargeable energy storage systems. Electrical energy storage thus becomes vital today than at any time in the human history. The recent evolution of nanotechnology may provide opportunities to meet the anticipated demand by improving the existing storage limits of battery system, especially Li-ion batteries, supercapacitors, and upcoming novel energy storage system namely Li–air rechargeable batteries. The engineering of new materials, and especially at the nanoscale, has become imperative to achieve enhanced energy and power efficiency to meet these challenges of energy storage systems. This book incorporates state-of-the-art understanding pertaining to nanoscale aspects of advanced energy storage devices such as Li-ion batteries, Li–air batteries, and electrochemical supercapacitors. It focuses on various fundamental issues related to the device performance of various positive and negative electrode materials, with special reference to their nanoscale advantages. It also includes fundamentals and processing techniques with regard to synthesis, characterization, physical and electrochemical properties, and applications of nanoscale materials pertaining to advanced electrochemical power sources. The goal of this book is to render an update of the current advancement in the field of electrochemical power sources with special reference to nanotechnology. A variety of advanced nanomaterials, such as transition metal oxides, phosphates, silicates, and conversion electrodes, together with carbonaceous materials that include carbon nanotubes, nanorods, and mesoporous carbons, are discussed by notable authorities of this exhilarating field. The book presents a balanced mix of theoretical and experimental approaches adopted over these years in the advancement of lithium-based power sources (Li-ion and Li–Air) and electrochemical capacitors. The book is a compact reference source for students, scientists, engineers, and specialists in various fields, including electron devices, electrochemistry, electrical engineering, nanotechnology, and solid state physics.

S. R. S. Prabaharan
M. S. Michael
Summer 2014

Chapter 1

Redox Reaction in Size-Controlled Li_xFePO_4

Atsuo Yamada

Department of Engineering, Building 5-607, The University of Tokyo, 7-3-1 Hongo, Bunkyo-ku, Tokyo 113-8656, Japan

yamada@chemsys.t.u-tokyo.ac.jp

Crystalline size has tremendous effect on the thermodynamics and kinetics in intercalation compounds. This includes diffusion/transport length, effective surface area for exchange current, surface energy, and interphase energy. We focus on Li_xFePO_4, where the phase changes by simple two-phase separation but with controllable miscible character by raising temperature and/or reducing crystalline size, thereby providing simple model system to rationalize thermodynamics and electrochemistry in electrode reaction. Our systematic experimental results on this issue are reviewed and summarized. Negative aspects of nano-particle including surface poisoning by air contact and its diagnosis will be also demonstrated.

Nanotechnology in Advanced Electrochemical Power Sources
Edited by S. R. S. Prabaharan and M. S. Michael

ISBN 978-981-4241-43-4 (Hardcover), 978-981-4303-05-7 (eBook)
www.panstanford.com

1.1 Background

The crystal chemistry of phospho-olivine Li_xFePO_4 ($0 < x < 1$) has received much attention ever since it was introduced as an alternative cathode material for new generation lithium ion batteries. The material provides several advantages over the conventional cathodes; lower cost, much safer performance, lower toxicity, and extremely flat charge–discharge profile at reasonably high potential of approximately 3.4 V *vs.* Li/Li^+.[1,2] Nevertheless, at an early stage of the development, its poor intrinsic electronic conductivity, on the order of 10^{-9} S/m, and small tap density had been addressed as major problems to be solved before it can be deployed on a commercial scale.[3–5] Now, at least the first problem has been overcome by using small particle of <100 nm and the efficient formation of composites with conductive phase, typically carbon. Surprisingly high specification for the power density of more than 5 kW/kg is reported on the commercial cells with $LiFePO_4$ cathode.

One of the essential reasons for the very low conductivity measured for the stoichiometric $LiFePO_4$ seems to be in the prefactor rather than in the activation energy. In addition, a key feature of the Li_xFePO_4 cathode is its extremely flat charge–discharge profile, typical of two-phase reaction systems. Conventional belief based on the Gibbs phase rule was that the reaction proceeds in a topotactic manner, varying the $FePO_4/LiFePO_4$ ratio throughout the range of x in Li_xFePO_4.[1] However, carrier (mobile electron or hole) densities in the monovalent stoichiometric end members should be negligible and this evidently accounts for the very low electronic conductivity of $<10^{-9}$ S/cm in $LiFePO_4$, despite its relatively small binding energy for polaron localization of 390–500 meV.

It is true that, within the polaronic conduction mechanism and the conventional $FePO_4/LiFePO_4$ two-phase model, the number of charge carriers in monovalent systems such as $LiFePO_4$ or $FePO_4$ is determined extrinsically by impurities. Defect chemistry studies of related $Li_{1-3x}Fe_xNiPO_4$ phases suggest that cation vacancies can be accommodated in the olivine structure.[6] However, without any external doping in $LiFePO_4$, a rapid electrochemical response as well as a good rate capability has been achieved in our laboratory with an optimized electrode composite of pure $LiFePO_4$. These motivated a closer look at the phase change of Li_xFePO_4 at room temperature to determine if the creation of charge career is spontaneous in non-doped Li_xFePO_4 during the electrochemical charge–discharge process (i.e., does lithium or lithium defect function as a carrier-doping impurity).

1.2 Binary Phase Diagram

As in many binary systems, thermodynamics suggests there are two solid solution regions outside the miscibility gap at finite temperatures. Generally, these two solid solution regions grow as a function of temperature finalized by the continuous solid solution over the whole composition between the two end members, as there is an increasing contribution of entropy to the total free energy at elevated temperatures. This is typically seen in spinodal decomposition in binary liquid systems with an inverse-parabolic phase boundary in the temperature–composition diagram. For intercalation solid electrode materials, this behavior was intensively studied in the 1980s using a lattice gas model based on mean field theory, where the attractive interaction between lithium ions is essential to induce phase separation. Onset temperature of the miscibility gap varies from approximately 270 K for metallic Sheverel $Li_xMo_6Se_8$ $(0 < x < 1)$[7,8] to approximately 1200 K for insulating NASICON-type $Na_{1+x}Ti_2(PO_4)_3$ $(0 < x < 2)$.[9] Analogous to these compounds, the intermediate phase in the Li_xFePO_4 cathode at room temperature may be dominated by the mixture of the $Li_\alpha FePO_4$ and $Li_{1-\beta}FePO_4$ phases with the narrow monophase regions, $0 < x < \alpha$ and $1-\beta < x < 1$, close to the stoichiometric end members of $LiFePO_4$ and $FePO_4$ at each end of the region.

This was first proved by the refinement of X-ray diffraction profiles measured for Li_xFePO_4 $(0 \leq x \leq 1)$ phases at room temperature.[10] In the bi-phase reaction of Li_xFePO_4 at room temperature, the orthorhombic lattice constants in each phase slightly deviate from those of the stoichiometric end components of $LiFePO_4$ and $FePO_4$. The deviation can be explained by assuming a model in which a most of the intermediate phase in the Li_xFePO_4

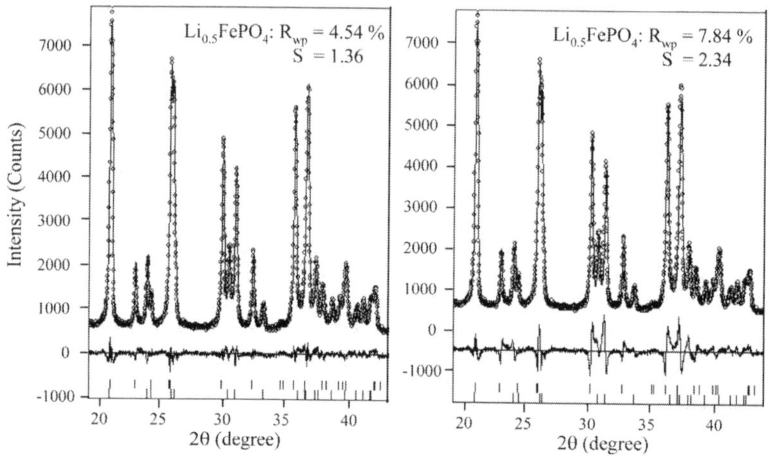

Figure 1.1 Difference in the goodness of fit in the Rietveld refinement of $Li_{0.5}FePO_4$ under the model with (left) a mixture of $Li_\alpha FePO_4$ and $Li_{1-\beta}FePO_4$ ($\alpha \approx 0.030$, $\beta \approx 0.036$) and (right) a mixture of the stoichiometric $FePO_4$ and $LiFePO_4$ phases.

cathode is dominated by the mixture of $Li_\alpha FePO_4$ and $Li_{1-\beta}FePO_4$ phases, as shown in Fig. 1.1.

1.3 Size Effect *versus* Surface Effect

Meethong *et al.* have suggested that the miscibility gap shrinks systematically with decreasing particle size and increasing temperature.[11] Such a size-dependent miscibility gap is a general phenomena observed in other two-phase intercalation reactions such as Li_xTiO_2 or MgH_x systems.[12,13] Thermodynamically, the dependence can be explained by the increasing contribution of the elastic energy induced along coherent two-phase interphase in the smaller particles by a factor of $r \sim A(x)/V(x)$, where r is the particle radius, $A(x)$ is the interface area, and $V(x)$ is the particle volume. The coexistence of two crystallographic phases within one particle leads to a domain wall related energy penalty, which is determined by the strain mainly due to a difference in lattice parameters. Such a strain induced energy penalty can destabilize the two-phase coexistence in smaller particles. As a result, the energy gain due to phase separation, represented by the free energy of mixing ΔG_{mix}, will decrease for smaller particle sizes. This will gradually close the miscibility gap for decreasing particle size, which explains the enhanced solubility of Li in the smaller particles. More recently, Gibot *et al.* reported the inherent crystalline disorder coupled with nano-crystallization below 40 nm leads to a peculiar situation where solid solution dominates most of the lithium compositional region in Li_xFePO_4 even at room temperature.[14] Such a thermodynamic concept have been verified in detail by Wagemaker *et al.*[15] including the effect of particle shape. Meethong *et al.*[16] related this intriguing findings to the kinetics of phase boundary movement, where the reduction in lattice strain caused by the reduction in lattice mismatches in smaller particles was claimed to be suitable to form coherent interface between $Li_\alpha FePO_4$ and $Li_{1-\beta}FePO_4$ so as to enhance the cooperative phase-boundary movement upon electrochemical reaction with minimum formation of dislocation or clacks. This logic was regarded to give comprehensive understanding of the enhanced electrochemical activity of the nano-sized $LiFePO_4$ particle as compared to the other slower two-phase electrode system with larger mismatch.

However, attention should be paid that the surface portion of the particle abruptly increases when the size is smaller than 100 nm. A simple estimation based on the sphere particle gives the surface (one unit cell) portions to be approximately 4%, 7%, and 17% for the particles with 100 nm, 50 nm, and 20 nm diameters, respectively. In such nanoscale regime, there is an increasing possibility to observe the simultaneous effects both from bulk crystal and from particle surface, which may lead to some confusion. For example, surface redox potential was calculated to be dependent on the crystallographic plane and was shown a wide dispersion over the range of 2.95–3.84 V.[17] Both of the surface impurities and the redox on the specific crystallographic facet plane would be important as the perturbation that may cause parasitic capacity below/above the bulk two-phase equilibrium potential in the voltage curve. The capacity outside the two-phase region for Li_xFePO_4 have been used as one of the indicator of the bulk solubility limit of lithium and lithium defect in $Li_\alpha FePO_4$ and $Li_{1-\beta}FePO_4$, and then may need careful reinvestigation.

More importantly, the side reaction upon ambient exposure is known to be pronounced for smaller particle leading to a spontaneous lithium extraction and surface oxidation.[18,19] Before going to any further discussions, such a side reaction should be removed, then, this phenomenon is explained in the next section.

1.4 Air Poisoning

X-ray diffraction (XRD) analysis has been carried out to investigate the effect of different atmosphere exposure conditions. Table 1.1 summarizes the lattice dimensions. In all cases, the diffractograms display a single phase with an ordered olivine structure indexed to the orthorhombic Pnmb space group. The Rietveld refinement permits to approximate the olivine crystallites size to 75 nm, which is verified on the SEM image of the powder presented.

Systematic decrease of the cell volume upon atmosphere exposure was clearly observed together with the corresponding changes in lattice parameters. It begins to be visible even after a short time (1 h) of ambient air exposure at room temperature, and become more evident after longer time (10 days) or higher temperature (120°C) exposure. Water and oxygen should be included in this side reaction, as the effect is significantly suppressed by using dry O_2 20% / N_2 80% and become negligible by using dry CO_2. These volume changes can be linked to the solid solution compositional domain in $Li_{1-x}FePO_4$ ($x < \beta$),[10,11,19,20] in which olivine can accommodate the corresponding lithium defect without first-order phase transformation.

Figure 1.2 demonstrates the linear relative dependence of the a and b lattice constants following the Vegard's law, in the hypothesis of a solid solution between $LiFePO_4$ prepared with no contact with air, and $FePO_4$. These results show that some lithium is certainly extracted when $LiFePO_4$ is put in contact with atmosphere. The Vegard's law gives 2.3% of missing lithium after ambient air exposure for 24 h at 120°C. It has to be noted that the lattice constants obtained by avoiding air contact are very

Table 1.1 Parameters and cell volumes of $LiFePO_4$ calculated from XRD refinement after various atmosphere exposure conditions

Strage condition		a(Å)	b(Å)	c(Å)	V(Å3)	x
Dry Ar (standard)	Room temperature	10.3308(2)	6.00813(11)	4.69444(10)	291.378(6)	0
Ambient air	Room Temperature, 1 h	10.3275(2)	6.00656(13)	4.69271(12)	291.1040(12)	0.007
	Room Temperature, 1 day	10.3254(2)	6.00573(11)	4.69329(11)	291.037(10)	0.011
	Room Temperature, 21 days	10.3227(2)	6.00428(15)	4.69346(14)	290.903(14)	0.017
	120°C, 1 day	10.3188(2)	6.00303(15)	4.69451(14)	290.79(4)	0.023
Dry air (N_2/O_2)	Room Temperature, 1 day	10.3294(4)	6.0072(2)	4.6932(2)	291.22(2)	0.004
	120°C. 1 day	10.3254(3)	6.0054(2)	4.69373(19)	291.043(19)	0.012
Dry CO_2	Room Temperature, 1 h	10.3302(4)	6.0075(2)	4.6936(2)	291.29(2)	0.001
	120°C. 1 day	10.3290(3)	6.00716(18)	4.696361(17)	291.231(16)	0.004

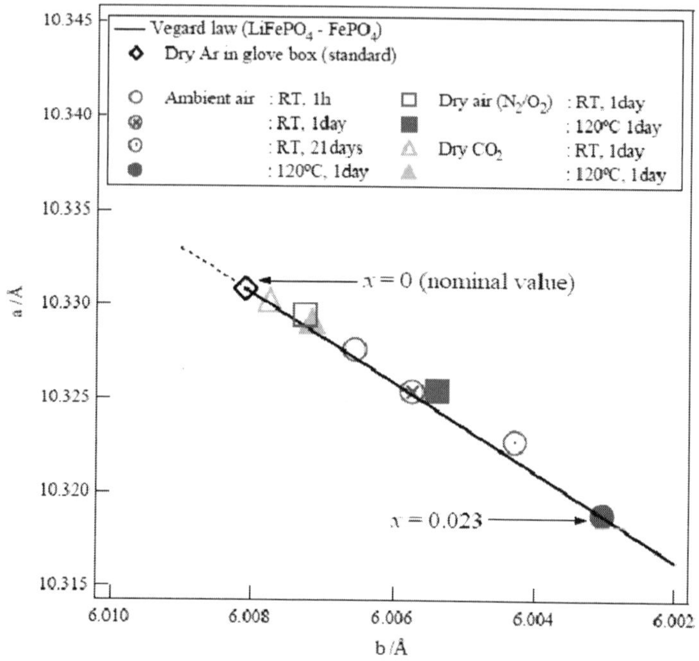

Figure 1.2 Lattice constants changes for $Li_{1-x}FePO_4$ after different air contact experiments compared to Vegard's law (arrow). The values for x are calculated using Vegard's law and $FePO_4$ cell parameters ($a = 9.8142$ Å and $b = 5.7893$ Å).

close to the ones reported for pure, much larger particles,[21] which present less possibilities of reactivity with air due to smaller solid solution domain and smaller exposed surface. It tends to prove that the sample has been well protected during our experiment and that possible air pollution during storage in glove box has been minimized.

For a subsequent study of size-dependent phase diagram, we took effort to eliminate such fluctuant factor in the pristine samples, because it would degrade the quality of any crystallographic and/or electrochemical data and hence mislead the subsequent quantitative discussions. The controlled pristine $LiFePO_4$ samples used in this study are free from such poisoning by air exposure and the particle size is the only varied parameter. The samples used in this study are

Table 1.2 Lattice dimensions of pristine and air-exposed $LiFePO_4$ with various particle sizes

No air exposure				
Particle size (nm)	a(Å)	b(Å)	c(Å)	v(Å^3)
200	10.33046(13)	6.00846(7)	4.69238(7)	291.257(7)
80	10.3303(2)	6.00764(12)	4.69313(11)	291.259(11)
40	10.3301(5)	60052(3)	4.6947(3)	291.23(2)
Air exposure; room temperature, 1 day				
Particle size (nm)	a(Å)	b(Å)	c(Å)	V(Å^3)
200	10.32839(15)	6.00767(8)	4.69228(8)	291.157(3)
80	10.3253(2)	6.00590(12)	4.69335(11)	291.047(11)
40	10.3223(5)	6.0015(3)	4.6955(3)	290.88(2)

free from any disorder as the lattice dimensions are almost identical to the single crystal values (a = 10.332(4) Å, b = 6.010(5) Å, c = 4.692(2) Å, V = 291.4(3) Å^3)[21] and showed negligible dependence on the particles size, while air-exposed samples suffered from significant volume contraction particularly on smaller particles, as in Table 1.2. With the same trend in variations of the lattice dimensions, Mössbauer spectra for the controlled pristine samples were analyzed by single Fe^{2+} doublet inherent to $LiFePO_4$, while parasitic Fe^{3+} sub-signal was always observed for the air-exposed samples increasing its intensity for smaller particles, as shown in Fig. 1.3.

1.5 Careful Verification of Phase Diagram

1.5.1 *Electrochemical Approach*

The first step for analysis is the accurate determination of the open-circuit voltage (OCV) profile *vs.* lithium composition x for three samples with different particle sizes of 40 nm, 80 nm, and 200 nm. Initially, the cell was charged galvanostatically at a rate of C/20 at 25°C followed by maintenance at 4.5 V for 24 h forming $FePO_4$. Capacity close to the theoretical capacity of 170 mAh/g was confirmed for all samples. Then, it was discharged under a rate of

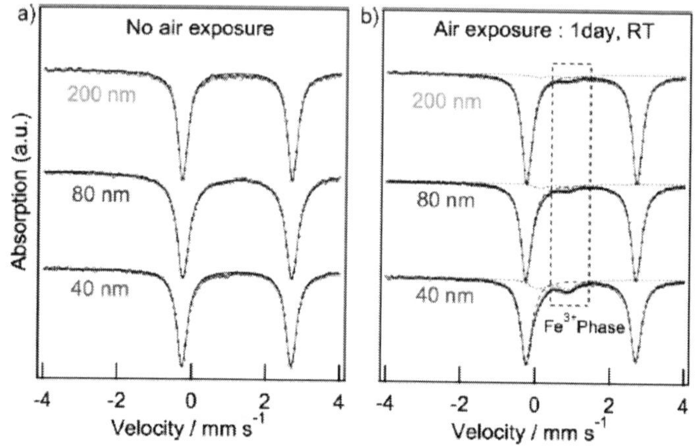

Figure 1.3 Mössbauer spectra measured for pristine (a) and air-exposed (b) $LiFePO_4$ with various particle sizes of 40 nm, 80 nm, and 200 nm.

C/20 for 12 min, which corresponds to approximately 1% of the theoretical capacity. The open circuit voltage at 25°C was measured after 24 h. Again, the cell was discharged under a rate of C/20 for 12 min. These procedures were repeated several times until the open circuit voltage at 25°C became constant at approximately 3.42 V in the bi-phasic region. A similar experiment was performed starting from $LiFePO_4$ toward the anodic direction and confirmed the identical OCV values at same lithium compositions. Size-dependent OCV curves for Li_xFePO_4 are given in Fig. 1.4, where the capacity observed below/above the two-phase equilibrium potential becomes larger as particle size decreases.

1.5.2 *Structural Approach*

Two target lithium compositions of x in Li_xFePO_4 ($x = 0.6, x = 0.93$) were chosen and were adjusted by the chemical way, allowing more homogeneous Li^+ incorporation over electrochemical reactions. Based on the room temperature phase diagram we reported previously,[10,19,20] the composition near an end member, $Li_{0.93}FePO_4$, was selected as an experimental criterion phase, since it would be around the phase transition zone between the two-phase region and the single-phase region in the $FePO_4$–$LiFePO_4$ binary phase

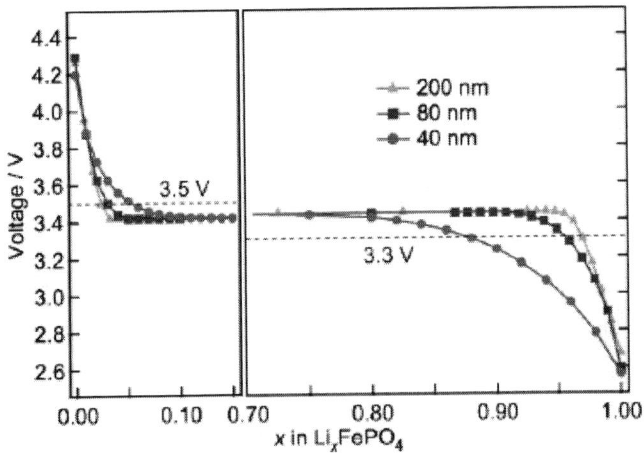

Figure 1.4 OCV curves measured for Li_xFePO_4 with various mean particles sizes of 200 nm, 80 nm, and 40 nm.

diagram, where the two-phase (larger particle) to single-phase (smaller particle) transformation is expected as a function of the particle size. The composition $x = 0.6$, $Li_{0.60}FePO_4$, was selected as a reference phase, because it should be composed of a mixture of two intermediate phases, analysis of which should lead to the definitive positions of the solid solubility limits, α and β, in $Li_\alpha FePO_4$ and $Li_{1-\beta}FePO_4$. In order to set the lithium composition in Li_xFePO_4 to $x = 0.6$ or $x = 0.93$, all of the lithium in pristine $LiFePO_4$ was first removed to form isostructural $FePO_4$ using a strong oxidizing agent NO_2BF_4 in acetonitrile. Then, $FePO_4$ was reacted with the specific amount of mild reducing agent LiI in acetonitrile to form Li_xFePO_4 with $x = 0.6$ or $x = 0.93$. The chemical treatment was performed for each pristine $LiFePO_4$ sample with three different particle sizes of 40 nm, 80 nm, and 200 nm for the subsequent high-resolution synchrotron X-ray diffraction.

In order to overview the change in the room-temperature phase diagram with decreasing particle size, the lattice parameters refined for $Li_{0.6}FePO_4$ (two-phase mixture of $Li_\alpha FePO_4$ and $Li_{1-\beta}FePO_4$) were analyzed. For all particle sizes, $Li_{0.6}FePO_4$ is evidently in two-phase state as typically seen in the two diffraction peaks clearly observed at around 13.5° and 14.5°, corresponding to the (200)

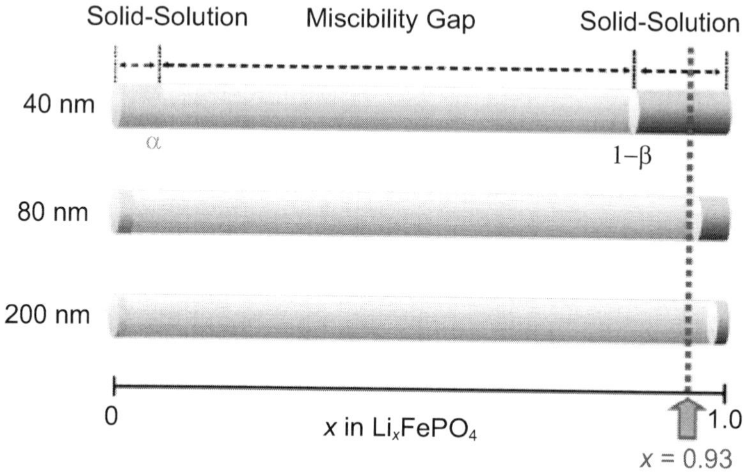

Figure 1.5 Schematic derivation of the size-dependent $FePO_4$–$LiFePO_4$ binary phase diagram using the XRD data for $Li_{0.6}FePO_4$.

reflection of $Li_{1-\beta}FePO_4$ and the (200) reflection of $Li_\alpha FePO_4$, respectively. All of the lattice parameters refined for the two separated phases in $Li_{0.6}FePO_4$ deviate from those of stoichiometric $FePO_4$ and $LiFePO_4$, and can be explained by assuming partial lithium occupancy α to $FePO_4$ and partial lithium deficiency β from $LiFePO_4$ as well as Vegard's law along $FePO_4$–$LiFePO_4$ solid solution.

Schematic derivation of the size-dependent $FePO_4$–$LiFePO_4$ binary phase diagram using the XRD data for $Li_{0.6}FePO_4$ is illustrated in Fig. 1.5, which highlights the fact that the solid solubility limits α, β, in $Li_\alpha FePO_4$ and $Li_{1-\beta}FePO_4$ systematically increases along with the reduction in particle size. Most probably, these systematic variations originate from the elastic thermodynamic effects on the miscibility gap as discussed in the various other binary systems. The solid–solid phase transitions are thermodynamically first order and involve a change in enthalpy (ΔH), which is highly related to the particle size as depicted by $\Delta H = \Delta H_{\text{intrinsic}} + \Delta H_{\text{D}}$, where ΔH_{D} represents the particle size contribution of opposite sign.[15] If the particle size decreases, the Gibbs free energy (ΔG) of mixing would be decreased by the increasing contribution of the penalty in elastic energy at the two-phase boundary, resulting in extension of the single phase regions near the end members on the composition phase diagram.

The extended solid solution has been claimed to be beneficial both for the charge movement and the subsequent cooperative phase boundary movement, because the diffusion/hopping prefactor along the one-dimensional diffusion path is increased while the lattice mismatch along the two-phase boundary is reduced. However, experimental evidences given so far for the existence of the small solid solution compositional domain are not direct ones and led some controversy whether they really originate from the bulk effect or from the artifact by the enhanced surface effect in nanoregime.[17] Then, isolation of the solid solution phase would give more definitive and straightforward experimental evidence for the size-dependent miscibility gap model.

Toward this end, we now focus on the sample with $x = 0.93$, which is around the phase transition compositional zone between the two-phase region and the single-phase region in the $FePO_4$–$LiFePO_4$ binary phase diagram (see Fig. 1.5). At the composition of $x = 0.93$, the two-phase (larger particle) to single-phase (smaller particle) transformation is expected as a function of the particle size as easily speculated from the phase behavior along the vertical compositional line at $x = 0.93$ in Fig. 1.5. The changes in the lattice parameter a are summarized in Fig. 1.6 together with the low-angle peaks of $Li_{0.93}FePO_4$ indexed as $(200)_T$ (lithium-rich triphylite phase) and $(200)_H$ (lithium-poor heterosite phase) at different particle sizes. The intensity of $(200)_H$ is gradually decreased when the particle size is reduced, and finally disappeared for the 40 nm particle. More importantly, the deviation of the lattice constant a at $x = 0.93$ from those of the phase-separated $Li_{0.6}FePO_4$ was observed only for 40 nm particle. Furthermore, it is noteworthy that the lattice constant a at $x = 0.93$ for 40 nm particle positions strictly on the line of Vegard's law assuming the complete solid solution of $FePO_4$ and $LiFePO_4$. The same behavior was identified also for the b-, c- axes. Clearly, the immiscible–miscible transition is induced by the particle size reduction at the identical lithium composition $x = 0.93$, and the solid solution in Li_xFePO_4 is isolated at $x = 0.93$ in the 40 nm particle, details of which are described below.

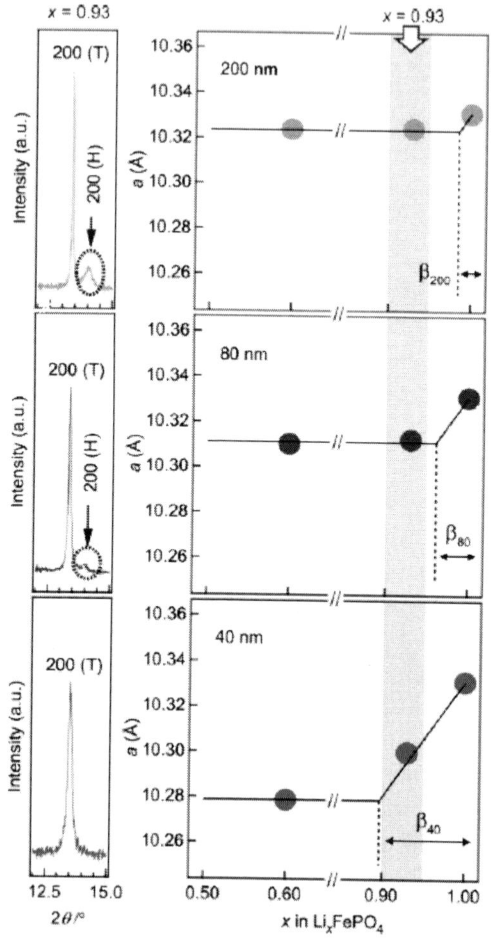

Figure 1.6 Left: magnified view of the 200 peak measured for $Li_{0.93}FePO_4$ with different particle sizes of 200 nm, 80 nm, and 40 nm at room temperature. Right: derivation of changes in a-axis lattice parameter as a function of lithium composition x for different particle sizes.

As for the 200 nm particle at $x = 0.93$, two independent peaks were clearly observed as indexed $(200)_T$ and $(200)_H$, respectively, which indicates that the $Li_{0.93}FePO_4$ phase consists of two intermediate phases, $Li_\alpha FePO_4$ and $Li_{1-\beta}FePO_4$. This is a reasonable result because the lithium composition for the immiscible–miscible transition was estimated to be $1 - \beta_{200} = 0.98$, which is much larger than the present lithium composition, $x = 0.93$. The lattice parameters are almost identical with those of $Li_{0.6}FePO_4$, which also reveal that the $Li_{0.93}FePO_4$ with a particle size of 200 nm is inside the miscibility gap at room temperature.

When the particle size is reduced to 80 nm, the phase at $x = 0.93$ is still in the two-phase separation, but the intensity of the $(200)_H$ peak has significantly decreased in the synchrotron XRD profile, which can be explained by the reduction of the $Li_\alpha FePO_4$ phase fraction in the two phase mixture of $Li_\alpha FePO_4$ and $Li_{1-\beta}FePO_4$, following the size-dependent miscibility gap model.[11] The estimated lithium composition for the immiscible–miscible transition, $1 - \beta_{80}$, has now decreased to 0.96, and becomes closer to the present lithium composition, $x = 0.93$. The $Li_{0.93}FePO_4$ phase with an average particle size of 80 nm is close to the transition zone from immiscible to miscible state and contains smaller amount of $Li_\alpha FePO_4$ at room temperature. The lattice constants are still identical with those of bi-phased $Li_{0.6}FePO_4$; the $Li_{0.93}FePO_4$ with a particle size of 80 nm is still inside the miscibility gap at room temperature but in closer access to the immiscible–miscible transition zone.

At 40 nm particle size with $x = 0.93$, the $(200)_H$ peak has disappeared, and the discrepancies in the lattice parameters from the two-phase values are prominent. Instead, the lattice constants are exactly on the line of Vegard's law assuming the complete solid solution between $FePO_4$ and $LiFePO_4$. This behavior proves that the $Li_{0.93}FePO_4$ phase with 40 nm size is in the single phase solid solution range and allows to conclude that a solid solution of Li_xFePO_4 could be isolated as a single phase with the solid solubility of $\beta_{40} > 0.1$.

For a direct comparison between the stoichiometric $LiFePO_4$ and the isolated solid solution of $Li_{0.93}FePO_4$ with an identical particle size of 40 nm, synchrotron X-ray diffraction profiles of the two samples are shown in Fig. 1.7 with some magnified presentations of the typical peaks. Peak shifts are visually evident and large enough to confirm the wide lithium compositional range of at least 7% for solid solution. The shift observations without appearing of any

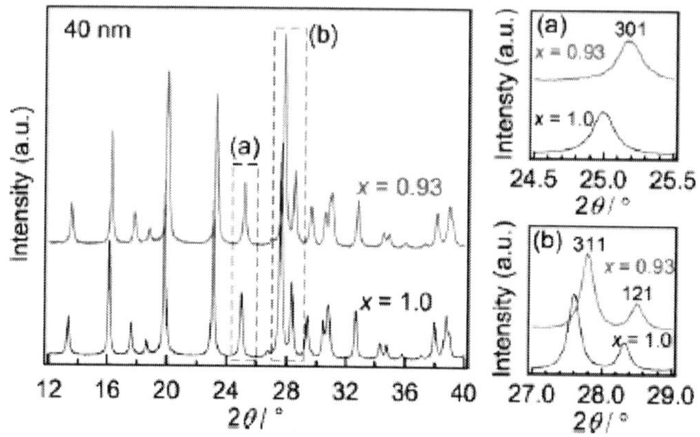

Figure 1.7 Comparison of the synchrotron X-ray diffraction profiles measured for stoichiometric $LiFePO_4$ and the isolated solid solution of $Li_{0.93}FePO_4$ with an identical particle size of 40 nm with magnified presentations of 301, 311, and 121 as the typical peaks.

second phase further support the room-temperature isolation of solid solution in Li_xFePO_4.

1.5.3 *Consistency between Electrochemical and Structural Data*

For a further confirmation of the inverse relationship between the solid solubility and particle size, the composition x has been electrochemically controlled by galvanostatically charging the cells in a potential range of 2.0 V to 4.5 V and then potentiostatically discharging them at a constant potential below the equilibrium potential at 3.30 V, which corresponds to $Li_{0.98}FePO_4$ (200 nm), $Li_{0.96}FePO_4$ (80 nm), and $Li_{0.88}FePO_4$ (40 nm) in OCV curves in Fig. 1.4. Similar experiment was performed with higher constant potential above equilibrium potential at 3.50 V, which corresponds to $Li_{0.02}FePO_4$ (200 nm), $Li_{0.03}FePO_4$ (80 nm), and $Li_{0.06}FePO_4$ (40 nm) in OCV curves in Fig. 1.4. The cell was dismantled in an Ar glove box, and then the cathode was washed in dimethyl carbonate (DMC) for subsequent XRD measurements.

Size-dependent peak shift is clear both for samples in equilibrium at 3.30 V and at 3.50 V as shown in Fig. 1.8. The lattice

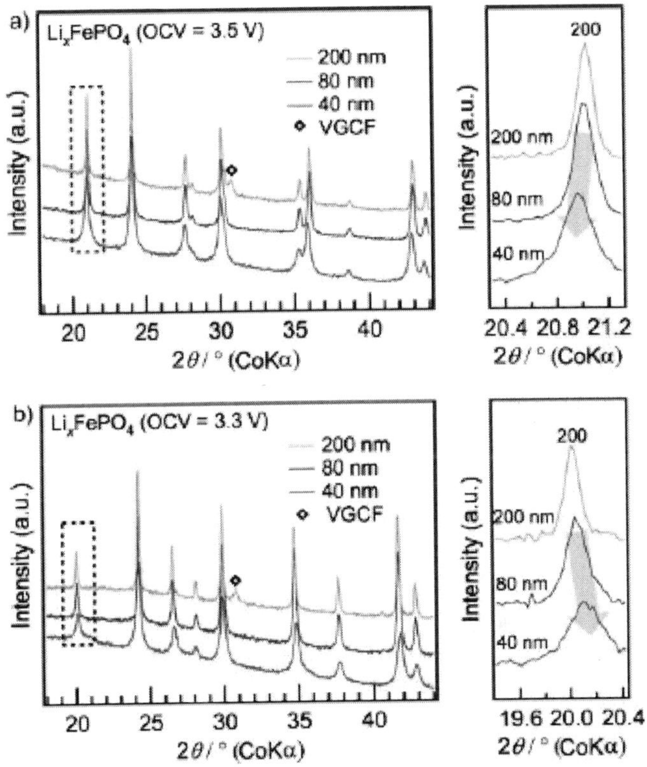

Figure 1.8 X-ray diffraction profiles for Li_xFePO_4 electrode with various mean particle sizes of 200 nm, 80 nm, and 40 nm, equilibrated at 3.3 V (a) and 3.5 V (b), with magnified presentation of the 200 diffraction peaks.

parameters were refined for all samples by Rietveld analyses and are summarized in Table 1.3. The extent of lithium nonstoichiometry x in Li_xFePO_4 were calculated for each sample assuming the Vegard's law between $FePO_4$ and $LiFePO_4$, showing excellent agreement with the electrochemical estimation of x, that is, x (Vegard's law) = x (OCV), for all of the six cases with three different particle sizes (40 nm, 80 nm, 200 nm) and two different OCV (3.30 V, 3.50 V). This provides firm experimental evidence that the capacity observed below/above the two-phase equilibrium potential is not caused by some impurities or the potential distribution dependent

Table 1.3 Lattice dimensions of Li_xFePO_4 electrode with various mean particle sizes of 200 nm, 80 nm, and 40 nm, equilibrated at 3.3 V (a) and 3.5 V (b). Lithium nonstoichiometry from $x = 1$ and $x = 0$ estimated by two different method, Vegard's law and OCV curves, show excellent agreement. The extent of the lithium nonstoichiomery is enhanced as mean particle size decreases

(a) Li-poor (Heterosite phase, OCV = 3.3 V)						
Particle size (nm)	a(Å)	b(Å)	c(Å)	v(Å)	x(Vegard law)	x(OCV)
200	9.8159(9)	5.7891(4)	4.7814(4)	271.70(4)	50.010	0.022
80	9.8292(19)	5.7994(5)	4.7792(11)	272.43(5)	0.030	0.029
40	9.8486(13)	5.8138(6)	4.7837(7)	273.90(6)	0.074	0.056
(b) Li-rich (Triphylite phase, OCV = 3.3 V)						
Particle size (nm)	a(Å)	b(Å)	c(Å)	v(Å)	x(Vegard low)	x(OCV)
200	10.3224(4)	6.0039(2)	4.6909(2)	290.67(2)	0.975	0.98
80	10.3184(7)	6.0016(14)	4.6904(4)	290.46(3)	0.965	0.96
40	10.2790(13)	5.9825(6)	4.7054(7)	289.35(6)	0.891	0.88

on the crystalline surface but largely due to the bulk lithium nonstoichiometry at least for particle sizes larger than 40 nm.

Further inspections on the solid solution compositional domain were performed for the electrochemically adjusted Li_xFePO_4 with a particle size of 40 nm, which allow largest miscibility in the present study. The potential was set to 3.8 V, 3.6 V, 3.5 V, 3.3 V, 3.2 V, 3.1 V, 3.0 V, and 2.8 V *vs.* Li^+/Li. They correspond to the lithium composition of 0.02, 0.04, 0.06, 0.88, 0.91, 0.93, 0.95, and 0.95, respectively, with an eye to the OCV data shown in Fig. 1.9, and are inside the solid solution domains. All the samples were subjected to the X-ray diffraction experiments with the subsequent Rietveld analysis, and the plots of lattice parameters *vs.* lithium compositions are given in Fig. 1.9. All the data are on the line of Vegard's law of $FePO_4$–$LiFePO_4$ solid solution without any systematic deviation, and further support the two main conclusions of this paper with no ambiguity: (i) validity of the size-dependent miscibility gap model in Li_xFePO_4 and (ii) bulk miscibility dominates the electrochemical behavior of Li_xFePO_4 at least for particle sizes larger than 40 nm.

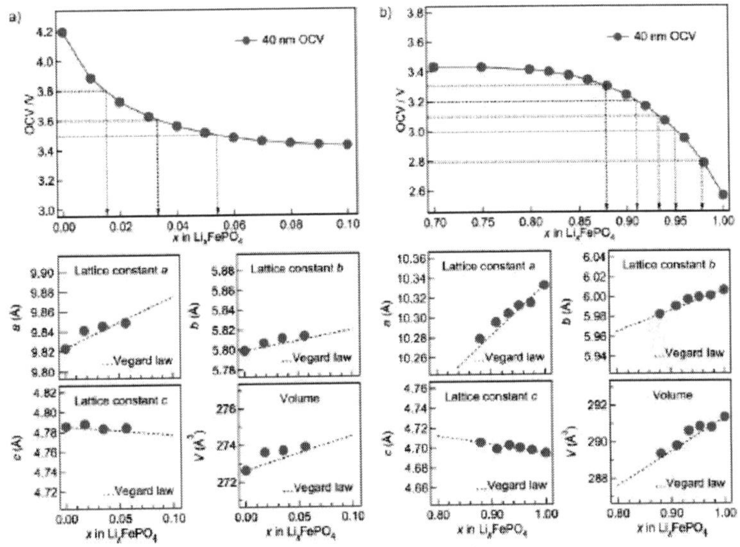

Figure 1.9 OCV curve and change in lattice dimensions above (a) and below (b) the two-phase equilibrium potential. Lithium composition was adjusted potentiostatically (equilibrium condition: <0.1 μA) and estimated by the OCV curves, as indicated by dashed lines. The linear lines along the plot of lattice constants show the Vegard's law between $FePO_4$ and $LiFePO_4$.

1.5.4 *Thermodynamic Approach*

We now address the thermodynamic experimental evidences for the solid solution outside of the miscibility gap.[20] Microcalorimetry can directly measure entropy changes *vs.* x at fixed temperature T during the electrochemical reaction. Observed heat flow P is related to entropy change ΔS by $P = -IT\Delta S/nF$, where I is the input current and F is the Faraday constant. The classical entropy of mixing $S_{\text{cf}}(x)$ can be calculated for a fully disordered lattice of lithium atoms and vacant sites, and heat flow varies linearly with $\Delta S_{\text{cf}}(x) = k_{\text{B}}\ln[1/x]$, where lithium atom—vacancy disorder when $0 < x < 1$. Although this expression is singular at $x = 0$ and $x = 1$, an important overall trend demonstrates that the heat flow changes symmetrically from negative at initial composition $x = 0$ to positive at final composition $x = 1$. Typical anomalies can be clearly seen at the very beginning of charging $LiFePO_4$, as shown

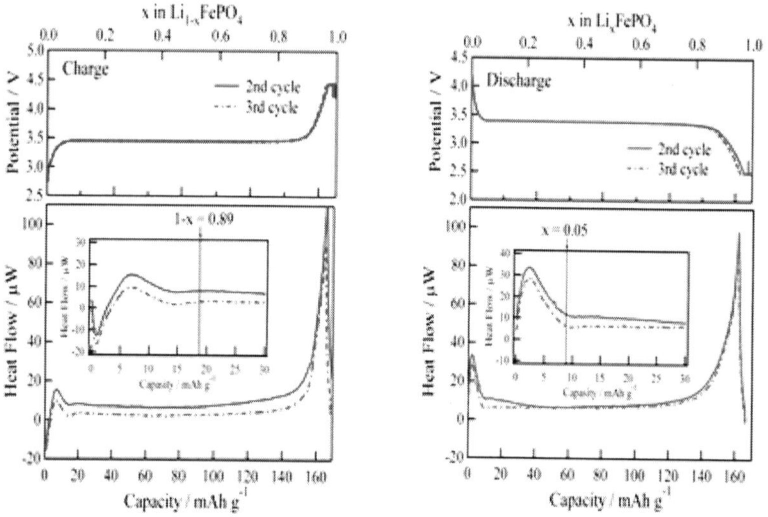

Figure 1.10 Calorimetric measurements for a 2032 type coin cell for charging and discharging processes. A clear anomaly exists at the beginning of both of charge and discharge. Vertical lines are the mono-phase/bi-phase boundary estimated from the structural data.

in Fig. 1.10. However, the endothermic region is narrow compared to the subsequent anomalous exothermic region. Similar anomalies were also detected at the very beginning of $FePO_4$ discharge, but again with a much narrower, negligible endothermic region. Such strong asymmetries for endo/exothermic behavior may be due to the dynamic character of calorimetric measurements. These two thermally anomalous compositional regions close to $LiFePO_4$ and $FePO_4$ in Li_xFePO_4 show good agreement with the values of α and $1-\beta$ in $Li_\alpha FePO_4$ and $Li_{1-\beta}FePO_4$ obtained by structural data also with the region in which the deviation of open circuit voltage from the constant value of $V_{oc} = 3.42$ V is evident (see Fig. 1.4). Furthermore, these initial anomalies in entropy upon charge–discharge were not observed when measurement was initiated at any point inside the miscibility gap. Therefore, it is likely they represent the change in configurational entropy in solid solution outside the miscibility gap in Li_xFePO_4.

Except for these two mono-phase regions, heat flow exhibited almost constant exothermic behavior with no significant anomalies, explained by a simple two-phase reaction scheme with a continuous phase-boundary movement without any structural or electronic phase transition. Abrupt increase of the exothermic heat flow at the end of charge–discharge processes in Li_xFePO_4 is due to the increasing polarization, and hinders the thermal anomaly for one of the two narrow mono-phase regions. It is noteworthy that the amount of heat flow during the charge–discharge reaction (<5 μW) is comparable or even smaller than those measured during similar experiments on conventional electrodes such as Li_xCoO_2 and $Li_xMn_2O_4$.

1.6 Summary

These systematic findings on nano-size effects for the phase diagram of important electrode material of Li_xFePO_4 redefine guidelines for understanding the polaronic transport, cooperative phase boundary movement, and its thermodynamics in bi-phasic intercalation processes. Polaronic insulators of the transition-metal polyanion family show a wide variety of framework structure and accommodate various transition metals and guest atoms. Recent advances in carbon coating technology have been applied to many of these compounds to shed light on their potential application as intercalation battery cathodes. There is, therefore, still plenty of scope for further investigations on this issue.

References

1. A. K. Padhi, K. S. Nanjundaswamy, and J. B. Goodenough, *J. Electrochem. Soc.*, 144, 1188 (1997).
2. A. Yamada, S. C. Chung, and K. Hinokuma, *J. Electrochem. Soc.*, 148, A224 (2001).
3. H. Huang, S.-C.Yin, and L. F. Nazar, *Electrochem. Solid-State Lett.*, 4, A170 (2001).
4. S. Y. Chung, J. T. Bloking, and Y. M. Chiang, *Nature Mater.*, 1, 123 (2002).

5. Z. Chen and J. R. Dahn, *J. Electrochem. Soc.,* 149, A1184 (2002).
6. A. Goñi, L. Lezama, M. I. Arriortua, G. E. Barberis, and T. Rojo, *J. Mater. Chem.*, 10, 423 (2000).
7. J. R. Dahn and W. R. McKinnon, *Phys. Rev.*, B32, 3003 (1985).
8. J. R. Dahn, W. R. McKinnon, J. J. Murray, R. R. Haering, R. S. McMillan, and A. H. Rivers-Bowerman, *Phys. Rev.*, B32, 3316 (1985).
9. C. Delmas, A. Nadiri, and J. L. Soubeyroux, *Solid State Ionics*, 28–30, 419 (1988).
10. A. Yamada, H. Koizumi, N. Sonoyama, and R. Kanno, *Electrochem. Solid-State Lett.*, 8, A409 (2005).
11. N. Meethong, H.-Y. S. Huang, W. C. Carter, and Y.-M. Chiang, *Electrochem. Solid-State Lett.*, 10, A134 (2007).
12. M. Wagemaker, W. J. H. Borghols, and F. M. Mulder, *J. Am. Chem. Soc.*, 129, 4323 (2007).
13. H. G. Schimmel, J. Huot, L. C. Chapon, F. D. Tichelaar, and F. M. Mulder, *J. Am. Chem. Soc.*, 127, 14348 (2005).
14. P. Gibot, M. Casa-Cabanas, L. Laffont, S. Levasseur, P. Carlach, S. Hamelet, J. M. Tarascon, and C. Masquelier, *Nature Mater.*, 7, 741 (2008).
15. M. Wagemaker, *Adv. Mater.*, (2009).
16. N. Meethong, H.-Y. S. Huang, S. A. Speakman, W. C. Carter, and Y.-M. Chiang, *Adv. Func. Mater.*, 17, 1115 (2007).
17. L. Wang, F. Zhou, Y. S. Meng, and G. Ceder, *Phys. Rev. B*, 76, 165435-1 (2007).
18. J. F. Martin, A. Yamada, G. Kobayashi, S. Nishimura, R. Kanno, D. Guyomard, and N. Dupre, *Electrochem. Solid State Lett.*, 11, A12 (2008).
19. G. Kobayashi, S. Nishimura, M.-S. Park, R. Kanno, M. Yashima, T. Ida, and A. Yamada, *Adv. Func. Mater.*, 19, 395 (2009).
20. A. Yamada, H. Koizumi, S. Nishimura, N. Sonoyama, R. Kanno, M. Yonemura, T. Nakamura, and Y. Kobayashi, *Nature Mater.*, 5, 357 (2006).
21. V. A. Streltsov, E. L. Belokoneva, V. G. Tsirelson, and N. K. Hansen, *ActaCrystallogr.*, B49, 147–153 (1993).

Chapter 2

Orthosilicate-Based Cathode Materials for Lithium-Ion Batteries

Robert Dominko

National Institute of Chemistry, Hajdrihova 19, SI-1000 Ljubljana, Slovenia ALISTORE-ERI, 80039 Amiens Cedex, France

robert.dominko@ki.si

This chapter deals with the studies carried out on the lithiated orthosilicate materials as potential cathode for Li-ion batteries. The only member of the orthosilicate family that, according to the present knowledge, shows potential usage in the commercial batteries is Li_2FeSiO_4. Coupled with graphite anode it can deliver a 2.4 V per cell. However, before its commercialization, the low activity at room temperature (RT) has to be solved. The main focus of this chapter is to enunciate the importance of size effects in the context of deliverable capacity. To attain the full capacity utilization at RT, it is necessary for the very small particles to have close contact with a conductive phase and structural arrangement in the composite electrode. Thus, further improvements of the existing synthesis methods and the introduction of even more advanced synthesis methods are a prerequisite for the final goal to prepare uniform particles of Li_2FeSiO_4 phase with particle size smaller than 20 nm.

Nanotechnology in Advanced Electrochemical Power Sources

Edited by S. R. S. Prabaharan and M. S. Michael

ISBN 978-981-4241-43-4 (Hardcover), 978-981-4303-05-7 (eBook)

www.panstanford.com

2.1 Background

In 2001, in a patent application[1] Armand proposed dilithium transition metal silicates (orthosilicates) as potential cathode materials for lithium ion batteries. The structure of this family of cathode materials is related to the structure of lithiophosphate or lithioarsenate (Li_3PO_4, Li_3AsO_4), and it was a subject of crystallographic studies three decades ago.[2–12] The renewed interest in this family of materials can be related to the original contribution of Goodenough's group on polyanionic cathode materials.[13–18] The discovery of electrochemical activity of $LiFePO_4$ opened a completely new branch of cathode materials, where oxygen is stabilized in the relatively strong P–O bond, and the structural stability and reasonably high redox potential for iron are provided by orthophosphate anion ($PO_4{}^{3-}$). The extensive work on $LiFePO_4$ revealed two major breakthroughs in the field of cathode materials for Li-ion batteries. First, polyanionic-based cathode materials are nowadays considered as safe cathode materials since the oxygen is covalently bonded to non-transition metal in the structure. This prevents oxygen loss during battery operation even at elevated temperatures. Another breakthrough that was achieved through the study of $LiFePO_4$ is realization of potential use of active materials with very low intrinsic conductivity.[19,20] Thus, using small particles and employing a proper electrochemically wiring, an excellent electrochemical behavior is observed at high current densities.[21] $LiFePO_4$ is today a serious candidate to replace more expensive and less safe $LiCoO_2$, even though the energy density of the former is higher.[22,23] All these breakthroughs and drawbacks of $LiFePO_4$ have inspired many researches to synthesize and characterize the members of orthosilicate family with a nominal chemical formula Li_2MSiO_4, where M is a transition metal of iron, manganese, cobalt, or vanadium oxide.[24–28] The strong Si–O bond is supposed to stabilize the oxygen in the polyanionic framework, while silicon, iron, or manganese—the major constituents of orthosilicate materials—are abundant, cheap, non-toxic, and relatively safe.[24] All mentioned facts make the orthosilicate materials attractive for large-scale batteries. The most attractive reason for utilization of orthosilicates as cathode materials in Li-ion batteries is their high theoretical capacity since the structure possesses two lithium atoms per formula unit. Utilization of the theoretical capacity would, in practice, lead to Li-ion batteries with much higher energy densities than we are witnessing today. A high theoretical capacity combined with improved safety is an especially critical precondition for large-scale batteries to be used in future electric vehicles, emergency power, and dispersed electric power storage concepts.

First full report[24] on Li_2FeSiO_4 electrochemical behavior is dated back to 2005, and soon after that the structural and electrochemical characterization of Li_2MnSiO_4 was published.[25] The common feature of those two papers is that they showed basic electrochemical characterization and they left a question mark on the structural characterization. Nyten *et al.* proposed an orthorhombic crystal structure with a space group $Pmn2_1$, however in their XRD pattern two XRD peaks remained non-fitted. In our work, we proposed a disordered orthorhombic cell with a space group $Pmn2_1$, but based on the electron diffraction patterns we left an option that the sample used in our refinement was actually a mixture of different polymorphs (*Pmnb* space group was proposed at the time when the Li_2MnSiO_4 sample was prepared at higher temperature). The first two publications were followed by papers from other groups.[23,28,29] Some of the early papers adopted the proposed structural parameters from the earlier publications showing the electrochemical behavior and were focused only on the electrochemical properties of $Li_2(Fe/Mn)SiO_4$ mixtures or the electrochemical properties of Li_2FeSiO_4. A separate work on Li_2CoSiO_4[27] clearly showed the differences in the XRD patterns between the samples prepared at different temperatures. The basic classification into low temperature polymorphs (the so-called β phase) and into high temperature polymorphs (the so-called γ phase) has been recoiled from the early crystallographical works on lithium phosphates and orthosilicates.[5,7,8,10–12] As mentioned earlier the structure of orthosilicates can be related to the structure of lithiophosphate (Li_3PO_4), which is known to crystallize at least in three different polymorphs. Two of them—the low temperature[30] and the high temperature form[31]—are structurally well characterized, while there is only a cubic unit cell without structural details proposed for the third one.[32] The tetrahedral

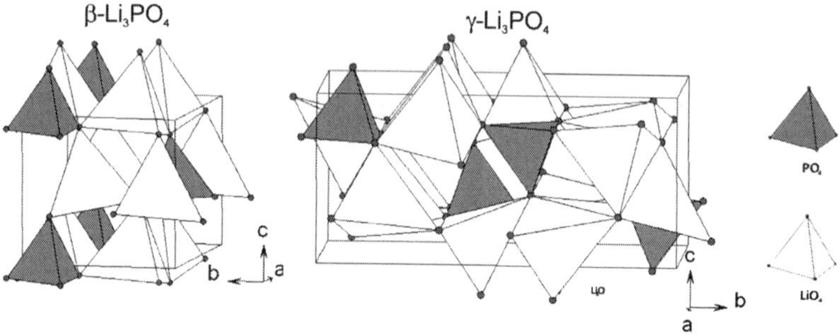

Figure 2.1 Structure of β-Li_3PO_4 and γ-Li_3PO_4 seen along the direction [100].

structures of Li_2MSiO_4 exhibit a rich polymorphism with more than eight polymorphs known,[27] basically divided into a low temperature group[30] and a high temperature group.[33] The structures of the β phases (i.e., Li_3PO_4[2] or Li_3AsO_4[34]) can be derived from wurtzite by slight distortion and transformation of the unit cell (transformation matrix 2 0 0, 1 2 0, 0 0 1) into an orthorhombic one and ordered occupation of the three different cationic sites. This results in the space group *Pmn2*$_1$, where Si, As, or P occupy the Wyckoff position 2a, Li is present on the position 4b in all cases. In the case of the low temperature Li_3PO_4 and Li_3AsO_4, another position 2a is occupied by another Li, while in the case of Li_2MSiO_4 (M = Fe, Mn, Co) the 2a position is occupied by transition metal. All tetrahedrons point in one direction—along *z*-axis (Fig. 2.1), which is coincident with the direction of the crystallographic twofold screw axis.

The high temperature γ forms of Li_3PO_4[2] and Li_3AsO_4[35] as well as Li_2CdSiO_4,[36] Li_2ZnSiO_4,[8,10] and Li_2MgSiO_4[37] also show cation ordering conforming to the unit cell with doubled *b* parameter, where the cations also order over the available tetrahedral sites, so that one half of the tetrahedra are pointing up and one half down (Fig. 2.1, γ-Li_3PO_4). The structure thus becomes centrosymmetric. It has been observed that the rate of cooling is an important variable that affects the isolation of polymorphs.[10] Recently, we performed extensive structural work on different Li_2FeSiO_4 samples. Their XRD patterns are shown in Fig. 2.2.

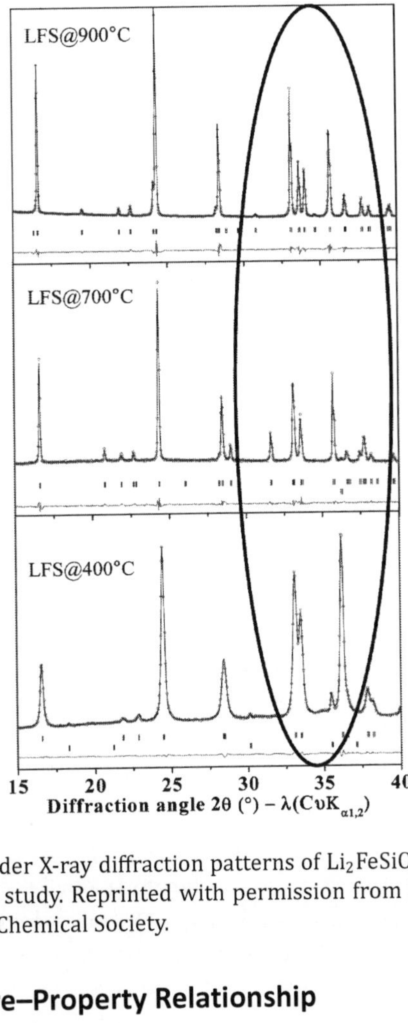

Figure 2.2 Powder X-ray diffraction patterns of Li_2FeSiO_4 samples used in crystallographic study. Reprinted with permission from Ref. 38. Copyright 2010 American Chemical Society.

2.2 Structure–Property Relationship

2.2.1 *X-Ray Diffraction Investigation*

X-ray powder diffraction patterns were refined using Rietveld refinement and, in this case when combined with Neutron diffraction and electron diffraction results.[38,39] Li_2FeSiO_4 samples were carefully prepared at selected temperatures, based on the knowledge obtained from a previous work.[40] With the aim to isolate different polymorphs we employed three different synthesis methods: a

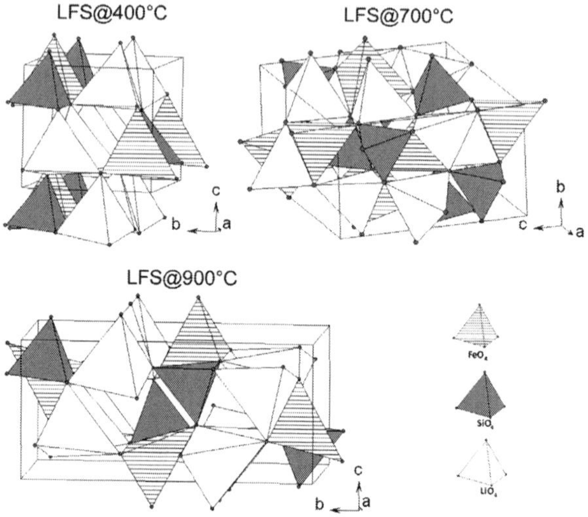

Figure 2.3 Structures of Li_2FeSiO_4 samples isolated from different temperatures seen along the direction [100].

ceramic synthesis procedure, a precipitation synthesis[41] developed by G. Quorin during his PhD work,[39] and a hydrothermal synthesis which was reported earlier.[25] The samples were prepared at 100°C (precipitated sample), 400°C (LFS@400°C), 700°C (LFS@700°C heated hydrothermal sample), and 800°C and 900°C (LFS@900°C quenched ceramic samples). The sample prepared at 900°C revealed a completely new polymorph of Li_2FeSiO_4 not known before. The samples obtained at 100°C and 400°C were fitted with the smallest cell, that is, the low temperature type of the structure.

More complicated structures can be derived from the samples prepared at 700°C and 800°C since in this temperature range at least one more polymorph exists. Indeed, a precise and systematic study of electron diffraction revealed the coexistence of two polymorphs in one particle which could explain all the problems with a crystal structure determination. Figure 2.3 shows three different polymorphs proposed in our recent work.[38,39]

The various possibilities of cation tetrahedral configuration in the slightly disordered hexagonal oxygen environment provide a lot of possibilities for the different arrangements of tetrahedra. Small energetic differences between polymorphs and their close structural relationships are the major reasons for complications in the structural characterization. From the uncertainties in the structural work in the earlier studies and from the recent detailed structural work on Li_2FeSiO_4, we can conclude that, most probably, some samples used for the structural determination were a combination of two or even more polymorphs.

2.2.2 *^{6}Li MAS NMR Investigation*

Besides the use of powder X-ray and neutron and electron diffraction, the characterization technique that helped us a lot in the understanding of structural diversity is lithium solid-state NMR. The use of this technique was introduced soon after we had succeeded to prepare the first crystallographically pure Li_2MnSiO_4 sample, for which we proposed a space group $Pmn2_1$ but with a certain disorder in the structure (Fig. 2.4a). The NMR spectra measured on this sample revealed a complexity but nevertheless we could extract two major characteristics typical for NMR spectra of silicates. First, all NMR signals measured on the silicates that include a paramagnetic transition metal exhibit a negative shift.[43,44] This phenomenon is not fully understood yet and it is a subject of debate between different groups working in that field. The second phenomenon observed during the extensive use of NMR is that all observed NMR shifts from various range of measurements in the different laboratories show a characteristic position of NMR shift in the measured spectrum. A possible explanation for different NMR shifts at several positions can be due to different numbers of neighbors in the samples which possess a certain amount of disorder.[44] Extensive work using different samples of Li_2MnSiO_4 clearly showed that NMR spectra can not only distinguish among polymorphs of Li_2MnSiO_4, but also correlate very well with the proposed structural models.[43] The correlation was established by calculating spinning-sideband patterns and by using first-principles calculations of hyperfine coupling constants.

Figure 2.4 shows four different NMR spectra. The first one (Fig. 2.4a) is from the sample used in the structural determination in our first publication and the rest three are from the samples that were

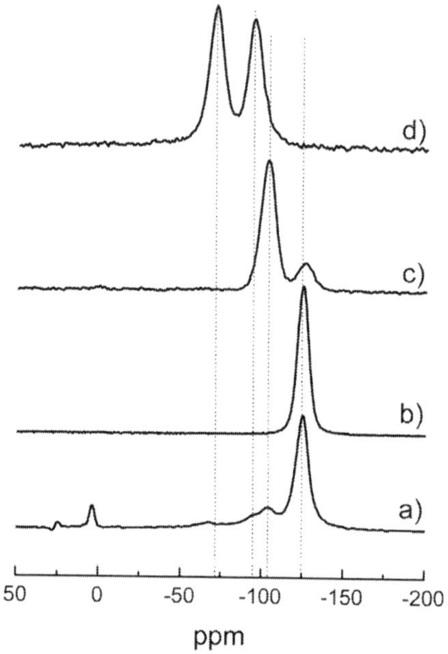

Figure 2.4 Patterns of the centerband regions of ^{6}Li MAS NMR spectra of (a) as-prepared sample, (b) HP Li_2MnSiO_4 sample, (c) sample prepared at 900°C, and (d) sample quenched from 1200°C.

carefully prepared with different approaches. The structure for the sample displayed in Fig. 2.4b has been refined in the $Pmn2_1$ space group,[45] the structure in Fig. 2.4c has been refined in the Pmnb space group, and the structure in Fig. 2.4d has been refined in the $P2_1/n$ space group. Based on these results we can propose the use of NMR technique as the most direct structure characterization technique. However, it should be noted that these relationships between the structure of polymorphs and the NMR shifts as well as the origin of the negative NMR shifts need to be confirmed by theoretical work.

Similar NMR shifts at different positions have also been observed in Li_2CoSiO_4 (Fig. 2.5a) and Li_2FeSiO_4 (Fig. 2.5b) samples that were prepared at different temperatures. Those unique and clear results suggest that NMR technique can be useful in the detection and isolation of new polymorphs.

The differences observed in the structural characterization can be correlated with the type of synthesis used for the preparation

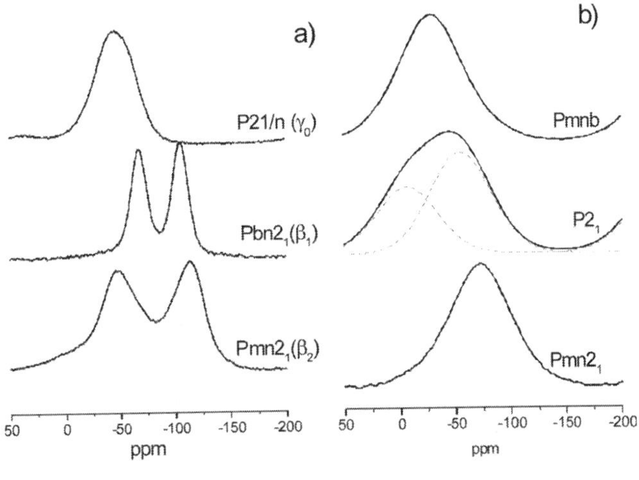

Figure 2.5 Patterns of the centerband regions of ^{6}Li MAS NMR spectra of (a) Li_2CoSiO_4 samples and (b) Li_2FeSiO_4 samples.

of samples. Several different synthesis methods have been used, and the existence of different polymorphs was detected. The diversity is coming from the temperature of synthesis used for sample preparation, from dynamic of cooling or heating samples, and sometimes also from the dwell time during the sample calcination management. Low temperature techniques, such as the precipitation or the hydrothermal synthesis, yield samples with a low symmetry (β type of polymorphs) and most of them can be fitted by using $Pmn2_1$ space group.[26,38] Solid-state synthesis, as by its nature, needs high temperature treatment to promote the solid-state reaction between precursors. At higher temperatures, the prepared samples have a higher symmetry (γ type of polymorphs). The γ type of polymorphs can be prepared by heating β type polymorphs.[26,38,39,46,47]

2.3 Synthesis and Microscopy

The type of synthesis and the temperature used are not only the determining factors for the structural properties but they also affect the electrochemical behavior of the prepared materials. A high temperature synthesis commonly ends up with materials having particles larger than 1 μm. Micrometer-sized particles are very convenient for commercialized cathode materials (for instance $LiCoO_2$ and $LiMn_2O_4$), but they are useless in terms of the electrochemical utilization of orthosilicates. The reason is the low conductivity of silicates which is in the range of insulators. We reported the intrinsic conductivity measured by ac impedance measurements in the range of $\sim10^{-14}$ Scm^{-1} for Li_2FeSiO_4 and $\sim10^{-16}$ Scm^{-1} for Li_2MnSiO_4 at room temperature with the activation energy close to 1 eV.[48] Most of the published electrochemical characterizations were performed at the elevated temperatures (from 55°C to 70°C) and in this case the intrinsic conductivity is approximately two orders of magnitude higher. Despite this improvement, it is still a fact that our materials have three to five orders of magnitude lower conductivity than the well-characterized $LiFePO_4$. Thus, the experience and the knowledge gained in the research of $LiFePO_4$[19] needs to be further upgraded if we wish to prepare silicates with reasonable power performance. Very small nanoparticles embedded in the carbon matrix and composite electrodes with good wiring (electronic and ionic) are prerequisite conditions for the potential use of silicates as an active cathode material. During several years of the work on silicates we optimized several different synthesis techniques, however until now the best samples of Li_2FeSiO_4 in terms of electrochemical behavior has been obtained from the hydrothermal synthesis, followed by heat treatment in the presence of organic precursor.[49,50] Figure 2.6 shows the TEM image of Li_2FeSiO_4 particles embedded in the matrix of carbon obtained after heat treatment of hydrothermally prepared sample.

2.4 Electrochemical Properties (*In situ* Structural Studies: XRD and XAS)

Using particles with the morphology shown in Fig. 2.6 and current density of 16 mAg^{-1} (corresponding to C/10 cycling rate), excellent capacity retention and full reversibility can be retained (Fig. 2.7). Typical electrochemical curve for Li_2FeSiO_4 shows that the first

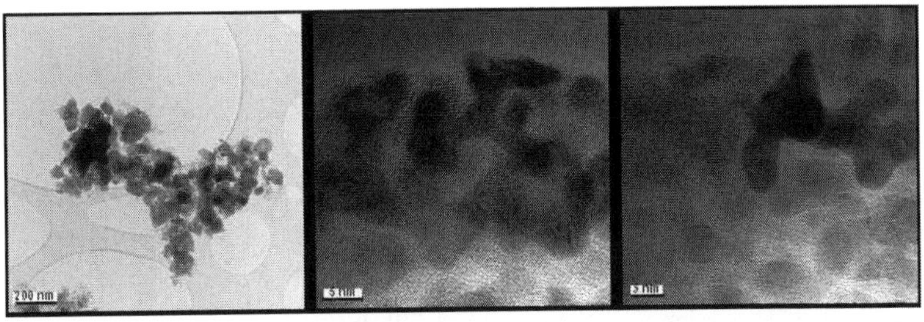

Figure 2.6 HR-TEM of carbon-coated Li_2FeSiO_4 particles obtained by hydrothermal synthesis.

oxidation goes through a plateau at 3.1 V *vs.* lithium reference, while the reduction proceeds through a plateau at approximately 2.76 V *vs.* lithium reference. The oxidation in the second and higher cycles occurs at 2.80 V *vs.* lithium reference.

This shift of the oxidation voltage plateau has been attributed to the structural rearrangements of the structure in the first cycle.[24,51] We need to stress out that this shift might occur through several cycles and it can be a function of the particle size and cycling rate. So far it is not known whether this phenomenon holds generally for all of the Li_2FeSiO_4 polymorphs or it is typical merely for

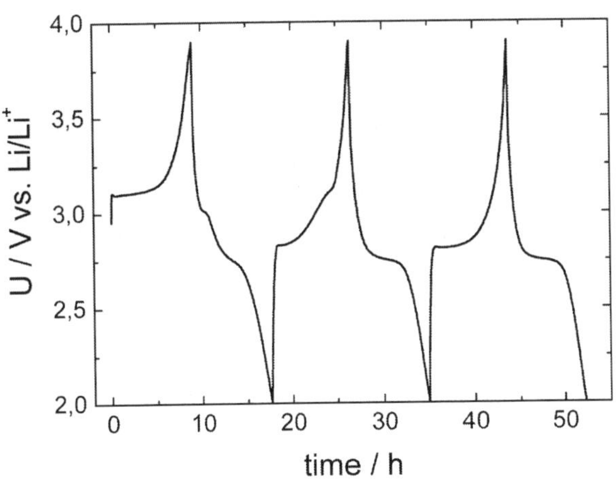

Figure 2.7 Typical shift of oxidation potential upon the cycling of Li_2FeSiO_4.

the samples prepared at 700°C which are typically a mixture of several polymorphs. We wish to emphasize that, in practice, the determination of differences in the electrochemical behavior of different Li_2FeSiO_4 polymorphs may be an extremely difficult task. There are many problems that prevent such an attempt, such as the difficulty of preparation of appropriate particle wiring at low temperatures or the difficulty to prepare small particles at elevated temperatures. To the best of our knowledge, all published results on the electrochemical activity of Li_2FeSiO_4 show the electrochemical characteristics of samples where a coexistence of different polymorphs can be observed with some minor impurities.[24,29,49,50,52–55] The work by Nyten *et al.* was the first published paper in which a complete electrochemical and structural characterization has been shown.[24] The preparation of Li_2FeSiO_4 sample in this work was based on an extensive grinding and milling in order to appropriately homogenize the slurry. This way they achieved a complete reaction of the precursors during solid-state synthesis and the well-distributed precursors of 10 wt% carbon gel[56] were a guarantee for the small particles. The sample was calcinated in a CO/CO_2 atmosphere which is essential when Li_2FeSiO_4 samples are prepared in the presence of organic (carbon) precursors. The obtained powder X-ray diffraction pattern was refined based on the structure of the low temperature Li_3PO_4 polymorph; however, they were not able to refine two diffraction peaks in the X-ray diffraction pattern. These non-identified peaks, however, disappeared after approximately five cycles, suggesting that they could be part of the active material. Based on the recent detailed structural characterization of Li_2FeSiO_4 samples[37,38] and their latter non-published work, we cannot take this sample as an example of electrochemical properties of $Pmn2_1$ phase. An important message from this paper[23] is related to the Li_2FeSiO_4 stability in the air. It was found that the electrodes exposed to air for a longer time showed a lower capacity. A possible explanation was that the sample was lithium deficient, that is, having a formula $Li_{2-x}FeSiO_4$. In any case, this pioneering work with a rather detailed structural and electrochemical characterization proposed a completely new family of active cathode materials. Later on, the complexity of the redox behavior during cycling of Li_2FeSiO_4 was a subject of a complete study in which the same authors used synchrotron X-ray diffraction and Mössbauer spectroscopy.[51] As indicated in the preliminary study[24] and confirmed later by Nyten,[50] and also by our group,[25] the electrochemical cycling of Li_2FeSiO_4 in the galvanostatic mode suggests a rearrangement of the structure during the first cycle as indicated by the shift in the potential plateau from 3.10 V during first oxidation to 2.80 V in the higher oxidation cycles. The complexity of the structural changes has been explained by the irreversible structural rearrangement. A proposal for the nature of the observed voltage shift to lower voltage was derived from XRD measurements. Rietveld refinements suggest the Li and Fe ion-site exchange and it is presumed that this led to structural stabilization, what indeed is reflected in lower chemical potential (lower operating voltage). The same group studied the observed voltage lowering by a density functional theory (DFT) calculations.[57] These calculations were in the relatively good agreement with the proposed model (the observed voltage lowering was 0.3 V, while the calculated one was 0.22 V). On the other hand, density of states (DOS) calculations suggested that Li_2FeSiO_4 was a semiconductor with a band gap of 0.15 eV. This calculated number does not fit with the electrical measurements where the activation energy was found to be close to 1 eV. The operation voltage for Li_2FeSiO_4 is in the range of thermodynamic stability of the typical electrolytes used for electrochemical characterization. Some concern is devoted to the use of $LiPF_6$ salt since it is known that it can hydrolyze and form HF which can attack the surface of Li_2FeSiO_4. However, the surface characterization by using XPS on carbon-coated samples, as expected, did not reveal any remarkable film formation, when a 1 M LiTFSI EC:DEC electrolyte was used.[58] More important information was obtained by the Uppsala group when they studied the surface and stability phenomena of a sample exposed to air. They found much bigger quantities of Li_2CO_3 on the pristine electrodes exposed to air than on electrodes stored in an inert atmosphere. They concluded that the possible mechanism was lithium withdrawal from the structure during the air exposure. The air exposed electrode delivered a low capacity during the first oxidation, while during the reduction process the obtained capacity was similar with the electrodes that were not exposed to the air. The lithium from the counter electrode was used in that case to substitute the missing lithium in the structure. Further work on the surface characterization revealed the importance of the electrolyte selection for long operation of Li_2FeSiO_4 cathodes. It was shown that even on the carbon-coated samples some degradation products of the reaction between Li_2FeSiO_4 and HF (formed from $LiPF_6$) could be detected.[59] No data about the surface chemistry on the non-carbon-coated Li_2FeSiO_4 samples is available till now.

Publications that were followed after the first work mainly adopted the $Pmn2_1$ space group for description of the structure of the prepared Li_2FeSiO_4 samples.[29,49,52,53] Most of these works focused on optimization of the preparation method in order to obtain a full capacity with current densities useful for commercial use. The common feature of all these papers is that they do not put special attention to structural characterization. The obtained capacities are being gradually improved from publication to publication; however, sometimes this improvement is merely due to a wider voltage window used in the experiment. We showed by the use of *in situ* techniques (*in situ* Mössbauer spectroscopy and *in situ* X-ray absorption spectroscopy) that galvanostatic cycling to high voltages (above approximately 4.2 V) triggers no change in the oxidation state of iron.[60] Our recent results based on a "GITT experiment" have shown that the processes occurring at high voltages during the oxidation exhibit much bigger voltage polarization compared to the polarization at the plateau close to 3 V *vs.* lithium reference where Fe^{II}/Fe^{III} oxidation occurs (Fig. 2.8).

The measurement was performed with time limitation corresponding to the charge transfer of 2 electrons per formula unit with a constant current corresponding to C/100. The battery was consecutively charged/discharged for 2 hours followed by 2 hours of relaxation. Even though we could reversibly exchange a charge corresponding to a 2 electron reaction, the processes occurring at the high-voltage plateau during oxidation and at the low-voltage plateau during reduction were obviously not connected with the oxidation/reduction of iron in the structure. Furthermore, the equilibrium potential during the first oxidation is close to 3 V *vs.* lithium reference. The theoretical predictions of the voltage profile for the lithium removal from Li_2FeSiO_4 are in agreement with

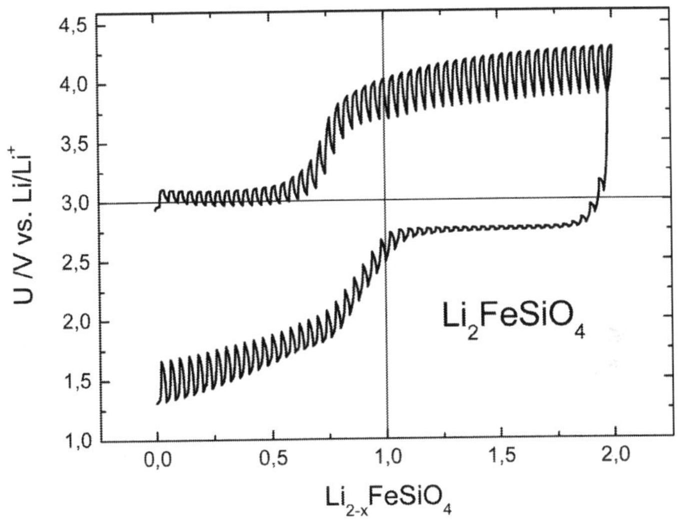

Figure 2.8 GITT measurement of carbon-coated Li_2FeSiO_4 sample.

the calculations[61] and the Fe^{III}/Fe^{IV} oxidation should happen, of course if possible, at voltages above 5 V. A similar scenario was calculated for Li_2MnSiO_4 where Arroyo *et al.*[60] suggested that both lithium atoms from the structure should be removed at very close potential slightly between 4.1 V and 4.2 V *vs.* lithium reference. Our own DFT calculations, however, have suggested that once 1 mol of lithium is removed, the second one should be removed even easier. In practice, this would mean that the oxidation to Mn^{4+} should proceed without a problem.[48,61] But the calculations also suggest that such a mechanism of lithium removal should lead to the materials degradation. Higher operating voltages and the possibility to have $>1e^-$ electrochemical reaction per transition metal are even nowadays a good motivation for deeper study of this class of materials, even though it was found that the stability and reversibility of Li_2MnSiO_4 were very poor. Similar as in the case of Li_2FeSiO_4, there are no systematic studies of the impact of polymorphs on the electrochemical behavior of Li_2MnSiO_4. In our first work,[25] we proposed that the crystal structure of our sample could be fitted with a *Pmn*2$_1$ polymorph, but with a consideration of two types of disorder: (a) exchange of Li and Mn over their primary sites and (b) partial migration of Li, Mn, and Si to alternate tetrahedral sites, just over the basal plane of the tetrahedron to the vicinal tetrahedral position. Although the refinement explained the X-ray powder diffraction pattern very well, the disagreement with the electron diffraction patterns exposed the need for further structural characterization. In our later work, we managed to isolate different polymorphs,[45] but we still do not have an answer if polymorphs have any impact on the electrochemical properties. In the first report we showed a preliminary electrochemical characterization in which we concluded that the poor electrochemical performance (low capacity retention and capacity fading upon cycles) was most likely due to a poor electrochemical wiring.[25] A capacity retention of approximately 0.5 mol of lithium in the voltage window from 4.5 V to 2.0 V *vs.* lithium reference was reported in that paper. In a further attempt to improve the electrochemical performance, we used composites prepared in different ways (composites with a different particle size—Fig. 2.9a) and we cycled our batteries with different current densities (Fig. 2.9b), in order to see the influence of sluggish kinetics due to the very low intrinsic conductivity.[62]

Despite the slight improvements in the reversible capacity and a slightly lower polarization, the cycling stability remained poor with a low cycling efficiency. Obviously, the strategy that has been proven efficient in the case of Li_2FeSiO_4 and $LiFePO_4$ does not work

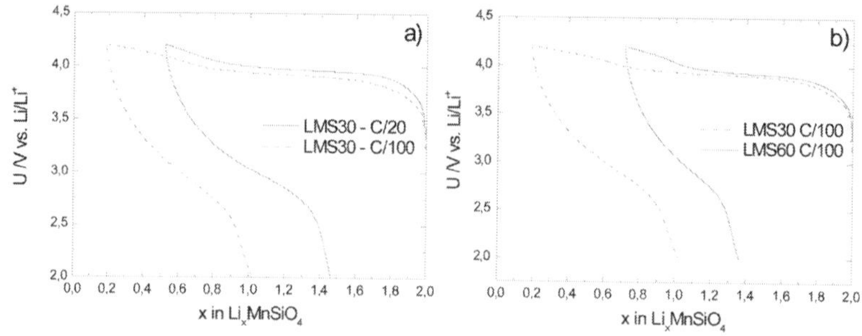

Figure 2.9 Electrochemical behavior of Li_2MnSiO_4 in the first cycle: (a) LMS30 sample (Li_2MnSiO_4 sample with average particle size of 30 nm) was cycled with current densities corresponding to C/20 and C/100 and (b) LMS30 and LMS60 (60 nm large particles) samples were cycled with a current density corresponding to C/100.

with Li_2MnSiO_4. After our initial publications, there has been no significant improvement of cycling stability of Li_2MnSiO_4.[63–67] In most cases, however, these follow-up works showed an improved discharge capacity in the first cycle. Furthermore, the improvement was mostly due to the use of a wider voltage window (with the upper cut-off voltage up to 4.8 V *vs.* lithium reference).

The electrochemical characteristics (the shape of oxidation/reduction curves and the cycling stability) reveal obvious differences in the mechanism of lithium exchange from both iso-structural materials: Li_2FeSiO_4 and Li_2MnSiO_4.[68] In Li_2FeSiO_4 the lithium exchange proceeds through a plateau which is in agreement with a two-phase mechanism (Fig. 2.10a). By the contrast, lithium exchange in Li_2MnSiO_4 involves the formation of an amorphous phase along the oxidation plateau at approximately 4.1 V *vs.* lithium reference since no diffraction peaks are detected in the X-ray powder diffraction after completing oxidation process (Fig. 2.10b). During the first reduction, this phase is not recovered and XRD patterns show mainly the non-destructed remainders of the initial material. The first low discharge capacity as well as the cycling instability has been explained by the loss of long range order. The two-phase system observed during the first cycle of Li_2FeSiO_4 shows a very small change of volume[51] between the oxidized and reduced state.

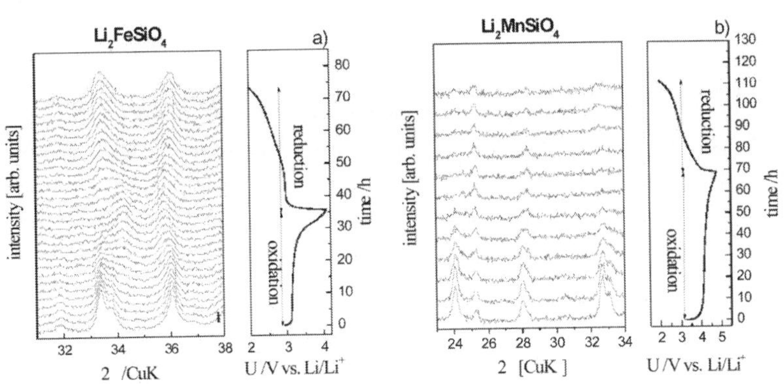

Figure 2.10 *In situ* XRD measurements in the first cycle: (a) for Li_2FeSiO_4 and (b) for Li_2MnSiO_4. Reprinted from Ref. 68, Copyright 2007, with permission from Elsevier.

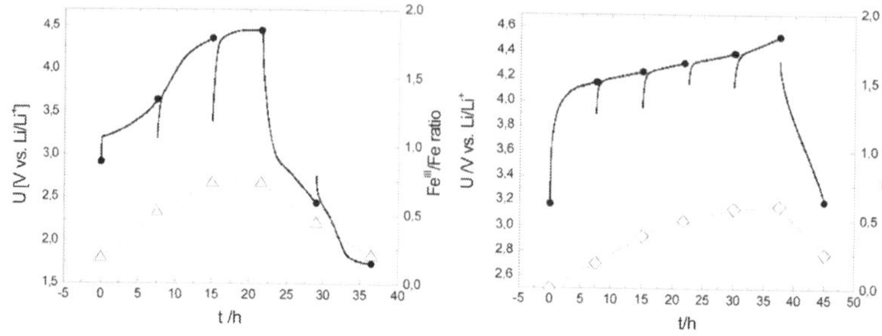

Figure 2.11 Semi-*in situ* XAS measurements of Li_2FeSiO_4 and Li_2MnSiO_4. Solid lines correspond to constructed oxidation/reduction curve, while the black circles are the points where we performed measurements. Dashed line with open symbols gives the ratio of transition metal (III) oxidation state in the material. Reprinted from Ref. 60, Copyright 2006, with permission from Elsevier.

Furthermore, the structural stability of Li_2FeSiO_4 and the loss of long range order in the case of Li_2MnSiO_4 have been carefully analyzed with an X-ray absorption spectroscopy (XAS) in the XANES and in the EXAFS part.[61] To better understand the observed differences in the electrochemical behavior and to elucidate the possible mechanism(s) during lithium extraction/insertion in Li_2FeSiO_4 and Li_2MnSiO_4 we performed *in situ* studies of the local environment and the changes in the oxidation state of Fe and Mn cations upon lithium exchange. XAS measurements were performed semi-*in situ* on batteries that were cycled with a current density corresponding to C/30, at room temperature. Figure 2.11 shows constructed electrochemical curves for the Li_2FeSiO_4 and Li_2MnSiO_4 samples used in the experiment. Black circles in the constructed curves show points where XAS measurements were performed. At each of this point, XAS spectra were measured and the changes in the shape of each measured spectrum indicate the changes in the local symmetry for both cations. We measured an additional Li_2FeSiO_4 cell pre-cycled five times. Comparison of the local symmetry before and after cycling showed no significant differences, suggesting that the local symmetry of Fe cations reversibly returns to the initial one after each cycle. In the case of Li_2MnSiO_4 sample, the shape of the Mn XANES spectrum after first reduction indicates that the process is not completely reversible. The changes in the oxidation state of Fe in Li_2FeSiO_4 and of Mn in Li_2MnSiO_4 are summarized in Fig. 2.11 (empty points). The changes of oxidation state of Fe show a completely reversible change and a linear dependence on the charge at low voltages; however, in the high-voltage region there is no change of oxidation state. This result puts under the discussion all the electrochemical results that were obtained at high voltages (above approximately 4.2 V *vs.* lithium reference) and it explains the higher polarization at these voltages observed in the GITT curve (see Fig. 2.8). The most important conclusions about the changes in the oxidation state of Mn in Li_2MnSiO_4 can be as follows: (a) a reversible change of oxidation state has been observed and (b) in all of our experiments we were not able to detect manganese in the oxidation state Mn^{4+}. From the point of local structure reversibility, the oxidation/reduction process is reversible for Li_2FeSiO_4, with a slight relaxation in the local environment, while the distortions of oxygen atoms in the MnO_4 tetrahedron are not reversible and those irreversible changes could be correlated with the observed amorphism using *in situ* XRD experiment during the first oxidation.

The third member of orthosilicate family of cathode materials, Li_2CoSiO_4, has not attracted a lot of attention[27,69] probably due to several reasons. The first could be due to expected much higher voltages of cobalt-containing polyanions, the second might be due to even worse intrinsic conductivity expected for Li_2CoSiO_4 and, last but not the least, cobalt-based orthosilicates cannot be justified as a cheap or environment-friendly material.

Irreversible changes in the local environment of Mn and the fact that we did not detect the change of oxidation state from Mn^{II} to Mn^{IV} are strong indicators that Li_2MnSiO_4 by itself is probably not suitable as a cathode material; however, those observations do not rule out the possible use of $Li_2(Fe,Mn)SiO_4$ solid solutions.[28,48] These preliminary works on mixed orthosilicates suggested the use of a $Li_2Fe_{0.5}Mn_{0.5}SiO_4$ solid solution. Besides the experimental results of Yang *et al.*,[63] our DFT calculations[48] predicted a relatively good stability of Mn/Fe orthosilicate solid solution upon delithiation. Although the experimental part of this work is still in progress, some latest electrochemical results[70] show cycling instability when a part of Fe is substituted with Mn. The idea of using a Li_2FeSiO_4 material partially substituted with Mn relies on the possibility

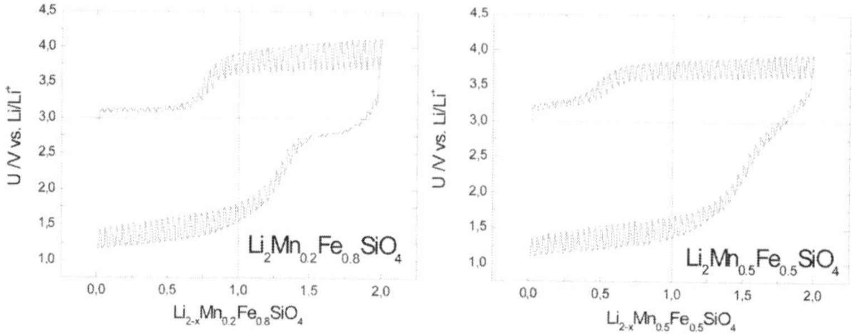

Figure 2.12 "GITT measurements" of carbon-coated $Li_2Mn_{0.2}Fe_{0.8}SiO_4$ and $Li_2Mn_{0.5}Fe_{0.5}SiO_4$ samples. Measurements were time limited during the oxidation and reduction.

that structurally stable Li_2FeSiO_4 will encompass the change in the local environment of manganese, which in that case should be able to exchange two electrons during the redox process. Solid solution $Li_2Mn_xFe_{1-x}SiO_4$ materials can be prepared using different synthesis methods. We prepared our samples with a hydrothermal synthesis. Figure 2.12 shows "GITT measurements" of carbon-coated $Li_2Mn_{0.2}Fe_{0.8}SiO_4$ and $Li_2Mn_{0.5}Fe_{0.5}SiO_4$ samples. Similar to GITT measurement of pure Li_2FeSiO_4 sample, also in the mixed Mn/Fe samples, time controlled oxidation and reduction show reversibility in the wide voltage window. A close comparison of the low-voltage plateaus shows that the oxidation plateau that correspond to the Fe^{2+}/Fe^{3+} oxidation is the lowest for Li_2FeSiO_4 but the voltage is in the correlation with the level of Mn in the sample; the higher the quantity of Mn, the higher equilibrium potential has been measured.

Although the GITT measurements show a high capacity, this capacity is at least in one part connected with additional oxidation/reduction processes that are occurring during the measurements. In order to find to what extent the observed capacity is connected with the change of oxidation state of iron and manganese, we performed *in situ* XAS study on a $Li_2Mn_{0.2}Fe_{0.8}SiO_4$ compound. Figure 2.13 shows the electrochemical curve obtained during the measurement carried out at a current density corresponding to C/15. Due to the rather fast kinetics (C/15), the experiment was stopped after 15 hours. Each XAS spectrum was measured within

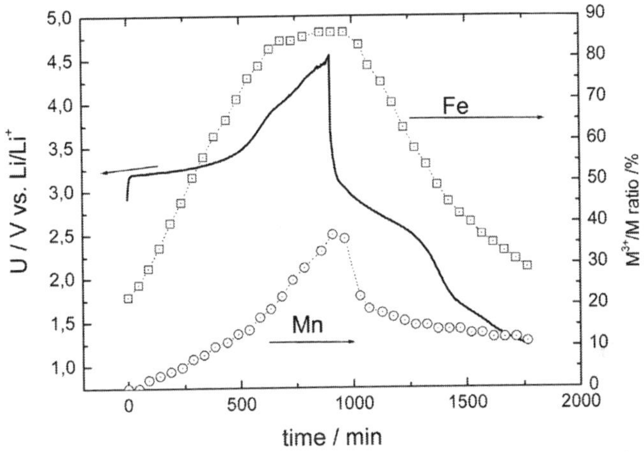

Figure 2.13 A set of *in situ* XANES measurements for a $Li_2Mn_{0.2}Fe_{0.8}SiO_4$ sample. Solid line is the experimentally measured oxidation/reduction voltage profile and empty symbols correspond to the determined ratio of M^{3+} oxidation state to overall quantity of Mn (circles) or Fe (squares).

55 minutes and for each one we calculated the average oxidation state of Fe and Mn. Empty squares and empty circles correspond to the ratio of Fe^{3+} or Mn^{3+} oxidation state according to the overall quantity of Fe or Mn. This experiment leads to a similar conclusion as did the XAS measurements with Li_2FeSiO_4 or Li_2MnSiO_4. Reversibility for iron is good, while the local environment of manganese is distorted and only part of manganese atoms were oxidized to Mn^{3+} oxidation state.

Thus, the strategy to use mixed silicates results in small improvement in the working potential but the oxidation state of Mn^{4+} was never achieved within the present experimental conditions. The use of mixed silicates even with small Mn concentration is not proved to be beneficial since the Mn oxidation is accompanied with sporadic reactions, while the existence of Mn^{3+} oxidation state in the tetrahedral coordination leads to some local structural deformations.

Further research on the impact of iron substitution with other transition metals and the electrochemical activity of non-stoichiometric orthosilicates or orthosilicates with a partial substitution of silicon with phosphorous or aluminum atoms[71] are directions which need to be explored in the future. Hopefully, materials with different electrochemical properties from those which are known from stoichiometric orthosilicates could be obtained. Use of differential theoretical approaches[45,48,57,61,71,72] can help in understanding and anticipating of different properties—from structural to electrochemical.

2.5 Effect of Carbon Coating

At the same time improvements of the rate capability of Li_2FeSiO_4 at the room temperature are required in order to attract potential commercial applications. General approach to obtain full capacity at useable current densities is minimization of particle size to few nanometers, in this case the particle size should be uniform. Furthermore, Li_2FeSiO_4 particles have to be in intimate contact with the conductive phase, preferably a high-quality carbon coating, which can protect the surface from parasitic reactions with the electrolyte. However, it seems that as important as the carbon coating, the morphological rearrangement of the particles in the final composites is imperative as well. The synthesis method that gives reproducible samples with a good particle size control is hydrothermal synthesis. Changing the synthesis parameters and conditions can lead to particles with different sizes and shapes. For instance, changing the Fe precursor from $FeCl_2$ to $FeSO_4$ leads to a change in the particle size and morphology of the Li_2FeSiO_4 particles (Fig. 2.14). Both precursors lead to particles with a mean size around 50 nm but particles prepared by $FeSO_4$ (Fig. 2.14a) as a precursor are more elongated while the particles prepared using the $FeCl_2$ precursor (Fig. 2.14b) have a rounder shape. Carbon-coated particles (citric acid has been used as a source of carbon) retain the shape and the morphology (Fig. 2.14c, d). Such small difference in the particle size has an impact on the electrochemical behavior of Li_2FeSiO_4-based electrodes (Fig. 2.15). This proves that not only the particle size but also the quality of the composite and the electrode morphology have a great impact on the electrochemical characteristics of orthosilicate-based cathode materials.

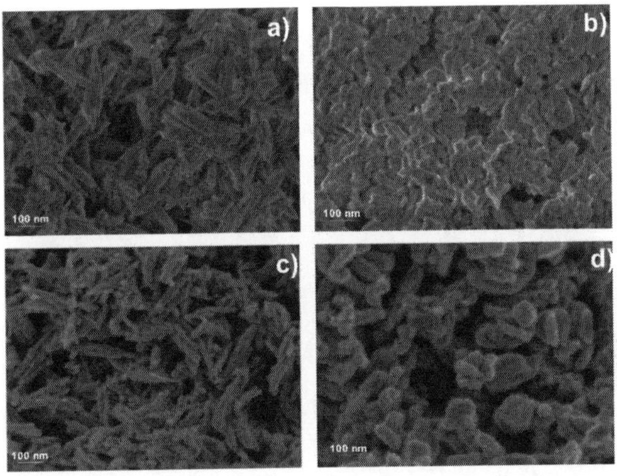

Figure 2.14 SEM micrographs of Li_2FeSiO_4 samples prepared by hydrothermal synthesis using two different types of Fe precursors: (a, c) Li_2FeSiO_4 samples prepared from $FeSO_4$ and (b, d) Li_2FeSiO_4 samples prepared from $FeCl_2$. Mircrographs of samples (a) and (b) are from prepared samples and micrographs of samples in (c) and (d) are from carbon-coated samples.

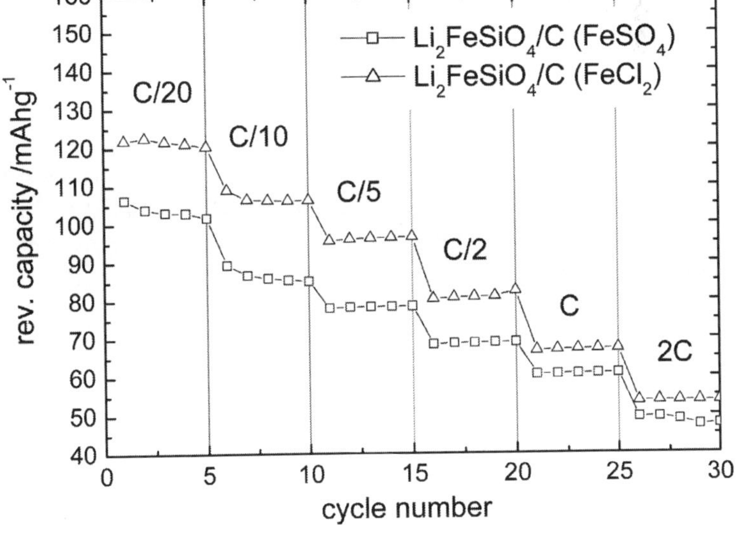

Figure 2.15 Cycling performance of Li_2FeSiO_4/C samples at different current densities prepared from $FeSO_4$ (empty squares) and from $FeCl_2$ (empty triangles).

Surprisingly, the described differences in the morphology show an impact on the electrochemical behavior at low current densities, while at moderate current densities the capacities of both samples are comparable. Nevertheless even at low current densities one cannot obtain the full capacity, so further particle size reduction and composite optimization are probably required.

2.6 Conclusions

The only member of the orthosilicate family that, according to the present knowledge, shows potential usage in the commercial batteries is Li_2FeSiO_4. Coupled with graphite anode it can deliver a 2.4 V battery. However, before its commercialization, the low activity at room temperature (RT) has to be solved. In order to attain the full capacity utilization at RT, unambiguously very small particles must be intact with a conductive phase and to be structurally arranged in the composite electrode. Thus, further improvements of the existing synthesis methods and the introduction of even more advanced synthesis methods are a prerequisite for the final goal to prepare uniform particles of Li_2FeSiO_4 phase with particle size smaller than 20 nm. Regardless of whether the room temperature cycling of Li_2FeSiO_4 is achievable or not, the exchange of the remaining 1 mol of lithium in the structure will be a challenge for the future. Several different approaches are possible such as the transition metal substitution, polyanion substitution, and non-stoichiometric samples, and they will be a subject of the research in the future.

Acknowledgment

I need to thank all my colleagues who worked with me on this topic. Particularly I would like to thank to Miran Gaberscek for the discussion and the motivation during the writing of this chapter, colleagues from the ALISTORE-ERI network for the fruitful discussion and Research Agency of Slovenia for the financial support for the work. Access to synchrotron radiation facilities of

ELETTRA (beamline XAFS) and HASYLAB at DESY (beamline C) is acknowledged.

References

1. Armand, M.; Michot, C.; Ravet, N.; Simoneau, M.; Hovington, P.; European Patent, EP 1 134 826 A1, 2001.
2. West, A. R.; Glasser, F. P.; *J. Solid State Chem.*, **1972**, *4*(1), 20.
3. West, A. R.; *Zeitschrift für Kristallographie,* **1975**, *141*(5/6), 422–436.
4. Tarte, P.; *Inorg. Nucl. Chem.*, **1967**, *29*(4), 915.
5. Tarte, P.; Cahay, R.; *Comptes Rendus Hebdomadaires Des Seances De L Academie Des Sciences Serie C,* **1970**, *271*(14), 777.
6. Okeeffe, M.; Hyde, B. G.; *Acta Crystall. Sec. B—Struct. Sci.*, **1978**, *34*(Dec), 3519–3528.
7. Monnaye, B.; *Revue De Chimie Minerale,* **1976**, *13*(5), 422–432.
8. Villafuertecastrejon, M. E.; West, A. R.; *J. Chem. Soc.—Faraday Trans. I,* **1979**, *75*, 374–384.
9. Plattner, E.; Vollenke, H.; Wittmann, A.; *Monatshefte Fur Chemie,* **1976**, *107*(4), 921–927.
10. Villafuertecastrejon, M. E.; West, A. R.; *J. Chem. Soc.—Faraday Trans. I,* **1981**, *77*, 2297–2307.
11. Yamaguchi, H.; Akatsuka, K.; *Acta Crystall. Sec. B—Struct. Sci.*, **1979**, *35*, 2678–2680.
12. Yamaguchi, H.; Akatsuka, K.; Setoguchi, M.; Takaki, Y.; *Acta Crystall. Sec. B—Struct. Sci.*, **1979**, *35*, 2680–2682.
13. Nanjundaswamy, K. S.; Padhi, A. K.; Goodenough, J. B.; Okada, S.; Ohtsuka, H.; Arai, H.; Yamaki, J.; *Solid State Ionics,* **1996**, *92*(1/2), 1–10.
14. Padhi, A. K.; Nanjundaswamy, K. S.; Goodenough, J. B.; *J. Electrochem. Soc.*, **1997**, *144*(4), 1188–1194.
15. Padhi, A. K.; Nanjundaswamy, K. S.; Masquelier, C.; Okada, S.; Goodenough, J. B.; *J. Electrochem. Soc.*, **1997**, *144*(5), 1609–1613.
16. Padhi, A. K.; Nanjundaswamy, K. S.; Masquelier, C.; Goodenough, J. B.; *J. Electrochem. Soc.*, **1997**, *144*(8), 2581–2586.
17. Masquelier, C.; Padhi, A. K.; Nanjundaswamy, K. S.; Goodenough, J. B.; *J. Solid State Chem.*, **1998**, *135*(2), 228–234.
18. Padhi, A. K.; Manivannan, V.; Goodenough, J. B.; *J. Electrochem. Soc.*, **1998**, *145*(5), 1518–1520.

19. Ravet, N.; Goodenough, J. B.; Besner, S.; Simoneau, M.; Hovington, P.; Armand, M.; Abstract#127, 196th ECS Meeting, Hawaii, 17–22 October, 1999.
20. Huang, H.; Yin, S. C.; Nazar, L. F.; *Electrochem. Solid State Lett.*, **2001**, *4*(10), A170–A172.
21. Gaberscek, M.; Dominko, R.; Jamnik, J.; *Electrochem. Comm.*, **2007**, *9*(12), 2778–2783.
22. Arico, A. S.; Bruce, P.; Scrosati, B.; Tarascon, J. M.; Van Schalkwijk, W.; *Nature Mater.*, **2005**, *4*(5), 366–377.
23. Armand, M.; Tarascon, J. M., *Nature*, **2008**, *451*(7179), 652–657.
24. Nyten, A.; Abouimrane, A.; Armand, M.; Gustafsson, T.; Thomas, J. O.; *Electrochem. Comm.*, **2005**, *7*(2), 156–160.
25. Dominko, R.; Bele, M.; Gaberscek, M.; Meden, A.; Remskar, M.; Jamnik, J.; *Electrochem. Comm.*, **2006**, *8*(2), 217–222.
26. Prakash, A. S.; Rozier, P.; Dupont, L.; Vezin, H.; Sauvage, F.; Tarascon, J. M.; *Chem. Mater.*, **2006**, *18*(2), 407–412.
27. Lyness, C.; Delobel, B.; Armstrong, A. R.; Bruce, P. G.; *Chem. Comm.*, **2007**, (46), 4890–4892.
28. Gong, Z. L.; Li, Y. X.; Yang, Y.; *Electrochem. Solid State Lett.*, **2006**, *9*(12), A542–A544.
29. Zaghib, K.; Salah, A. A.; Ravet, N.; Mauger, A.; Gendron, F.; Julien, C. M.; *J. Power Sources*, **2006**, *160*(2), 1381–1386.
30. Keffer, C.; Mighell, A.; Mauer, F.; Swanson, H.; Block, S.; *Inorg. Chem.*, **1967**, *6*(1), 119–125.
31. Wang, B.; Chakoumakos, B. C.; Sales, B. C.; Kwak, B. S.; Bates, J. B.; Synthesis, *J. Solid State Chem.*, **1995**, *115*(2), 313–323.
32. Zhu, B.; Mellander, B. E.; Chen, J. X.; *Mater. Res. Bull.*, **1993**, *28*(4), 321–328.
33. Zemann, J.; *Acta Crystall.*, **1960**, *13*(11), 863–867.
34. Elfakir, A.; Wallez, G.; Quarton, M.; *Phase Trans.*, **1993**, *45*(4), 281–288.
35. Abrahams, I.; Bruce, P. G.; David, W. I. F.; West, A. R.; *J. Solid State Chem.*, **1994**, *110*(2), 243–249.
36. Riekel, C.; *Acta Crystall. Sec. B—Struct. Sci.*, **1977**, *33*(Aug 15), 2656–2657.
37. Jousseaume, C.; Kahn-Harari, A.; Vivien, D.; Derouet, J.; Ribot, F.; Villain, F.; *J. Mater. Chem.*, **2002**, *12*(5), 1525–1529.
38. Sirisopanaporn, C.; Boulineau, A.; Armstrong, A. R.; Bruce, P. G.; Hanzel, D.; Budic, B.; Dominko R.; Masquelier, C.; *Inorg. Chem.*, **2010**, 49, 7446–7451.

39. Boulineau, A.; Sirisopanaporn, C.; Dominko R.; Armstrong, A. R.; Bruce, P. G.; Masquelier, C; *Dalton Trans.*, **2010**, 39, 6310.
40. Quoirin, G.; Taulelle, F.; Dupont L.; Masquelier, C.; 211th ECS Meeting, 2007, Abstract #98.
41. Quoirin, G.; Tarascon, J. M.; Masquelier, C.; Delacourt, C.; Poizot, P.; Taulelle, F.; World Patent, WO2008/107571 A2.
42. Quoirin, G.; PhD thesis, Universite de Picardie Jules Verne, 2007.
43. Mali, G.; Meden, A.; Dominko, R.; *Chem. Comm.*, **2010**, *46*(19), 3306.
44. Armstrong, A. R.; Lyness, C.; Menetrier, M.; Bruce, P. G.; *Chem. Mater.*, **2010**, *22*, 1892–1900.
45. Politaev, V. V.; Petrenko, A. A.; Nalbandyan, V. B.; Medvedev, B. S.; Shvetsova, E. S.; *J. Solid State Chem.*, **2007**, *180*(3), 1045–1050.
46. Nishimura, S. I.; Hayase, S.; Kanno, R.; Yashima, M.; Nakayama, N.; Yamada, A.; *J Am. Chem. Soc.*, **2008**, *130*(40), 13212–13213.
47. Kokalj, A.; Dominko, R.; Mali, G.; Meden, A.; Gaberscek, M.; Jamnik, J.; *Chem. Mater.*, **2007**, *19*(15), 3633–3640.
48. Dominko, R.; Conte, D. E.; Hanzel, D.; Gaberscek, M.; Jamnik, J., *J. Power Sources*, **2007**, *178*, 842–847.
49. Nadherna, M.; Dominko, R.; Hanzel, D.; Reiter, J.; Gaberscek, M.; *J. Electrochem. Soc.*, **2009**, *156*(7), A619–A626.
50. Nyten, A.; Kamali, S.; Haggstrom, L.; Gustafsson, T.; Thomas, J. O.; *J. Mater. Chem.*, **2006**, *16*(23), 2266–2272.
51. Gong, Z. L.; Li, Y. X.; He, G. N.; Li, J.; Yang, Y.; *Electrochem. Solid State Lett.*, **2008**, *11*(5), A60–A63.
52. Li, L. M.; Guo, H. J.; Li, X. H.; Wang, Z. X.; Peng, W. J.; Xiang, K. X.; Cao, X.; *J. Power Sources*, **2008**, *189*, 45–50.
53. Peng, Z. D.; Cao, Y. B.; Hu, G. R.; Du, K.; Gao, X. G.; Xiao, Z. W.; *Chinese Chem. Lett.*, **2009**, *20*(8), 1000–1004.
54. Zhang, S.; Deng, C.; Yang, S. Y.; *Electrochem. Solid State Lett.*, **2009**, *12*(7), A136–A139.
55. Abouimrane, A.; Ravet, N.; Armand, M.; Nyten, A.; Thomas, J.O.; Abstract #350, IMLB 12, Nara, Japan, 27 June–2 July, 2004.
56. Larsson, P.; Ahuia, R.; Nyten, A.; Thomas, J. O.; *Electrochem.Comm.* **2006**, *8*(5), 797–800.
57. Nyten, A.; Stjerndahl, M.; Rensmo, H.; Siegbahn, H.; Armand, M.; Gustafsson, T.; Edstrom, K.; Thomas, J. O.; *J. Mater. Chem.*, **2006**, *16*(34), 3483–3488.

58. Ensling, D.; Stjerndahl, M.; Nyten, A.; Gustafsson, T.; Thomas, J. O.; *J. Mater. Chem.*, **2009**, *19*(1), 82–88.
59. Dominko, R.; Arcon, I.; Kodre, A.; Hanzel, D.; Gaberscek, M.; *J. Power Sources*, **2008**, *189*, 51–58.
60. Arroyo-de Dompablo, M. E.; Armand, M.; Tarascon, J. M.; Amador, U.; *Electrochem. Comm.*, **2006**, *8*(8), 1292–1298.
61. Dominko, R.; Bele, M.; Kokalj, A.; Gaberscek, M.; Jamnik, J.; *J. Power Sources*, **2006**, *174*, 457–461.
62. Li, Y. X.; Gong, Z. L.; Yang, Y.; *J. Power Sources*, **2006**, *174*, 528–532.
63. Yang, Y.; Fang, H. S.; Li, L. P.; Yan, G. F.; Li, G. S., Rare Met. Mater. Eng., **2008**, *37*(6), 1085–1088.
64. Liu, W. G.; Xu, Y. H.; Yang, R.; *J. Alloy Compd.*, **2009**, *480*(2), L1–L4.
65. Ghosh, P.; Mahanty, S.; Basu, R. N.; *J. Electrochem. Soc.*, **2009**, *156*(8), A677–A681.
66. Belharouak, I.; Abouimrane, A.; Amine, K.; *J. Phys. chem. C*, **2009**, *113*, 20733–20737.
67. Dominko, R.; *J. Power Sources*, **2007**, *184*, 462–468.
68. Gong, Z. L.; Li, Y. X.; Yang, Y., *J. Power Sources*, **2006**, *174*, 524–527.
69. Deng, C.; Zhang, S.; Yang, S. Y.; *J. Alloy Compd.*, **2009**, *487*, L18–L23.
70. Kuganathan, N.; Islam, M. S.; *Chem. Mater.*, **2009**, *21*(21), 5196–5202.
71. Frayret, C.; Masquelier, C.; Villesuzanne, A.; Morcrette, M.; Tarascon, J. M.; *Chem. Mater.*, **2009**, *21*(9), 1861–1874.
72. Wu, S. Q.; Zhu, Z. Z.; Yang, Y.; Hou, Z. F.; *Comput. Mater. Sci.*, **2009**, *44*(4), 1243–1251.

Chapter 3

Nanoscale Conversion Materials for Electrochemical Energy Storage

Maximilian Fichtner
Institute of Nanotechnology, Karlsruhe Institute of Technology (KIT), P.O. Box 3640, D-76021 Karlsruhe, Germany
m.fichtner@kit.edu

3.1 Introduction

Batteries for future portable and automotive applications must meet a variety of requirements in order to be acceptable for the market. Their energy density, the charge–discharge rates, and the cyclic life must be high, and the systems must be safe and cost-effective. Moreover, with increase in demand for electrochemical energy storage, it is necessary to develop systems which are based on sustainable element compositions. A particular challenge is the development of batteries for future electric cars [battery electric vehicles (BEV)] because the energy density has to be very high and cost is an important factor.

A state-of-the-art battery system has an energy density in the order of 120 Wh/kg which is low compared to the energy density

Nanotechnology in Advanced Electrochemical Power Sources
Edited by S. R. S. Prabaharan and M. S. Michael

ISBN 978-981-4241-43-4 (Hardcover), 978-981-4303-05-7 (eBook)
www.panstanford.com

of a hydrogen storage tank (1100 Wh/kg) or a diesel tank (9000 Wh/kg). Hence, the challenge is to develop new materials which allow for a considerable increase of the storage capacity while other factors, such as power density, cost, and safety must not be neglected.

Nowadays, the systems with the highest gravimetric energy density on the market are Li-ion batteries (LIBs) which are based on insertion materials. Here, lithium is intercalated in the host structure of the cathode (e.g., lithium cobalt oxide or lithium iron phosphate) or anode material (graphite). Due to the limited number of lattice sites for the lithium ions or lithium atoms, the storage capacity of such systems is in the order of the already achieved values. New materials with unexpected properties need to be discovered, so that the goal of at least one lithium cycling per transition metal at the rate of 10 mA/cm^2 can be achieved with pulse charging and discharging at still higher rates.

There are strategies for improving the storage density, which include the search for new intercalation compounds which allow for higher packing densities of lithium and the development of new materials and electrolytes to reach higher operation voltages.

An alternative approach having the potential to considerably increase energy densities is making use of several oxidation or reduction steps so that more than one electron can be transferred by an active electrode element. This is possible by solid-state conversion reactions in reactive nanocomposites based on metal nitrides, metal chalkogenides, or metal fluorides.

Working on these systems showed that the performance of such materials may suffer from their low electronic conductivity of active components and the large volume changes in the electrodes. In this context, metal nano/mesoscale structures gained considerable attention because of their novel physical and chemical properties that are not present in their corresponding macroscale structures.[1] The mesoscale structure of the battery material seems to be vital to further improvement of the electrochemical performance and first promising results will be presented for both the negative and the positive electrode of this type of batteries. The chapter will also provide examples for anode and cathode systems which are based on the conversion principle and on the use of nanoscale and mesoscale structures to ameliorate the properties of the system. Strategies will be outlined and discussed for improving the performance and cyclic life time of batteries based on conversion electrodes and basic principles of solid-state reactions.

3.2 Current Electrode Materials

Current commercial battery systems contain anodes and cathodes which are almost exclusively based on the insertion principle of lithium or hydrogen. Although these systems already exhibit a greatly improved performance when compared to batteries of 10 or 20 years ago, their principle is also subject to intrinsic limitations leading to comparably low energy system densities.

A challenging application for batteries in terms of energy and power density is future electric vehicles [hybrid electric vehicles (HEVs) and BEV]. In the technical environment of cars, energy density of current LIBs still is two orders of magnitude below that of liquid organic carriers, such as gasoline, or one order of magnitude smaller than the energy density of a hydrogen storage system. Hence, in order to be competitive in terms of weight, size, and cost, new concepts are necessary to increase the energy density by a factor of 3 to 5 and at least be competitive with hydrogen-driven systems, the efficiency of which, however, suffers from a >50% efficiency loss caused by the fuel cell. A recent tank-to-wheel comparison that includes the efficiency of the drivetrain is shown in Fig. 3.1.

It is obvious that the nature of the active materials in a battery is of primary importance to the resulting cell energy density. The main features which determine cell energy are the number of electrons which the materials can store per unit volume or weight (volumetric capacity or specific capacity) and the electrochemical potential they produce. Using a cell construction model, it is possible to calculate the energy density as a function of the intrinsic material properties. Figure 3.2 illustrates the impact of specific capacity of the materials on the complete cell-specific energy using an "average" design of Li-ion electrodes and cell stack technology, as can be found in many of the commercial cells today.

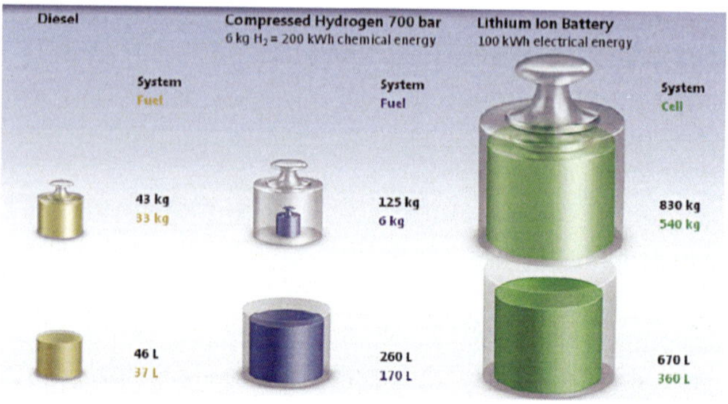

Figure 3.1 The weight and volume of various fuels and tank systems required for a 500 km range vehicle. Reprinted with permission from Ref. 2.

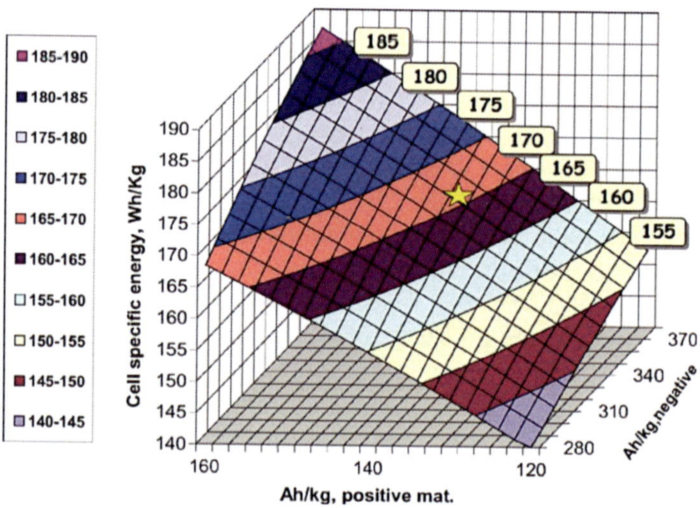

Figure 3.2 Impact of material properties (negative and positive electrodes) on cell-specific energy. Star indicates a $LiCoO_2$/graphite example. Reprinted with permission from Ref. 3.

Various attempts have been made to improve gravimetric and volumetric energy density, cycle life, safety, and rate capability of LIBs. On the anode side, the first alternative to the pure Li metal anode were carbon-based Li-ion intercalation materials, which were adapted to overcome safety and reversibility problems.

This development gave rise to the concept of a "rocking-chair", a type of rechargeable battery in 1985. Lithium ions are reversibly stored between the carbon layers, which thereby develop an electrochemical potential which are low enough to act as negative electrode. There have been numerous improvements in the Li storage capacity of carbon materials, as a result of which the stoichiometric limit of Li-ion intercalation in graphite (LiC6), 372 mAh/g, was exceeded. Deep doping of Li is one possibility. However, this may cause the formation of hazardous metallic Li (dendrite) on the electrode surface, such that the safety of the anode cannot be guaranteed.

On the cathode side, LIBs use lithium cobalt oxide, $LiCoO_2$ (or variants) as cathode material. The compound produces a high potential of 4.3 V *vs.* Li^0/Li^+ when charged. Due to the materials' ability to release about half (0.5) a lithium atom per mole and oxidize half of the cobalt to the tetravalent stage, the resulting specific capacity is close to 140 Ah/kg and capacity density 690 Ah/l. $LiCoO_2$ has the advantage of a long cycle life, but the cost of the cobalt is high. This may be acceptable for small items, such as mobile phones, but not for larger items, such as electric vehicles. Ideally, the cobalt should be replaced by a cheaper metal, like iron.

Other materials, such as $LiNiO_2$ or the spinel oxide, $LiMn_2O_4$ may have advantages in terms of a higher specific capacity (200 mAh/g, $LiNiO_2$) or lower cost and better stability on overcharge ($LiMn_2O_4$), but suffer from other problems and are still under development. The so-called "5 V class materials" are rather promising, as they significantly increase the cell working voltage with no or small sacrifice in capacity. A well-investigated system is $Li[Ni_{1/2}Mn_{3/2}]O_4$ with an average working voltage of about 4.7 V *vs.* a lithiated graphite negative electrode. With a specific capacity of 120 Ah/kg, an approximate energy density increases to 15% is obtained at the cell level.

3.3 Conversion Electrodes

A principle limitation of the established Li-ion concept is that typically only one electron per transition metal atom is exchanged or stored. Furthermore, lattice sites for lithium in the host matrix are limited, not densely packed, and cannot always be filled completely.

Conversion reactions may circumvent these limitations and calculated theoretical capacities are above those of lithium cobalt oxide, lithium iron phosphate, or graphite by a factor of 5 to 10. This is due to a complete electrochemical reduction with the transfer of several electrons per 3d transition metal ion typically leading to metal nanoparticles which are embedded in a host matrix of the lithium compound.

While the theoretical capacities are highly attractive, it is evident from the first work in this field that there are also drawbacks associated with the currently investigated conversion systems. The following problems have been identified and are subject of research:

(1) Limited reversibility, which is often caused by the large volume changes that may be associated with conversion reactions or by irreversible chemical reactions of the electrode with the electrolyte.
(2) Large polarization between charge and discharge curves leading to lower energy efficiencies.

While promising results have already been obtained in increasing the cycle life by tailored nanostructured electrodes, the polarization seems to depend on the particular ionic nature of the M–X bond (X = halide, oxide, nitride sulfide, phosphide) which makes new battery chemistries necessary in order to circumvent that problem.

3.3.1 *Concept and Examples*

Irrespective of the question whether the material acts as anode or cathode, the following simplified and general reaction can be used to describe the overall process in a conversion electrode

$$M^{n+}X + ne^- + nLi^+ \leftrightarrows M^0 + nLiX \tag{3.1}$$

M is a metal, preferably a 3d transition metal, X is an anion, such as H^-, N^{3-}, S^{2-}, O^{2-}, Se^{2-}, F^-, etc.

3.3.2 *Anode Materials*

3.3.2.1 Metal–air batteries

Metal–air batteries have reached a much higher theoretical specific energy than most currently available primary and rechargeable batteries. These batteries are unique, as the cathode active material, O_2 does not have to be stored in the battery. It is taken from ambient air. In the battery, the oxygen is reduced to O^{2-} or O_2^{2-} at a catalytic surface in the air electrode during the discharge of the battery. The oxide or peroxide ion further reacts with cationic species in the electrolyte.

The design of the battery and the properties of the electrolyte are of crucial importance in this case and a number of systems have been investigated, which are based on Zn/O_2, Al/O_2, Ca/O_2, Li/O_2, and Fe/O_2.[4] Early attempts focused on utilizing aqueous solutions of KOH, NaCl, NH_4Cl, $(NH_4)_2SO_4$, KNO_3, and a conventional electrode design. Table 3.1 summarizes several interesting metal/O_2 systems, their idealized cell reaction, the open circuit voltage (OCV), and the specific energy. It should be mentioned that most of the current systems are primary batteries, or as secondary batteries, still suffer from insufficient reversibility. However, due to the very high theoretical capacities of these systems, it would be very attractive to find ways to enable or improve the reversible behavior. The work in this field shows that an elaborate design of the micro- and mesoscale structure of the electrode may be of crucial importance in such a case. Embedding the solid products in a manner that the material is accessible and can be reduced by electrons under ambient conditions may be a key issue for the reversible properties of these systems.

Selected examples of metal–air batteries where nanomaterials are relevant already to the performance of the system will be presented below.

3.3.2.1.1 Li–air batteries

From Table 3.1, Li–air secondary batteries may be particularly attractive due to their high theoretical energy density which mainly results from the light weight of lithium metal.

Table 3.1 Characteristics of metal/O_2 couples

Metal/O_2 couple	Idealized cell reaction	Calculated OCV [V]	Theoretical specific energy [Wh/kg][a]
Li/O_2	$4Li + O_2 \rightarrow 2Li_2O$	2.91	13,000
Al/O_2	$4Al + 3O_2 \rightarrow 2Al_2O_3$	2.73	8100
Ca/O_2	$2Ca + O_2 \rightarrow 2CaO$	2.99	4600
Zn/O_2	$2Zn + O_2 \rightarrow 2ZnO$	1.65	1300
Mg/O_2	$2Mg + O_2 + H_2O \rightarrow 2Mg(OH)_2$	3.1	6800
Fe/O_2	$4Fe + 3O_2 \rightarrow 2Fe_2O_3$	1.3	1200

[a] without O_2

It has already been mentioned that the chemical discharge reaction in the battery may lead to the formation of peroxides and oxides. Due to the altered thermodynamic situation, this has consequences for the OCV and the specific capacity. In the case of the Li–air battery, the Gibbs free energy of formation is −606.1 kJ/mol for Li_2O and −560.1 kJ/mol for Li_2O_2. Thus, the following OCVs can be calculated for the two reactions:

$$4Li + O_2 \rightarrow 2Li_2O \qquad E^o = 2.91\,V \tag{3.2}$$

$$2Li + O_2 \rightarrow Li_2O_2 \qquad E^o = 3.10\,V \tag{3.3}$$

The ratio of Li_2O_2 to Li_2O is a function of the discharge rate and electrolyte formulation. This can be understood because the lower oxygen content in the electrolyte results in the preferential formation of Li_2O.

If ambient air is used, the presence of water vapor leads to a side reaction forming LiOH ($2Li + 2H_2O \rightarrow 2LiOH + H_2$, theoretical potential 2.08 V). This reaction may lead to irreversible processes in the anode and a decreasing performance of the system.[5] Hence, attempts have been made in order to exclude water vapor from the electrode. By using a non-aqueous electrolyte, the parasitic reactions can be avoided and alleviate the safety concerns.

Abraham and Jiang were the first to report a lithium–oxygen organic electrolyte battery.[6] The battery was shown to have an open circuit potential close to 3 V, an operating voltage of 2.0 to 2.8 V, and due to a cobalt phthalocyanine-catalyzed air electrode, a good Coulombic efficiency upon recharge over several cycles. The discharge mechanism was determined to be primarily the deposition of Li_2O_2 in a carbon-based air electrode. Specific energies were in the order of 250–350 Wh/kg.

Later, several different air cathode formulations were studied by Read.[7] The air electrode's physical properties, its thickness, porosity, and volume fraction of carbon black were found to be significant factors in determining cell performance. The discharge capacity per gram of carbon is affected directly by the ratio of pore volume to carbon black weight. The Brunauer, Emmett, and Teller (BET) surface area of the carbon black does not appear to be an important factor in determining discharge capacity. However, a wetted surface area may be important and the ability of an electrolyte to wet a particular carbon is an important factor in determining cell capacity. A particular issue when using organic electrolytes is the fact that both O_2^{-2} and O^{-2} are insoluble in an organic electrolyte. Consequently, Li_2O_2 or Li_2O deposits near where the O_2^{-2} and O^{-2} is generated, namely, in the cathode.[8]

Dobley *et al.*[9] presented a way of how to increase the energy density of the Li–air battery through improved air cathodes with a mesoscale structure. The reactivity of oxygen was increased by incorporating nanosized transition metal catalysts into the carbon electrode. As a result, both the oxygen reduction kinetics and the specific capacity of the cathode were increased. The anode was composed of a lithium metal foil pressed into a nickel mesh current collector that was welded to a nickel tab. The separator used was Setela, which is an organic polymer film of 20 microns in thickness. The liquid electrolyte used was 1 M LiPF6 in 1:1:1 EC/DEC/DMC. The air cathode was a carbon composite made by combining carbon, the transition metal catalyst, and a binder deposited on a metal current collector. The binder used was Teflon combined with latex. A thin Teflon film between the air cathode and the atmosphere provides the cathode with hydrophobicity to repel atmospheric water and creates channels for oxygen diffusion.

The air cathodes demonstrated very high specific capacities of 3471 mAh/g (at 0.1 mA/cm^2) and 1850 mAh/g (at 0.2 mA/cm^2) depending on the weight of the active carbon material in the

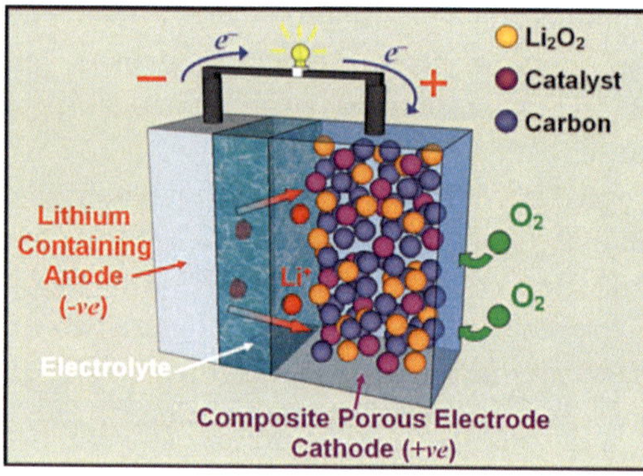

Figure 3.3 Schematic representation of a rechargeable Li–O_2 battery. Reprinted with permission from Ref. 10.

cathode. However, no information was given on the cyclability and lifetime of the system.

In another study it was shown that the nature of the catalyst plays a key role in the Li–air system. By using α-MnO_2 nanowires as catalyst in an oxygen cathode, capacities of 3000 $mAhg^{-1}$ of carbon or 730 and 505 $mAhg^{-1}$ were achieved based on total electrode masses excluding and including O_2, respectively.

The reason why the α-MnO_2 nanowires perform better than other closely related manganese oxides can be explained by their particular crystal structure. The hollandite structure consists of 2×2 tunnels formed by edge- and corner-sharing MnO_6 octahedra, with the tunnels being sufficiently large to incorporate Li_2O. The ability to accommodate both Li^+ and O^{2-} within the tunnels suggests the possibility of incorporating also Li^+ and O_2^{-2}. Such incorporation is not possible in the other MnO_2 polymorphs or other MnO_x materials. Since the major role of the catalyst is to promote the reversibility of Li_2O_2 formation, the ability of the crystal structure to incorporate Li^+ and O^{2-} is believed to be a cause of the good performance of this catalyst[10] as depicted in Fig. 3.3.

3.3.2.1.2 Mg–air battery

Batteries based on Mg are attractive due to their sustainable composition, low cost, and the electrochemical reactivity.[11] Compared to other systems, for example, the standard potential of Mg (-2.37 V) is more negative than that of zinc (-0.76 V) and the electrochemical equivalent of magnesium (0.45 $gA^{-1}h^{-1}$) is less than that of zinc (1.22 $gA^{-1}h^{-1}$). The electrochemical reactions of the Mg–air battery can be expressed by the following equations:

$$\text{Cathode: } O_2 + H_2O + 4e^- \rightarrow 4OH^- \quad P^\circ = 0.40\,\text{V} \tag{3.4}$$

$$\text{Anode: } 2Mg + 4OH^- \rightarrow 2Mg(OH)_2 + 4e^- \quad P^\circ = -2.69\,\text{V} \tag{3.5}$$

$$\text{Overall: } 2Mg + O_2 + H_2O \rightarrow 2Mg(OH)_2 \quad E^\circ = 3.09\,\text{V} \tag{3.6}$$

The major problem in using Mg–air primary batteries is the low Coulombic efficiency of the traditional Mg plate electrode and the high level of polarization during the discharge process. Moreover, the electrolyte for Mg–air batteries must satisfy the objectives of low anodic polarization, a uniform corrosion rate of Mg, and fast sedimentation of the final anodic product ($Mg(OH)_2$) in the electrolyte.

To circumvent these problems, magnesium nanoparticles were synthesized by a vapor transport process and spheres, plates, rods, and sea urchin-like structures were prepared by varying the process parameters,[12] (Fig. 3.4). The electrochemical activity of vapor-deposited Mg structures was found to be significantly higher than that of commercial Mg powder. The energy densities of the system were up to 565 Wh/kg at a discharge current density of 5 mA/cm^2 and 305 Wh/kg at a discharge current density of 50 mA/cm^2. These values are almost an order of magnitude higher than those obtained with a commercial Mg powder.

The best characteristics among the nanostructures were obtained with the sea urchin-like structures, the lowest values were obtained with sphere structures; both materials have particle diameters in the order of 400–500 nm. The high specific surface area of the Mg sea urchin-like nanostructures obviously helps to reduce the real current density for the electrode during the discharge process and leads to a relatively low level of polarization for the Mg electrode, which is favorable for improving the energy density and the high rate discharge capability of the battery.[12]

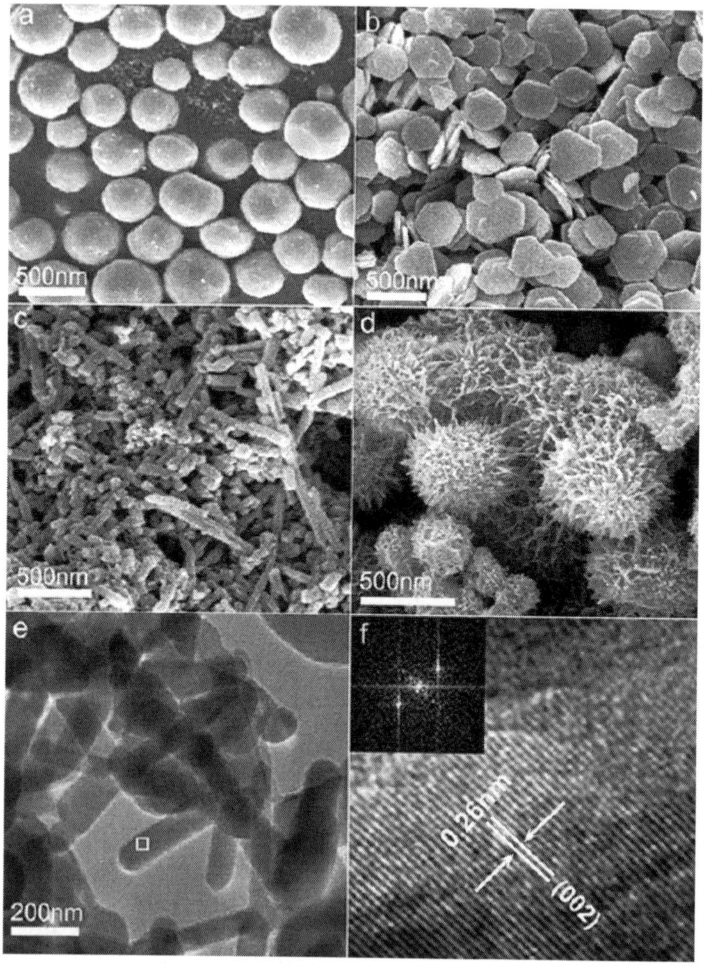

Figure 3.4 (a–d) Scanning electron microscopy (SEM) images of the Mg nanostructures; (e) transmission electron microscopy (TEM) image; (f) high-resolution transmission electron microscopy (HRTEM) image of the white square marked in (e); the inset in (f) shows the corresponding fast Fourier transform (FFT) pattern. Reprinted with permission from Ref. 12.

3.3.2.1.3 Fe–air battery

Similar to the Mg–air batteries, Fe–air batteries are attractive due to their sustainable composition and low cost of the active material. In this system the anodic oxidation of iron proceeds in two main steps:

$$Fe + 2OH^- \leftrightarrows Fe(OH)_2 + 2e^- \quad (3.7)$$

$$Fe(OH)_2 + OH^- \leftrightarrows FeOOH + H_2O + e^- \quad (3.8)$$

Although the theoretical capacity (1200 mAh/g) is 5–6 times below that of the Mg–air battery, the iron system has a certain advantage of showing a better reversible behavior. Moreover, the system offers good resistance against mechanical shocks, vibrations, overcharge, and deep discharge.[13]

Nevertheless, the porous iron electrode in practice suffers from a low charge efficiency, since hydrogen evolution at around −1.2 V from the aqueous KOH electrolyte competes with the discharge reaction and a passivation of the electrode occurs, probably due to the formation of an $Fe(OH)_2$ layer at the surface.

It was found that the charge–discharge performance can be improved by an electrode design where the active material has a high surface area and is in intimate contact with conductive carbon. Various types of carbons, including natural graphite, acetylene black, and several types of graphitic nanofibers (CNF) were loaded with Fe_2O_3 by wet impregnation of the carbon with Fe nitrate and subsequent calcination of the material. Comparison of the carbons employed for the preparation of Fe_2O_3-loaded electrodes indicated that the tubular CNF, acetylene black, and graphite exhibited larger redox currents than vapor-grown CF and platelet CNF. However, although the use of Fe nanoparticles and nanocarbons led to a certain improvement, a strong fading of the redox capacity was still observed. This was attributed to an increase in iron particle size and the dissipation of soluble $HFeO_2^-$ species in the electrolyte during cycling. Cyclic voltammetry revealed that the Fe(II)/Fe(III) redox couple was predominant during cycling, while the Fe/Fe(II) couple became less significant.

In order to circumvent particle growth, Fe_2O_3 was synthesized as nanoparticles inside multi-walled carbon nanotubes (MWCNT) and electrodes were prepared by mixing and rolling 90% of the MWCNT/Fe_2O_3 and 10% polytetrafluoroethylene (PTFE).[14] However, due to the incorporation of the active component in the nanotubes, its accessibility by the electrolyte is hampered. Hence,

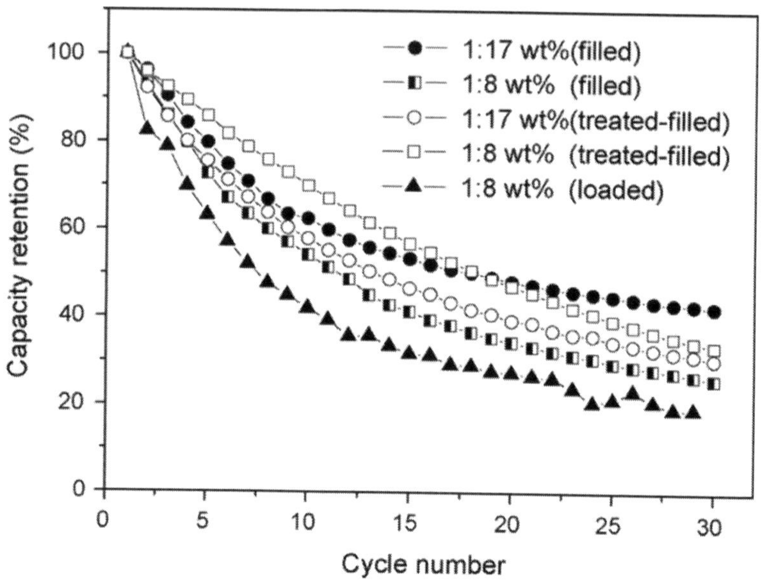

Figure 3.5 Capacity retention of Fe_2O_3-filled CNT and Fe_2O_3-loaded CNT composite electrodes. Reprinted with permission from Ref. 14.

the electrolyte diffusion was improved by forming pores in an oxidative pre-treatment of the nanotubes. The experiments showed that the re-distribution of Fe occurred mainly inside the nanotubes, the particle size was limited, and the dissipation of soluble species was reduced. Comparison with the cycle performance of Fe_2O_3-loaded onto CNTs shows that the capacity of Fe_2O_3-filled CNTs was significantly improved (Fig. 3.5). The data also indicate that the microstructure of the electrode is not only the issue which affects reversibility but also it will be essential to develop an electrolyte with a reduced solubility for $HFeO_2^-$.

3.3.2.2 In Li-ion batteries

3.3.2.2.1 Transition metal oxides

Simple 3d-metal oxides with Fe, Co, Ni, and Cu as metal have long been disregarded as possible reversible electrode materials for rechargeable Li-based batteries because of the absence of interstitial sites within their crystallographic structure for Li insertion–de-insertion and of the non-existence of Li–M alloys. Fuji Photo Film's announcement of the development of lithium batteries containing convertible metal oxides in the mid-1990s caused a renewed interest in lithium alloys as alternative materials for use in the negative electrode of rechargeable lithium cells. Since then, a great deal work has been done on the potential use of TMOs based on Co, Cu, Fe, Ni, and Sn as active electrode material.

A major advantage of TMO-based lithium ion cells over commercial $LiCoO_2$/C cells is that TMO may have about twice the capacity of carbon per unit mass and three times its density. TMO therefore, has about six times the capacity of carbon per unit volume. Prolongation of cycle life of anode materials is another goal. Prominent reversible electrode capacities up to approximately 800 mAhg^{-1} at voltages around 1 V were reached, together with a good prolonged cycling and acceptable rate performance.

Various TMOs, such as cobalt, iron, and copper oxides with different M:O ratios, were examined first and it was found that both the transition metal and lithium undergo a reversible redox reaction, while the oxygen atom is exchanged.[15] In the case of cobalt oxide, the following reversible redox reaction is observed:

$$\mathrm{CoO} + 2\mathrm{Li}^{+}2e^{-} \leftrightarrows \mathrm{Co} + \mathrm{Li_2O} \tag{3.9}$$

$$2\mathrm{Li} \leftrightarrows 2\mathrm{Li}^{+} + 2e^{-} \tag{3.10}$$

$$\text{Total: } \mathrm{CoO} + 2\mathrm{Li} \leftrightarrows \mathrm{Co} + \mathrm{Li_2O} \tag{3.11}$$

Lithium reactivity process in the oxides mainly consists of two steps. The first occurs during the first reduction at a voltage plateau in the 1 V range corresponding to a capacity of two lithium ions. It corresponds to a Li-driven decomposition of the metal oxide into a composite matrix made of metallic particles (M^0) dispersed into Li_2O. The origin of the second step, visualized by a sloping part of the voltage-composition discharge curve down to 0 V, involves an organic-type layer growth that disappears upon the subsequent charge.

The size of the cobalt-containing particles is greatly reduced when the material is cycled. Starting from initial particle sizes in the order of 100–200 nm, the reduced metallic Co particles are in the order of 1–2 nm only and embedded in a Li_2O matrix. Nanoscale effects due to the very small particle size of the cobalt in the Li_2O and a catalytic function of the nanoparticles were considered responsible for the observation of an electrochemically active and reversible Li_2O behavior which had not been reported before.

The reversibility of such conversion systems was explained by an acido-basic concept which involves O^{2-} anion transfer (Lux–Flood model) and is based on the acidity or basicity of oxides (Smith scale). All oxides that react reversibly with Li are basic as is the Li_2O itself.[16]

The concept can be generalized and applied to other systems which are based on nitrides, sulfides, and selenides. Thus, the general reaction can be written as follows, with M = metal and X = basic anion.

$$M^{n+}(X) + ne^- + nLi^+ \leftrightarrows M^0 + nLi(X) \quad (3.12)$$

The proton affinity of the anion is considered to be the decisive quantity in this case and it was shown that the intrinsic proton affinity measured in the gas phase (Table 3.2) is useful to basically explain the reactive behavior of the inorganic compounds. A reversible reaction is possible when metal reduction is thermodynamically feasible ($\Delta_r G < 0$, Table 3.3) and the anion can be exchanged easily due to its basicity.

3.3.2.2.1.1 SnO_2 electrodes

Various efforts have been made to implant Sn-based anode into a Li-secondary battery, since it was first reported by Idota *et al.* in 1997.[17] The theoretical capacity of the SnO_2 anode is 781 mAh/g and can be

Table 3.2 Values of the intrinsic proton affinity for chalcogenide and nitride ions given by Waddington

Anion	Proton affinity in the gas phase (kJ/mol)
N^{3-}	3084
O^{2-}	2318
S^{2-}	2300
Se^{2-}	2200

Table 3.3 Calculated Gibbs free energy change ($\Delta_r G$) and emf (E) values *vs.* Li^+/Li^0 for the reaction $M_xO_y + 2yLi^0 \rightarrow yLi_2O \ldots \Delta_r G$

Compound	$\Delta_r G$ (kJ mol^{-1})	$E = E_{eq}$ (V)
TiO_2	−233.53	0.605
MnO	−199.17	1.032
FeO	−317.10	1.643
NiO	−346.24	1.794
CoO	−347.01	1.798
Cu_2O	−414.18	2.146
CuO	−428.26	2.219

Note: Data obtained with permission from Ref. 16

explained by the following mechanism:

$$SnO_2 + 4Li \rightarrow Sn^0 + 2Li_2O \tag{3.13}$$

The reduced Sn^0 intercalates Li:

$$Sn + xLi \rightarrow Li_xSn \ldots 0 \leq x \leq 4.4 \tag{3.14}$$

The first electrodes exhibited a higher capacity (>600 mAh/g) than the conventional graphite anode (372 mAh/g), but showed fast capacity fading upon cycling. A drastic volume change between Sn and $Li_{4.4}Sn$ of 358% occurs, which has been deemed the main reason for the poor cycling stability of Sn oxide anodes. As a consequence of the volume change, the particles become detached and electrically inactive. This was confirmed by diffraction studies where small and active Sn particles were found to aggregate into larger and inactive Sn clusters during cycling.[18]

Efforts were made afterwards to circumvent the problem of particle detachment. With respect to cycle capacity and charge–discharge rate, SnO_2 nanowires were found to have a clear advantage over SnO_2 thin films and bulk materials.[19] Figure 3.6 shows that a high reversible capacity and Coulombic efficiency of 97% could be obtained even at high charge–discharge rates of 8°C. The nanowires were prepared by immersion of $SnCl_2$ solution in a porous membrane made of polycarbonate. The solvent was evaporated and the polymer was burned away in oxygen plasma. Then, the sample was annealed in air. In this way, cylindrical SnO_2 nanofibers were obtained, which were attached to a Pt collector as substrate.

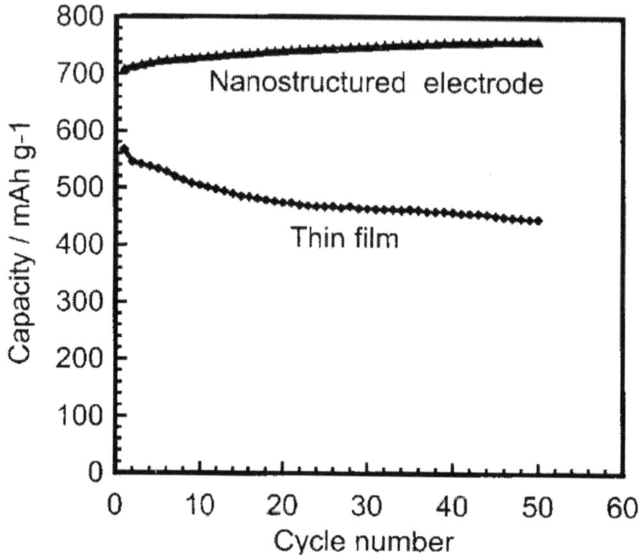

Figure 3.6 Capacity *vs.* cycle number for the nanostructured SnO_2 and thin-film control electrodes at a charge–discharge rate of 8°C over the potential window of 0.2–0.9 V. Reprinted with permission from Ref. 19.

It has also been demonstrated that the size of the SnO_2 nanoparticle has a significant influence on the cycling behavior. Only 3% of the capacity was retained after 60 cycles for an 8 nm-sized anode material. Using a 4 nm-sized anode material, 73% were retained. Smaller particles in the order of 3 nm size showed both a very high initial capacity of 740 mAh/g and negligible capacity decay after 60 cycles even at high charge–discharge rates. The smallest Sn nanoparticles undergo a reversible volume change without aggregation into larger Sn clusters during cycling, which might explain the superior electrochemical properties of the smallest particles.[20]

Improved cycling behavior was also observed when SnO_2 nanoparticles were infiltrated into and precipitated onto multi-walled carbon nanotubes.

It can also be observed with larger crystallites when the embedding matrix prevents agglomeration of the SnO_2 nanoparticles. For example, a carbon matrix produced *in situ* in a spray pyrolysis process of $SnCl_2$ may have a stabilizing effect on the system. However, the amount of carbon has to be high, >40%, in order to lead to significant improvements.

3.3.2.1.2 Metal oxysalts

Recently, it was found that not only oxides, but also transition metal oxysalts, such as carbonates and oxalates, can be electrochemically active and a conversion reaction takes place where Li reacts reversibly with the complex anion. As compared to other displacement reactions with metal oxides and fluorides, the use of a low molecular weight salt does not penalize the capacity, while giving extra stability due to the formation of lithium carbonate as the main side product.[21]

The electrochemical reactions of $MnCO_3$ and FeC_2O_4 can be written as

$$MnCO_3 + 2Li \leftrightarrows Li_2CO_3 + Mn \tag{3.15}$$

$$FeC_2O_4 + 2Li \leftrightarrows Fe + Li_2C_2O_4 \tag{3.16}$$

The theoretical capacities of the two reactions are 372 and 466 $mAhg^{-1}$, respectively.

Both systems exhibit an electrochemical behavior (Fig. 3.7) which is different to that of the corresponding oxide. The voltage profile of the carbonate shows a small plateau at 0.9 V and a second extended plateau at 0.4 V. Whereas, the voltage profile of the first discharge is characterized by a sharp voltage drop and the initial reversible capacity of a carbonate is lower compared with the oxide, capacity retention is significantly better for $MnCO_3$. The large irreversible capacity of the carbonate can be ascribed to irreversible reactions with the electrolyte during the first discharge until Mn nanoparticles are protected by a coating layer of lithium carbonate. Moreover, electrolyte degradation was assumed to affect the reversible capacity of the system.

For FeC_2O_4 a first extended plateau at about 1.2 V *vs.* Li is observed, followed by a smooth decrease in voltage to 0 V. The difference between Fe and Mn compounds is presumed to be a consequence of a different catalytic effect of manganese as compared with iron metal nanoparticles.

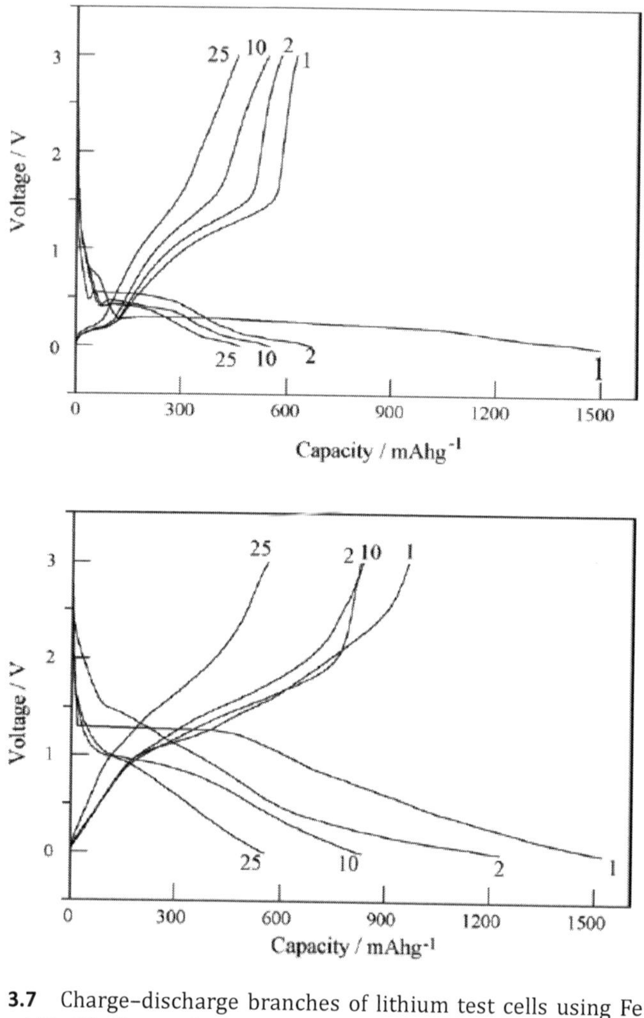

Figure 3.7 Charge–discharge branches of lithium test cells using FeC_2O_4 (top) and $MnCO_3$ (bottom) as initial active cathode material. Reprinted with permission from Ref. 21.

The rate performance of iron oxalate is strongly influenced by the charge–discharge rates. While the initial capacities are significantly higher at low rates (880 mAh/g at C/5 *vs.* 350 mAh/g at 5°C), the difference shrinks to 450 mAh/g *vs.* 350 mAh/g after 60 cycles.

3.3.2.1.3 Metal hydrides

So far, metal hydrides have only been considered for use in nickel-metal hydride batteries where the anode consists of a $LaNi_5H_x$-based metal hydride. During discharge, the hydride ions are oxidized and react with OH^- to H_2O. At the cathode, NiOOH is reduced to $Ni(OH)_2$. The charge transfer reaction in the system therefore is based on protons.

Recently, an electrochemical reactivity of a metal hydride with lithium electrolyte was discovered and it was demonstrated that metal hydrides can also play a role in LIBs.[22] Several binary (MgH_2, NaH, TiH_2) and ternary ($LaNi_{4.25}Mn_{0.75}H_5$, $Mg_2NiH_{3.7}$) hydrides were tested and each was found to be electrochemically active toward lithium.

Magnesium hydride (MgH_2) is regarded as particularly attractive due to its high theoretical gravimetric (2038 mAh/g) and volumetric (2878 Ah/L) capacity. Moreover, magnesium is abundant, available at low cost, and environmentally benign. The electrochemical reaction results in the formation of a composite containing Mg embedded in a LiH matrix, which on charging converts back to MgH_2. Compared to other conversion systems, the Mg particles produced in the discharge reaction are large, in the order of 20 nm to 60 nm.

$$\mathrm{MgH_2} + 2\mathrm{Li}^+ + 2e^- \leftrightarrows \mathrm{Mg} + 2\mathrm{LiH} \tag{3.17}$$

The reaction is spontaneous, as is apparent from its Gibbs free energy of reaction, $\Delta G_r = -108.2$ kJmol^{-1}. The free energy corresponds to an equilibrium potential of $E_{eq} = 0.560$ V *vs.* Li^+/Li^0 from Nernst law ($E = -\Delta G_r/2F$) and is similar to the value found in the experiment at a discharge rate of 1 Li in 10 h (0.537 V) (Fig. 3.8).

The voltage-composition trace of the MgH_2–Li system discharged at very low currents and between 3 V and 0.005 V shows an initial abrupt voltage drop down to a first plateau which is located at around 0.44 V. The plateau extends to around Li/Mg = 2 and can be explained by the conversion reaction above. Another fast decay to a lower voltage plateau (near 0.095 V) extends up to Li/Mg = 2.5 and is associated with a shift of the position of the Bragg peaks of Mg in the XRD, while those of LiH remain unaffected. The observation is

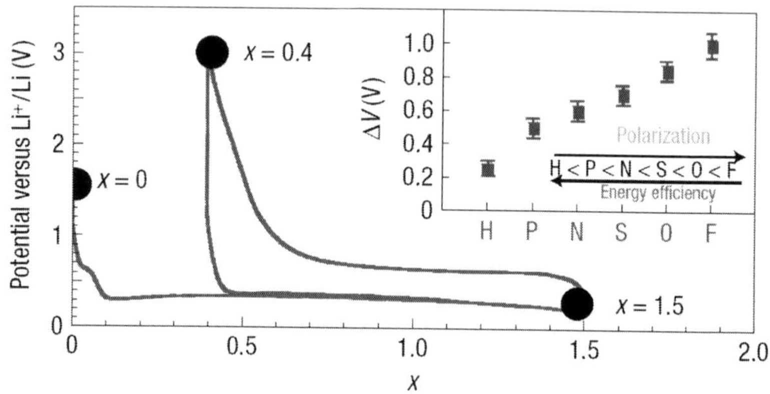

Figure 3.8 Evolution of the potential (V) as a function of x for a Li/MgH_2 cell that was cycled down to $x = 1.5$ at a rate of one lithium in 10 h. (Inset: The evolution of the polarization in volts (1 V) for various Li/M_yX_z cells having X = H, P, N, S, O, F and run under similar current rate conditions; M = metals. Reprinted with permission from Ref. 22.

therefore explained by the onset of a solid dissolution of Li in Mg. At Li/Mg >2.5, the X-ray reveals a two-phase process corresponding to the coexistence of Mg hcp- and Li bcc-type solid solutions and one single phase.[22]

Good reversibility in the first 50 cycles was obtained when the electrode was prepared by pressing the powder into copper foam and limiting the capacity to 520 mAh/g.

An interesting feature of the system is the very low polarization compared to other conversion electrodes (Fig. 3.8, inset). This has been attributed to the mass transport kinetics where the mobility of the ionic species plays an important role. Hydride ions are supposed to move faster than O^{2-} and F^- and it is believed that there are lower kinetic barriers in the case of hydride systems. But as the role of anion diffusion in the rate-determining step has not been elucidated in detail, the statement remains to be verified.

3.3.3 *As Cathode Materials*

The development of efficient cathode materials may be of particular importance, because current cathode materials have energy densities which are smaller than those of the anode by a factor of two. An improvement of the energy density of the cathode will therefore have a large impact on the overall capacity of the cell.

3.3.3.1 Sulfides

It was established in the early days of lithium rechargeable batteries that transition metal sulfides, such as titanium or molybdenum sulfides, did cycle well and these were used in lithium metal rechargeable batteries. For a low-cost system, iron sulfide was preferred and first studies using a room-temperature secondary Li–FeS_2 battery were performed in the 1970s.[23]

FeS_2 has been studied and used in Li-Al–FeS_2 molten salt rechargeable batteries at 400°C and in Li/FeS_2 primary batteries at ambient temperature for the last two decades. Li/FeS_2 cells have an excellent reversibility at 400°C, but only a limited reversibility has been obtained at ambienttemperature. There has been renewed research interest and efforts were undertaken to improve the rechargeability of the Li–Fe–S system in lithium polymer batteries at moderate temperatures around 55–135°C for potential applications in electric vehicles.

$$FeS_2 + 4Li \rightarrow Fe + 2Li_2S \tag{3.18}$$

The chemical and structural details of intermediates are still discussed controversially and there has been a debate of whether $Li_3Fe_2S_4$, Li_2FeS_2-type, and/or $Fe_{1-y}S$-type phases are present as intermediate phase. One of the reasons is that the chemistry during the electrochemical process is complex and may vary depending on the electrolyte properties and the material properties of the iron sulfide (e.g., natural pyrite or synthetic FeS_2). Different intermediates of $LiFe_xS$-type were obtained depending on the nature of the sulfide. A one-step reduction process at 1.5 V was observed at room temperature for natural FeS_2 electrodes and synthetic FeS_2 electrodes exhibited a two-step reaction at 1.7 and 1.5 V (Li anode, 3 mA/g).[24] In both cases, x was 3.3. Higher current densities decreased the efficiency of the electrode due to the internal cell resistance, and a value of $x = 2.3$ was obtained at 213 mA/g.

It was shown that lithium reduction of FeS_2 pyrite at 1.5 V and higher rates led to the direct formation of X-ray amorphous Fe and Li_2S platelets with thicknesses in the order of 20 nm. Similar plate-like crystals of Li_2S were found to cover the surface of larger fractured FeS_2 particles which were discharged at higher rates.

In order to circumvent safety problems and to reach a longer cycle life, lithiated iron sulfide, Li_2FeS_2, was proposed as electrode material.[25]

The sulfide is supposed to react according to the following reversible reaction

$$Li_2FeS_2 \leftrightarrows 2Li + FeS_2 \tag{3.19}$$

Two electrons are exchanged in the process; the specific capacity is 400 mAh/g (calculated) at 2.3 V (observed), which gives an energy density of 921 Wh/kg. A major problem is the synthesis of the Li_2FeS_2 which needs high temperatures and long reaction times, followed by extraction and purification of the product.

3.3.3.2 Fluorides

Metal fluorides are an interesting class of compounds which may be used for conversion electrodes. Due to the highly ionic nature of their metal-fluoride bond, they should theoretically exhibit a much higher output voltage and lithium storage capacity than any other systems. However, many of the metal fluorides have a limited electrochemical activity with lithium due to their poor electronic conductivity brought about by their large band gap.

The first report of metal fluorides as cathode materials for rechargeable Li batteries was published in 1970 by Seiger *et al.* where NiF_2 was shown to be discharged at low potential with a limited degree of reversibility. Two decades later, Arai *et al.* reported an electrochemical activity of TiF_3, VF_3, and FeF_3.[26] Regardless of their high theoretical capacities of up to 767 mAh/g for TiF_3 (Table 3.4), these compounds reached only a moderate reversible capacity of 80 mAh/g in the potential range between 4.5 V and 2.0 V, which corresponds to the intercalation portion of the M^{3+}/M^{2+} redox couple.

Table 3.4 EMF of the electrochemical reactions between various metal fluorides and lithium and their corresponding Li storage capacities (data from Ref. 27)

MF_n	ΔG_f [kJ]	E° [V]	Capacity [mAh/g]	MF_n	ΔG_f [kJ]	E° [V]	Capacity [mAh/g]
AgF	−187	4.156	211	FeF_3	−972	2.742	712
CoF_3	−718	3.617	694	ZnF_2	−713	2.404	518
CuF_2	−491	3.553	528	MnF_2	−807	1.919	577
SnF_2	−601	2.984	342	VF_3	−1226	1.863	745
NiF_2	−604	2.964	554	TiF_3	−1361	1.396	767

The redox potential of isolated Fe(III) in strong acids is +0.770 V *vs.* the hydrogen electrode. Together with the Li/Li^+ potential of −3.045 V, the cell voltage is 3.815 V for the reaction

$$Fe^{3+} + Li \leftrightarrows Fe^{2+} + Li^+ \tag{3.20}$$

Recently, also heterogeneous conversion reactions of TiF_3 and VF_3 at room temperature were reported.[27] In this work LiF nanoclusters were found to be electrochemically active in M/LiF composites in which X-ray amorphous LiF and metal are dispersed on an atomic or lower nanometer scale. The reversible Li storage capacities of TiF_3 and VF_3 observed were in the order of 500–600 mAh/g between 4.3 V and 0.02 V. The electrochemical activity of LiF and the extractable capacities varied strongly with the preparation of the Ti/LiF composite and were found to depend on the microstructure and crystallite size of the material.

A detailed study of the TiF_3/Li system revealed a first plateau above 1.4 V during discharge, with a capacity of about 120 mAh/g corresponding to storage of 0.5 Li in $Li_{0.5}TiF_3$. The second region is a wide plateau at 0.95 V which narrows after the first cycle. The plateau is an indication of a two-phase region and can be assigned to the following reaction:

$$MF_n + nLi^+ + ne^- \leftrightarrows M/LiF \tag{3.21}$$

A third region is a low-voltage slope region with an additional capacity of 300 mAh/g in the first cycle. This feature is highly reversible and was also found for TMO systems. The origin of the sloped region is believed to be an interfacial interaction of Li within the nanocomposite's matrix, which possibly leads to a distinct local

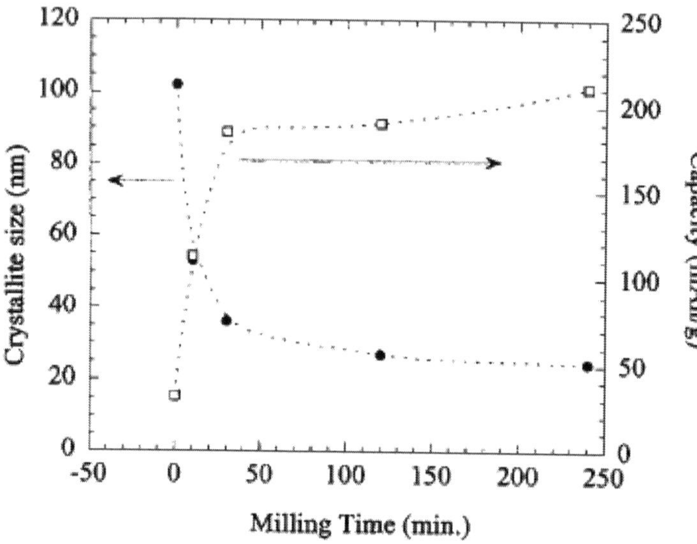

Figure 3.9 Crystallite size and specific capacity plotted as a function of composite milling time. Reprinted with permission from Ref. 28.

charging, in addition to the formation of a solid–electrolyte interface (SEI).

Badway *et al.*[28] demonstrated at the same time that an FeF_3/C (85/15 wt%) nanocomposite can reach a specific capacity of about 200 mAh/g for the Fe^{3+}/Fe^{2+} couple in the region between 4.5 and 2.5 V at room temperature.

The composites were prepared by high-energy ball milling under an inert gas atmosphere of iron trifluoride and carbon in the form of expanded graphite, carbon black, or activated carbon. Nanodomains of FeF_3 (<30 nm) were formed in the mechanical synthesis, which obviously were able to support appreciable electron conductivity. This was attributed to the shortening of bulk path lengths and augmentation of electron tunneling. In addition, the nanodomains can electrically connect to each other by the use of highly conductive carbon. The correlation of the FeF_3 crystallite size with the specific capacity measured in the 2.8–3.5 V range (Fig. 3.9).

In order to increase the kinetics of the reaction, the temperature level was increased to 70°C and cycle tests were performed. It was found that under these conditions, three-electron reduction

(Fe^{3+}/Fe^{0}) can be achieved (4.5 to 1.5 V) with a specific capacity in the first cycle of 770 mAh/g FeF_3, which is close to the theoretical value and in accordance with the full conversion of iron fluoride.

In this case a two-step reaction was observed. The Fe^{3+} to Fe^{2+} reaction mechanism was found to consist of a two-phase transformation followed by a topotactic insertion reaction as suggested by XRD and electrochemical characterization:

$$Li^{+} + e^{-} + FeF_3 \rightarrow LiFeF_3 \tag{3.22}$$

Reducing the Fe^{2+} in $LiFeF_3$ to Fe^{0} would necessitate the breaking of all Fe–F bonds and result in significant structural rearrangement. In fact, *ex situ* XRD of the reaction products of the fully lithiated material showed that only LiF and Fe were present. Hence, the second step proceeds according to

$$2Li^{+} + 2e^{-} + LiFeF_3 \rightarrow 3LiF + Fe \tag{3.23}$$

Although the systems seem to be very interesting and specific capacities were exceptionally high and efficiency was good, a noticeable fade in capacity with cycle number existed and the full capacity was available at low rates and elevated temperatures only.

To overcome the stability problem, it is obvious that an alternative approach to material synthesis would be needed. In this respect a recent report suggested that iron oxide encapsulated in meso- and macroporous carbon can be used as anode in Li batteries and reaches a greatly improved reversibility and rate performance.[29] At the same time, a similar structure for a cathode material based on iron and lithium fluoride was synthesized and investigated. It was demonstrated that encapsulation of transition metal–metal fluoride in nanocarbon might be an effective strategy to improve the cycling performance of such a cathode material.[30]

The synthesis of a Fe/carbon/LiF nanocomposite was performed by pyrolysis of a pre-milled ferrocene/LiF mixture in an autoclave. The treatment led to the formation of 5–20 nm sized Fe nanoparticles which were intimately embedded in multi-walled nanocarbon structures. HR-TEM studies showed that the particles were encapsulated in onion-like graphite layers which were partly defective (Fig. 3.10). LiF was in close vicinity to the iron and physisorption analysis revealed that the composite possesses an

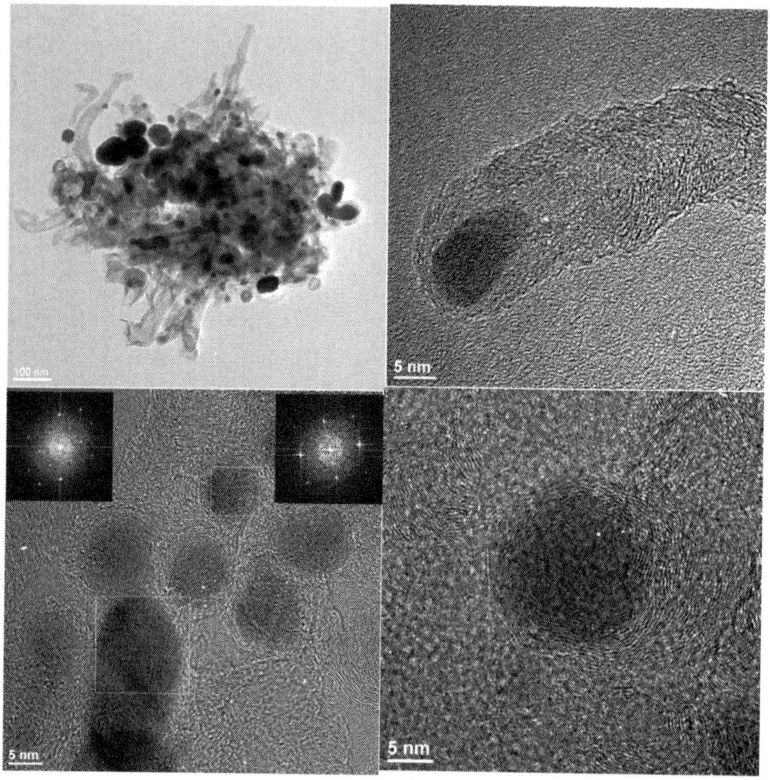

Figure 3.10 TEM characterization of the as-prepared carbon Fe–LiF composites. Top left: BF-TEM overview image. Top right: Iron-rich nanoparticle inside a tubular graphitic structure. Bottom: Iron-rich particles confined by onion-like graphite. Based on the observed lattice fringes, the two particles indicated were identified to be α-Fe [001] and Fe_3C [001].

interesting porosity for materials transfer (mesoporosity together with the macroporosity given by the tubular structure).

The particular microstructure of the material is likely to provide for a protection against grain growth, segregation of active materials, etc., but also allows for a contact to the electrolyte, materials transfer, and can maintain microstructural stability and electronic conductivity. In this respect, encapsulation of metal–metal fluoride in various carbon materials, such as carbon nanotubes (CNTs),

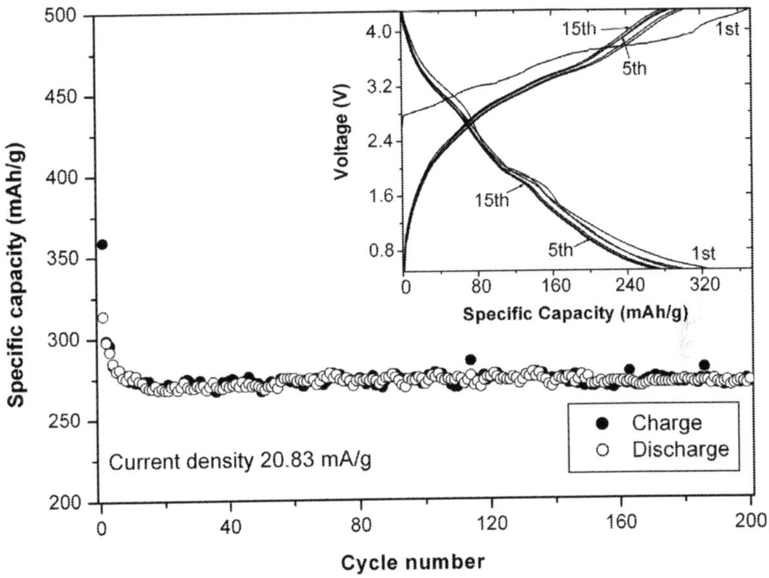

Figure 3.11 Specific capacity of the LiF-Fe–C composite as a function of cycle number, measured at room temperature. Inset: Galvanostatic charge–discharge time *vs.* voltage.

carbon fibers, and porous carbon materials, may lead to improved properties.

Cycling studies at room temperature indicate that after an initial capacity decrease in the first 10–15 cycles the material was stable for the next several hundred cycles (the first 200 cycles are shown in Fig. 3.11). Interestingly, a conversion between Fe and FeF_3 was detected by Mossbauer spectroscopy, also at room temperature. The reason why only a part of the material seems to be active is not clear yet and will be subject of further investigations. It is likely that the materials transfer is prohibited for a part of the sample by the defect-free encapsulation of the iron nanoparticles.

3.3.4 *Solid-State Reactions in Conversion Electrodes*

Chemical reactions of solid phases with reactants in the gas, in the liquid or in the solid phase are the basis of the conversion-based electrochemical storage systems described above. The factors which may influence reaction kinetics and thermodynamics therefore are of particular importance. Knowledge of the underlying thermodynamics, the influence of the particle size, the rate-limiting processes at interfaces, phase transformations, and materials and electronic transport properties is the prerequisite for understanding and tuning the materials properties in order to be able to design and develop an efficient electrochemical storage system.

3.3.4.1 Thermodynamics

The thermodynamics of a redox reaction may be regarded a basic parameter set which defines the theoretical or optimal properties of a system. The theoretical potential of an electrode can be calculated on the basis of the Nernst equation and the Gibbs free energy of the reaction. Such estimations are often done based on literature data of standard enthalpies and entropies of formation of reactants and products. However, there may be deviations in the electrochemical experiment, which may be due to the following reasons:

(1) The solid-state reaction leads to the formation of intermediates, the properties of which may have to be taken into account.
(2) There may be parallel reactions leading to an ambiguous mixture of products and energetic situations.
(3) There may be size and interface effects in case nanoparticles are involved in the reaction.
(4) Energetic barriers may lead to over potentials which reduce the system's efficiency.

The first three points influence the thermodynamics and the last point is a kinetic argument.

3.3.4.1.1 Intermediates and parallel reactions

If a reaction system runs through an intermediate state, this should basically not change the overall values for the electrochemical potential and the capacity. However, the situation may be different, if the intermediate state has properties which prevent the system from proceeding further along the reaction path or puts other restrictions on it.

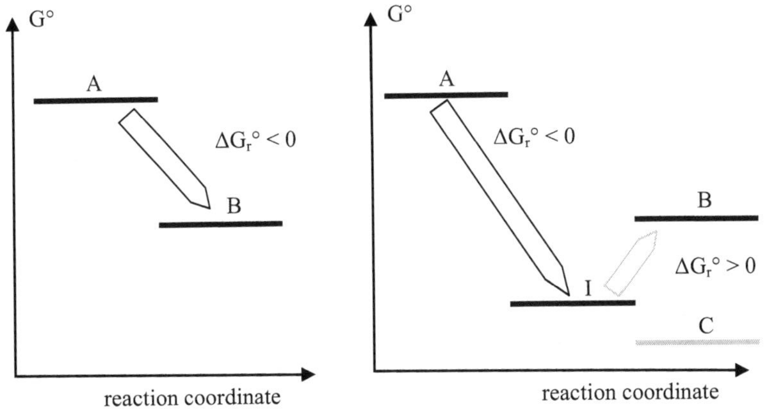

Figure 3.12 Standard free energy of an exergonic chemical reaction A → B without intermediate (left) and with a stable intermediate I (right).

A simplified representation of the energetic situation where a stable intermediate may form is depicted in Fig. 3.12. In both cases, the reaction A → B exhibits the same overall standard free energy of reaction. However, if a stable intermediate (I) forms, the second step of the reaction I → B is endergonic and therefore thermodynamically unfavorable. As a result, the reaction stops at I and the electrochemical data are correlated with the transition of A → I rather than A → B. Such processes can occur, if there is an element in the reaction mixture, which forms very stable compounds with one of the elements of the electrode material so that, at least gradually, A → B becomes less and less available and the system degrades or alters its properties. The capacity may fade, if reaction A → I is electrochemically not reversible.

For example, another problem may occur, if the intermediate has structural properties which prevent any further reaction of I with Li. One possibility is that the first transition leads to the formation of an intermediate which cannot incorporate Li in a following intercalation step. However, the reaction systems (e.g., of oxides, fluorides) investigated so far rather show an intercalation step first, followed by a conversion step.

Table 3.5 Instrumental methods for analysis of structural parameters and chemical composition

Parameter	Investigation method
Phase transformation temperature, decomposition temperature	Differential scanning calorimetry (DSC) Thermogravimetric analysis (TGA) Thermal desorption spectroscopy (TDS)
Crystal structure/long-range order	Single crystal and powder X-ray diffraction (XRD) Neutron diffraction (ND)
Molecular environment	Infrared spectroscopy (IR) Raman spectroscopy
Local order	X-ray absorption spectrometry (XANES, EXAFS) Nuclear magnetic resonance (NMR) Inelastic neutron scattering (INS)
Specific surface area	Physisorption method (BET)
Particle size	Electron microscopy (HR-TEM) Powder X-ray diffraction Small-angle X-ray or neutron scattering (SAXS, SANS)

The particular properties of the intermediate may also alter the reaction path and force it along a reaction $A \rightarrow I \rightarrow C$, if this is thermodynamically more favorable.

It is quite possible that a series of intermediate phases forms during the electrochemical process, their exact nature being controlled by the differences in the kinetics of the diffusion of the different ions or atoms in the system. Parallel reactions to products with similar thermodynamic stability may lead to a degradation of the reversible properties, if one of the products is not electrochemically reversible. Hence, knowledge of the real reaction partners and of their properties is the key for understanding the electrochemical processes in the system and elucidating the reaction mechanism. This is possible by making *ex situ* and/or *in situ* experiments with methods that supply information about the chemical composition, structural, and thermal properties of the compounds in the reaction mixture. A number of examples have been presented in literature based on the various methods as listed in Table 3.5.

3.3.4.1.2 Size and interface effects

Particle size and the nature of interfaces are factors which may influence the thermodynamic properties of a system due to an increasing significance of interfaces, edges, and corners. The smaller the particle is, the more the energy from the surface or interface of the particle can contribute to the total energy.

In a simple example where a small metal particle M and a species X react to a compound MX, the excess surface energy of both reactants and products must be taken into account. Thus, the molar free energy of formation of a particle MX may be described in a simplified manner on the basis of the Gibbs concept in which excess quantities are addressed:

$$\Delta G\,(r) = \Delta G_0(r) + RT\ln\left(\frac{a_{MX}}{a_M a_X}\right) + \frac{3V_M E_{M\rightarrow MX}(\gamma, r)}{r} \qquad (3.24)$$

where, V_M is the molar volume, r particle radius, a chemical activity, and γ the surface free energy or surface tension. $E_{M\rightarrow MX}$ denotes a surface energy term which depends on the surface free energies of the metal compound and the metal particle, on the molar volumes of the two solid reaction partners, and on an additional energy term E_{ads} (Eq. 3.25).

$$E_{M\rightarrow MX}(\gamma, r) = \left[\gamma_{MX}(r)\cdot\left(\frac{V_{MX}}{V_M}\right)^{2/3} - \gamma_M(r)\right] + E_{ads} \qquad (3.25)$$

The additional term E_{ads} takes into account that binding of X at the surface of both the metal and MX will reduce the respective surface energy by minimizing the excess energy γ arising from unsatisfied bonds of surface atoms. It should be mentioned that the particular combination of the various parameters in Eq. 3.25 may shift the reaction to the right or to the left side depending on the actual system and its interaction with its local environment or interface. Thus, E_{ads} is a term which may increase or decrease the excess energy difference of both sides of the reaction.

For a more detailed and comprehensive overview of interfacial thermodynamics and the thermodynamics of charge carriers (nano-ionics), the reader is referred to a recent review and references therein.[31]

3.3.4.2 Kinetics

The overall kinetics or macrokinetics of an electrochemical process with a conversion reaction is determined by the kinetics of several subsequent processes (microkinetics). These involve

- Lithium transport in the electrolyte.
- Lithium transfer from the electrolyte to the particle interface.
- Lithium transfer from the interface to the bulk.
- Mass transport due to phase segregation.

In parallel, electronic transport occurs:

- Electron transfer from the collector to conductive parts of the electrode material.
- Electron conduction in the conductive part of the material.
- Electron transfer to the non-conductive, active material.
- Electron transport in the active material.

In both the ionic and the electronic part of the process, there are steps which may be fast or slow depending on intrinsic properties of the materials and/or the design of the electrode. Consequently, charging and discharging can be greatly improved by targeted acceleration of the slowest kinetic step. A recent example is the ultra-fast charging of $LiFePO_4$ by creating a fast ion-conducting particle surface phase through controlled off-stoichiometry.[32]

Charge carriers in nanosized systems in which the spacing of adjacent interfaces is on that scale typically exhibit interfacially dominated transport properties or even mesoscopic transport properties. The kinetics is normally changing heavily when the grain size of a solid reaction partner is reduced down to the nanoscale. Measuring and modeling of kinetic data can be useful to learn about the influence of reduced diffusion lengths, increasing fraction of grain boundaries or other factors, and thus, obtain mechanistic insight in the materials transformation. The observed kinetics reflects the mechanism of the rate-limiting step in the whole materials transformation which normally is a sequence of events. Faster reaction steps are not visible in the macrokinetics, but may also be less important in case the investigation aims at recognizing the slowest step and at finding ways to increase the rate of it.

In principle, there are several methods for determining the kinetics of a solid-state reaction. It is possible to conduct either an isothermal experiment by keeping the thermal status of the sample and detecting the charge state or the phase composition with time or different temperature ramps are used for heating up and decomposing the sample. Generally, the transformation of solids can be described by a reaction rate $\mathrm{d}f/\mathrm{d}t$ which depends on the temperature T and the reacted fraction f in the following way.

$$\frac{\mathrm{d}f}{\mathrm{d}t} = y(f) \cdot k(T) \tag{3.26}$$

The temperature dependence is generally assumed to follow an Arrhenius law

$$k = k_0 \cdot \exp\left(-\frac{E_\mathrm{A}}{RT}\right) \tag{3.27}$$

where E_A is an activation barrier, R the gas constant, and k_0 a constant pre-exponential factor. In an ideal isothermal experiment, $k(T)$ is a constant and the function $y(f)$ can be determined easily. E_A can be determined in a straightforward manner by measuring k at different temperatures.

For modeling purposes, care must be taken that the thermodynamic driving force is constant throughout the experiment. The transformation reactions can possibly be described by any of several kinetic models. These include diffusion, moving boundary and nucleation, and growth. To determine which, if any, of these kinetics models describe these reactions, it is necessary to construct plots corresponding to the theoretical equations. The equations are summarized below.

$$(1-f)^{1/3} = 1 - \frac{\sqrt{kt}}{R} \tag{3.28}$$

$$f = 1 - \exp(-kt^n) \tag{3.29}$$

$$(1-f)^{1/3} = 1 - \frac{k}{R}t \tag{3.30}$$

Equation 3.28 corresponds to a diffusion-controlled process; Eq. 3.29 depicts a process that is limited by a reaction at a moving boundary; and Eq. 3.30 represents a nucleation- and growth-controlled process. If diffusion was controlling the rates, then, according to Eq. 3.28, a plot of $(1-f)^{1/3}$ *vs.* time$^{1/2}$ should be linear.

3.4 Summary

In summary, this chapter outlines various concepts involved in the nano-aspects of conversion materials for electrochemical energy storage.

References

1. C. Burda, X. B. Chen, R. Narayanan, M. A. El-Sayed, *Chem. Rev.* 105, 1105–1102 (2005).
2. U. Eberle, M. Felderhoff, F. Schüth, *Angew. Chem. Int. Ed.* 48, 2–25 (2009).
3. M. Broussely, G. Archdale, *J. Power Sources* 136, 386–394 (2004).
4. D. Linden, T. B. Reddy (Eds.), *Handbook of Batteries*, 3rd ed., McGraw-Hill, New York (2002).
5. E. L. Littauer, K. C. Tsai, *J. Electrochem. Soc.* 124, 850 (1977).
6. K. M. Abraham, Z. Jiang, *J. Electrochem. Soc.* 143(1), 1–5 (1996)
7. J. Read, *J. Electrochem. Soc.* 149(9), A1190–A1195 (2002).
8. J. Read, K. Mutolo, M. Ervin, W. Behl, J. Wolfenstine, A. Driedger, D. Fostera, *J. Electrochem. Soc.* 150(10), A1151–A1156 (2003).
9. A. Dobley, C. Morein, R. Roark, K. M. Abraham, *42nd Power Sources Conference*, Philadelphia, PA (2006).
10. A. Débart, A. J. Paterson, J. Bao, P. G. Bruce, *Angew. Chem.* 120, 4597–4600 (2008).
11. S. Sathyanarayana, N. Munichandraiah, *J. Appl. Electrochem.* 11, 33–39 (1981).
12. W. Li, Ch. Li, Ch. Zhou, H. Ma, J. Chen, *Angew. Chem. Int. Ed.* 118, 6155–6158 (2006).
13. C. A. Caldas, M. C. Lopes, I. A. Carlos, *J. Power Sources* 74, 108–112 (1998).
14. B. T. Hang, H. Hayashi, S.-H. Yoon, S. Okada, J. Yamaki, *J. Power Sources* 178, 393–401 (2008).
15. P. Poizot, S. Laruelle, S. Grugeon, L. Dupont, J.-M. Tarascon, *Nature* 407, 496 (2000).
16. P. Poizot, S. Laruelle, S. Grugeon, J.-M. Tarascon, *J. Electrochem. Soc.* 149, A1212–A1217 (2002).
17. Y. Idota, T. Kubota, A. Matsufuji, Y. Maekawa, T. Miyasaka, *Science* 276, 1395–1397 (1997).

18. I. A. Courtney, W. R. McKinnon, J. R. Dahn, *J. Electrochem. Soc.* 146, 59 (1999).
19. N. Li, Ch. Martin, B. Scrosati, *Electrochem. Solid State Lett.* 3(7), 316–318 (2000).
20. Ch. Kim, M. Noh, M. Choi, J. Cho, B. Park, *Chem. Mater.* 17, 3297–3301 (2005).
21. M. J. Aragon, B. Leon, C. Perez-Vicente, J. L. Tirado, *J. Power Sources* 189, 823–827 (2009).
22. Y. Oumellal, A. Rougier, G. A. Nazri, J.-M. Tarascon, L. Aymard, *Nature Mater.* 7, 916–921 (2008).
23. R. A. Sharma, *J. Electrochem. Soc.* 123, 448–453 (1976).
24. Y. Shao-Horn, S. Osmialowski, Q. C. Horn, *J. Electrochem. Soc.* 149(12), A1547–A1555 (2002).
25. A. G. Ritchie, P. G. Bowles, D. P. Scattergood, *J. Power Sources* 136, 276–280 (2004).
26. H. Arai, S. Okada, Y. Sakurai, J. Yamaki, *J. Power Sources* 68, 716–719 (1997).
27. H. Li, G. Richter, J. Maier, *Adv. Mater.* 15, 736–739 (2003).
28. F. Badway, N. Pereira, F. Cosandey, G. G. Amatucci, *J. Electrochem. Soc.* 150, A1209–A1218 (2003); F. Badway, N. Pereira, F. Cosandey, G. G. Amatucci, *J. Electrochem. Soc.* 150, A1318–A1327 (2003).
29. P. Adelhelm, Y. S. Hu, M. Antonietti, J. Maier, B. M. Smarsly, *J. Mater. Chem.* 19, 1616–1620 (2009).
30. German Patent Application 102009017262.9; R. Prakash, A. K. Mishra, A. Roth, Ch. Kübel, T. Scherer, M. Ghafari, H. Hahn, M. Fichtner, *J. Mater. Chem.* 20, 1871–1876 (2010).
31. J. Maier, *Adv. Mater.* 21, 2571–2585 (2009).
32. B. Kang, B. Ceder, *Nature* 458, 190–193 (2009).

Chapter 4

Nanoengineered Lithium–Air Secondary Batteries: Fundamental Understanding and the Current Status of Development

M. S. Michael,[a] Junichi Kawamura,[b] Naoaki Kuwata,[b] and S. R. S. Prabaharan[c]

[a] *Chemical Sciences Research Centre, SSN College of Engineering, Anna University Affiliated, Kalavakkam, Chennai, India*
[b] *Institute of Multidisciplinary Research for Advanced Materials (IMRAM), Tohoku University, Katahira, Aobaku, Sendai, Japan*
[c] *School of Electronics Engineering, VIT University, Chennai Campus, Chennai, Karnataka, India*
prabaharan.srs@vit.edu.in

Lithium–air (Li–air) batteries are regarded as a transformational energy storage technology and categorized under advanced green energy power sources for the future. In Li–air technology, oxygen from ambient air is consumed at the cathode during discharge and the pure oxygen (O_2) is released during recharge of the battery. In this advanced conversion technology, elemental lithium metal and atmospheric oxygen form an electrochemical couple yielding the highest theoretical energy density of 11,680 $Whkg^{-1}$, the value very close to the energy released on combustion of liquid gasoline (13,000 $Whkg^{-1}$). Li–air battery system involves the reduction

Nanotechnology in Advanced Electrochemical Power Sources
Edited by S. R. S. Prabaharan and M. S. Michael

ISBN 978-981-4241-43-4 (Hardcover), 978-981-4303-05-7 (eBook)
www.panstanford.com

of nonportable O_2 via $2e^-$ transfer process in aprotic medium, while in aqueous medium, it exhibits either a $2e^-$ or $4e^-$ transfer. Accordingly, in aprotic medium, ORR (oxygen reduction reaction) results in the formation of Li_2O_2 during charge and the OER (oxygen evolution reaction) process promotes the decomposition of Li_2O_2 during discharge. What is considered paramount in Li–air system is the construction of an efficient gas diffusion electrode (GDE) made of engineered nanoporous carbons (activated carbons, carbon blacks, carbon monoliths, carbon foams, graphene, and so on) embedded with bifunctional nanocatalysts (metal oxides for instance) to felicitate both ORR and OER. This chapter presents great challenges which exist at the cathode side (air) of Li–air system, the importance of nano electro-catalysts, and how it helps achieve high reversibility, high energy density, and round trip efficiency in Li–air batteries.

4.1 Introduction

Although the concept of Li–air battery dates back to 1976,[1] it has recently emerged as a potentially transformational energy storage technology, exhibiting very high specific energy much needed for transportation as well as stationary energy storage applications. At present, the status of Li–air is at its very infant stage; however, large R&D efforts show promising advancements and the interests to develop the exhilarating battery system is growing in a rapid fashion. In recent years, Li–air battery research has gained popularity among researchers in the universities and industries (IBM, Toyota, for instance) owing to its high theoretical energy density in competition with fossil fuels.[2,3] Moreover, Li–air cell reactions are often hyped as the technological heir to Li-ion batteries. Li–air battery system may be thought of as primary as well as secondary version in which the latter can be recharged. Hence, in pursuit of next-generation energy storage systems, Li–air systems may pose a great challenge to Li-ion batteries as the energy density of the latter is now approaching its theoretical limit, set by the energies of cathode and anode materials used. In Li–air batteries, lithium metal anode (protected

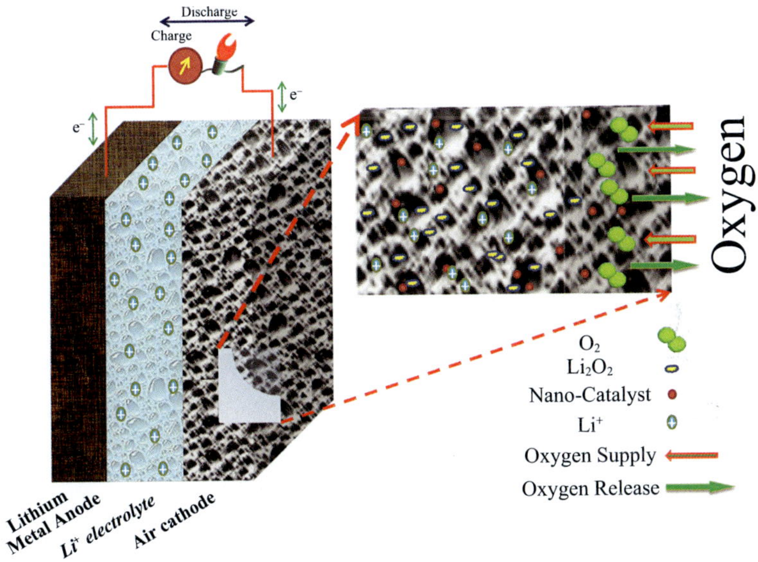

Figure 4.1 Exploded view of Li–air battery system arranged in a cut-away view.

or unprotected) is electrochemically coupled to atmospheric oxygen through an air cathode as shown in Fig. 4.1. It is fundamentally considered a secondary battery system owing to its galvanic nature, as the assembled Li–air cell is ready for discharge without any prior charging unlike Li-ion.

The Li–air battery combines the unique advantages of fuel cell and battery. This means that oxygen supplied from the atmosphere at the cathode side is being a nonportable cathode, as in the fuel cell and lithium metal being the anode provides the galvanic nature of a secondary battery. Fundamentally, it has no 'fuel infrastructure issue', as do hydrogen fuel cells.[4–7] Li–air system is regarded as the best environment-friendly, high-energy density having highest theoretical energy density next to gasoline as shown in the Ragone plot (refer Fig. 4.2). In general, Ragone plot is a very useful means of comparing various energy storage systems, which helps to identify an optimum system based on power requirement. The ratio of energy to power is time and it is typically plotted on

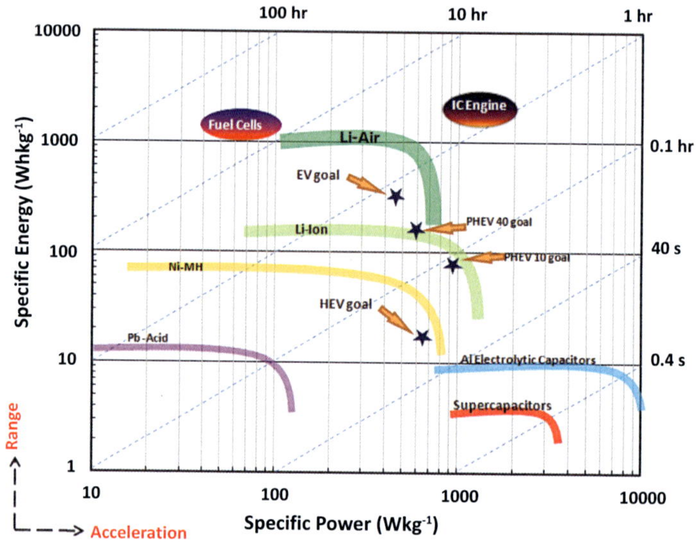

Figure 4.2 Ragone plot describing various electrochemical power sources linked with present and future electric mobility options (*Source:* USABC PHEV Goals, US defense logistics agency, Product data sheets).

the plots as cross lines and time is referred to as the time of discharge or charge. The plot shows that Li-ion batteries is the best choice for applications that require discharge time greater than 5 s, and for shorter times of discharge (e.g. start-stop hybrid cars), electrochemical capacitors (ECs) are more appropriate. Nevertheless, in terms of theoretical energy density, Li–air shows promise compared with IC engines.

4.2 Why Li–Air Battery?

The major demand for Li–air battery is justified by the fact that it is able to dispense extremely high-energy density, an amount of energy a battery can store for a given volume, which competes that of conventional gasoline powered engines. Li–air batteries gain this advantage in energy density because they use a nonportable cathode, oxygen available lavishly from the environment, no need to be stored in the battery itself. They are attractive for many

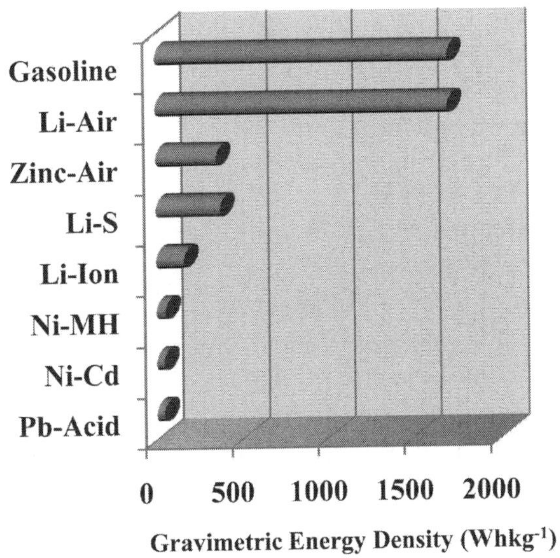

Figure 4.3 The gravimetric energy density ($Whkg^{-1}$) of various types of rechargeable battery systems compared with gasoline. In the practical Li–air battery, the value is just an estimate. For gasoline, the value includes the average tank-to-wheel efficiency of cars.

applications including electrified transportation in which battery's energy-density rating and the amount of energy that can be delivered for a given weight are important. One of the biggest drawbacks of today's electric vehicles that run with Li-ion batteries is the limited range between recharging, <100 miles, that is much lower than the acceptable driving range of 300 miles per charge.[8] There is no other battery that has a potential to offer up to 10 times the energy density of Li-ion batteries except Li-air. Hence, Li-air battery would be the better alternative for fossil fuels in plug-in electric vehicles (EVs) if it is possible to unlock all the theoretical energy of it (see Fig. 4.3).[9]

4.3 Other Metal–Air Battery Systems

Several metal–air batteries have been studied hitherto owing to their energy densities, for instance zinc–air (Zn–air), magnesium–air (Mg–air), aluminum–air (Al–air), and iron-air (Fe–air) batteries.[10–14] Each system has its own merits and demerits. Fe–air batteries exhibit long cycle life, but they have low voltage and specific energy compared with other metal-air batteries. Considering Al–air batteries, aluminum is one of the most abundant resources in the earth and has a high specific energy. The disadvantage of the system is its low columbic efficiency. Zn–air systems, on the contrary, deliver high-energy densities and are relatively inexpensive to produce in different sizes ranging from very small button cells for hearing aids, to very large batteries used for electric vehicle propulsion. Even though the energy density is high (1084 $Whkg^{-1}$), it still cannot meet the requirements of many high energy applications. Moreover, the high working voltage of nonaqueous Li–air battery (3 V) compared with other metal/air batteries using aqueous electrolyte as well as the light weight of Li anode result in high theoretical energy density, 11680 $Whkg^{-1}$, and makes it a promising replacement to traditionally utilized Li-ion batteries in EVs. Apart from that, the non-aqueous Li–air batteries can be recharged while other metal–air batteries are not reversible. Yet, another equal contender for Li–air is Li–S (lithium–sulfur) batteries,[15] which is out of scope of this chapter and hence will not be discussed.

4.4 Rechargeable Li–Air Battery System

4.4.1 *Li–Air Battery Fundamentals*

Recently, despite the continuous investigations on advanced Li-ion batteries, there have been great efforts in new transformational energy storage battery technology "beyond Li-ion" battery technologies aimed at providing more energy storage capacity for practical EVs[4]. Li–air batteries have created rush owing to its projected theoretical energy density values, which are on par with gasoline. It has brought a new impetus to the use of lithium metal in Li–air (O_2) battery. Lithium metal possesses the highest theoretical capacity of 3600 $mAhg^{-1}$ next to non-metal hydrogen. Use of lithium metal (Li metal) in Li–metal oxide battery, the first generation of lithium rechargeable batteries, has been hampered by the growth of dendrites during charge making the battery system vulnerable to catastrophic reaction. Hence, the concept of using Li^+ ion in metal oxide electrode against intercalable graphite anode led to the discovery of Li-ion batteries.

Ever since the possibility of fabricating a practical nonaqueous Li–air battery as reported by Abraham *et al.*,[16] a new renewed interest has been triggered by world battery community to bring back lithium's true advantage in the form of Li–air secondary battery system.[17–20]

The fundamental principle of Li–air batteries is that a lithium metallic anode is coupled to atmospheric oxygen through a porous carbon-based air cathode instead of the heavy conventional compounds found in Li-ion batteries. It is imperative that efficient oxidation/reduction of oxygen (O_2) is vital to realize the high theoretical energy density of this system. In order to achieve this, different types of Li–air batteries configuration are being investigated: nonaqueous system, aqueous system, and the combinations of both (in the form of hybrid).[4,21] Although a practical version is still a long due, efforts are tremendously underway in both academic and industrial sectors.

The Li–air cell can be thought as being a battery-fuel cell hybrid or a semifuel cell kind owing to O_2 cathode reactant in the form of air from the atmosphere and the only fuel for reaction, lithium metal is kept inside the cell. It is anticipated that the large increase in theoretical specific energy on migrating from Li-ion to Li–air (O_2), arises because of Li_2O_2, the discharge product formed at the cathodes, can store more Li per unit mass, than any metal oxide cathodes of Li-ion batteries and Li metal as anode stores more charge per unit mass than graphite (Li_xC_6), the anode material currently used in commercial Li-ion batteries. The discharge product of Li–air system, Li_2O_2, predominantly involves $2e^-$ transferreaction with the formation of intermediate superoxide, O_2^-, LiO_2 as described below:[22–24]

[ORR]

$$O_2(g) + e^- + Li^+ \rightarrow LiO_2 \text{ [via } O_2 + e^- \rightarrow O_2^-] \quad (V_o = 2.87 \text{ V } vs. \text{ Li/Li}^+)$$

$$2LiO_2 \rightarrow Li_2O_2 + O_2 \quad (V_o = 2.96 \text{ V } vs. \text{ Li/Li}^+)$$

[OER]

$$Li_2O_2 \rightarrow 2Li^+ + O_2 + 2e^- \qquad (V_o = 2.91\,V\,vs.\,Li/Li^+)$$

It should be understood that the OER process does not involve LiO_2, instead the Li_2O_2 is directly discomposed into Li^+.

The partial formation of Li_2O as a discharge product has been reported by a few research groups, suggesting that either O_2 can be discharged down to Li_2O possibly involving $4e^-$ transfer or the major discharge product Li_2O_2 could further reduce into Li_2O according to the following reaction[23–25]

$$O_2(g) + 4e^- + 4Li^+ \rightarrow 2Li_2O\ (V_o = 2.91\,V\,vs.\,Li/Li^+)$$

$$Li_2O_2 + 2Li^+ + 2e^- \rightarrow 2Li_2O\ (V_o = 2.87\,V\,vs.\,Li/Li^+)$$

The formation of Li_2O is not favored kinetically, as it requires the full cleavage of O–O bond. The rechargability of Li–air cell has been accomplished by electrochemical (oxidation) decomposition of Li_2O_2 with or without catalyst in the porous carbon matrix, the so-called gas diffusion electrode (GDE) at the air cathode side.

It is a great challenge to make an efficiently rechargeable Li–air battery, owing to the involvement of semimetal Li_2O_2. This is somewhat surprising that Li_2O_2 exhibits a half metallic (surface level) nature (duly explained by the first principle calculations using surface state analyses) despite the fact that Li_2O_2 is a bulk insulator.[25] Generically, it is thought that Li_2O_2 is a catalyst by itself rather than a reactant, which has been demonstrated by Cannon *et al.*[26] These authors have proven the catalytic effects of Li_2O_2 in the decomposition of sodium chlorate. However, when Li_2O_2 is added to sodium chlorate containing 1% Co_3O_4, the catalytic effect of cobalt oxide is partially suppressed or deactivated. Recently, Hun-Gi Jung *et al.*[27] demonstrated that they could achieve the capacity as high as 5000 mAhg^{-1} carbon without having catalysts at the air cathode side of Li–air employing a glyme-based electrolyte. These findings are an additional proof that Li_2O_2 is oxidizable without the help of an external catalyst meaning that Li_2O_2 by itself is a catalyst, although many authors claim that it does require external catalyst to induce the electrocatalytic decomposition of Li_2O_2 upon recharge in Li–air cell.[28–33]

As discussed elsewhere in this chapter, it may be Li_2O_2 or Li_2O, discharge product holds up the process in Li–air system, limiting the number of charge–discharge cycles to a mere handful. It is difficult to get back all lithium from these discharge products and only partially possible even when assisted by special catalysts. Essentially, it appears that OER catalysts are mandatory to decompose the discharge product reversibly to enhance the roundtrip efficiency. Moreover, it has been the trend in any O_2 involved electrochemistry for instance, fuel cells, ORR kinetics is normally slow and electro-catalysts are mandatory to speed up the ORR kinetics at the cathode side. In the Li–air being considered to be battery-fuel cell hybrid, the need for catalyst for O_2 reduction (ORR) and Li_2O_2 decomposition (OER) processes during discharge and charge, respectively, becomes inevitable.

Before Li–air batteries can find use in hybrid and electric vehicles, it must be able to recharge thousands of such cycles, which becomes a huge challenge for battery scientist. Worldwide rush thus has been created to speed up and develop a fully useable battery system. A number of reviews have appeared;[34–38] all these reviews describe a variety of niche issues viz., air cathode porosity, stable electrolyte, electrode polarization, electro-catalysts, ways to protect the lithium metal and cell formation, and other engineering design aspects of Li–air system. Among which, one of the vital issues being considered is the selection of efficient bi-functional electrocatalysts suitable for ORR and OER at the air cathode. This chapter focused thus far on the recent achievements made hitherto in developing Li–air cell with a specific reference to authors' achievements in development of cell engineering and development of nanoengineered bi-functional (discharge/charge) catalysts embedded in air cathodes. It also reviews the effective use of nonprecious (no-noble) metal oxide nano-catalysts *vs.* precious (noble) catalysts (Pt, Au, and Pd) in terms of achievable capacity of mesoporous air-cathode and yet diminishing the electrode polarization during charge and discharge cycles in the context of improving the round trip efficiency of the system.

Table 4.1 Thermodynamic electrode potentials of O_2 reduction[39,40]

Electrolyte	ORR reaction	Electrode potential at standard conditions (V)
Aqueous-Acidic	$O_2 + 4H^+ + 4e^- H_2O \rightarrow$	1.229
	$O_2 + 2H^+ + 2e^- H_2O_2 \rightarrow$	0.70
	$H_2O_2 + 2H^+ + 2e^- 2H_2O \rightarrow$	1.76
Aqueous-Alkaline	$O_2 + H_2O + 4e^- 4OH^- \rightarrow$	0.401
	$O_2 + H_2O + 2e^- HO_2^- + OH^- \rightarrow$	−0.065
	$HO_2^- + H_2O + 2e^- 3OH^- \rightarrow$	0.867
Nonaqueous aprotic	$O_2 + e^- [O_2]^- \rightarrow$	
	$O_2^- + e^- [O_2]^{2-} \rightarrow$	2.96 *vs.* Li/Li^+
	$[O_2]^{2-} + 2e^- 2O^{2-} \rightarrow$	2.91 *vs.* Li/Li^+

4.4.2 *Principle of Operation of Li–Air Battery*

As shown in the schematic in Fig. 4.1, the Li–air battery consists of lithium metal as anode, Li-ion conducting organic/aqueous electrolyte, and a porous cathode composed of carbon, catalyst, and binder. During discharge, lithium metal is oxidized to lithium ions and releases electrons as described by the following half-cell reaction

$$Li \rightarrow Li^+ + e^-$$

The electrons follow the external circuit to the cathode and the lithium ions migrate across the electrolyte to reduce oxygen at the cathode. The oxygen is reduced in either a two or four electron process by the half-cell reaction as shown in Table 4.1.

4.4.2.1 Non-aqueous Li–air system

The aprotic Li–air battery consists of a lithium metal anode, a liquid organic electrolyte, and a porous carbon cathode. Electrolytes can be any organic solvent capable of solvating lithium salts [$LiPF_6$, $LiAsF_6$, $LiN(SO_2CF_3)_2$, $LiSO_3CF_3$], which typically consist of carbonates, ethers, and esters.[41–43] The carbon cathode is usually high surface area carbon material with a nanosized metal oxide catalyst (commonly MnO_2 or Mn_3O_4). A major advantage of aprotic electrolyte is that the spontaneous formation of a barrier the so-called SEI (solid electrolyte interface) between the anode and electrolyte (much like the barrier formed between electrolyte and carbon-lithium anodes in conventional Li-ion batteries), which protects the lithium metal from further reaction with the electrolyte. However, it has a major drawback that Li_2O_2 produced at the cathode is generally insoluble in the organic electrolyte leading to build up along the cathode/electrolyte interface. This makes cathodes in aprotic electrolyte-based Li-air batteries prone to clogging and volume expansion, which reduces conductivity and degrades battery performance over time.[44]

4.4.2.2 Aqueous Li–air system

The aqueous Li-air battery consists of a lithium metal anode (protected), an aqueous electrolyte, and a porous carbon cathode. The aqueous electrolyte is prepared by simply dissolving lithium salts in water. The aqueous Li–air battery avoids the issue of cathode clogging experienced in aprotic batteries because the reaction products are water soluble, which allows aqueous Li–air batteries to maintain performance over time.[45] On the contrary, lithium metal reacts vigorously with water and thus the aqueous design requires a solid electrolyte interface between the lithium metal and aqueous electrolyte. Normally, a lithium ion conducting ceramic or glass is used, but their conductivities are generally low (on the order of 10^{-3} S/cm at ambient temperatures).

4.4.2.3 Aqueous/nonaqueous hybrid electrolyte

Hybrid Li-air battery consists of two parts (one part aqueous and another part nonaqueous electrolyte) connected by a lithium-conducting ceramic such as LISICON, LiPON, LAGP, and LTAP.[46,47] A lithium metal anode is in contact with the nonaqueous side of the electrolyte, whereas the porous cathode is in contact with the aqueous side. A lithium-conducting ceramic is typically employed as the membrane joining the two electrolytes.

During discharge, the electrode reaction

$$O_2 + 2H_2O + 4e^- \rightarrow 4OH^-$$

4.4.2.4 Solid-state electrolyte

The solid-state battery eliminates problem at the anode/cathode interfaces associated with liquid electrolyte. Currently, solid-state Li–air battery is composed of a lithium anode, a ceramic or glass, or a glass-ceramic electrolyte, and a porous carbon cathode. The anode and cathode are typically separated from the electrolyte by polymer-ceramic composites that enhance charge transfer at the anode and electrochemically couple the cathode to the electrolyte. The polymer–ceramic composites serve to reduce the overall impedance of the solid-state Li–air battery. The main drawback of the solid-state battery design is the low conductivity of most glass-ceramic electrolytes. Lithium aluminum germanium phosphate has been found to be an effective electrolyte, but the ionic conductivity of current lithium fast ion conductors is still lower than liquid electrolyte.[48]

4.5 Current Problems Encountered in Li–Air Rechargeable System

4.5.1 *Non-aqueous Solvents*

Rechargeable Li–air system is based on nonaqueous electrolyte, introduced in 1996 by Abraham *et al.* that consisted of polyacrylonitrile-based gel electrolyte,[16] and they identified Li_2O_2 as the discharge product. The other two products, Li_2O and LiO_2 have been found by Tarascon *et al.* to be formed during the reduction of O_2.[22] The formation of these discharge products could be explained by the formation of superoxide, O_2. during the reduction of O_2 in a nonaqueous electrolyte. The half-life of the superoxide depends on the nature of cation present in the electrolyte. As the superoxide has a very short life, it disproportionates into Li_2O_2 and O_2.

Peng *et al.* have studied the mechanism of oxygen reduction reactions (ORRs) in a nonaqueous electrolyte, CH_3CN, by employing *in situ* Raman spectroscopic technique. Their findings are an additional proof confirming the formation of LiO_2, an intermediate on O_2 reduction, which disproportionates into Li_2O_2. During charge, in-situ spectroscopic studies revealed that Li_2O_2 decomposed directly in one-step reaction and released O_2 without the formation of intermediate product LiO_2.[48]

Many solvents have been explored for use in Li–air battery, including carbonate, ethers, ionic liquids, and acetonitrile, but no solvent has demonstrated highly reversible cycling yet. The organic solvents used hitherto include carbonates (EC, PC, DMC), ethers (tetrahydrofuran, dioxolane, Crown ethers, especially 12-crown-4, 15-crown-5, 1,2-dimethoxyethane, methyl nonafluorobutyl ether, tris(2,2,2-trifluoroethyl) phosphate, tetraethylene glycol dimethyl ether), and esters (γ-butyrolactone), and commonly used lithium salts are $LiClO_4$, $LiPF_6$, $LiSO_3CF_3$, and $LiN(SO_2CF_3)2$.[49–51] Organic carbonate electrolytes have been used widely in Li–air batteries to date. It has been reported that during discharge, the carbonate electrolytes got decomposed rather than forming Li_2O_2. The product formed on decomposition of organic carbonate electrolytes has been found to be mainly carbonates species such as $C_3H_6(OCO_2Li)_2$, Li_2CO_3, HCO_2Li, CH_3CO_2Li, CO_2, and H_2O at the cathode.[52–55] The formation of CO_2 was supported by Xu *et al.*, analyzing the gas during the charge/discharge process in carbonate electrolytes by using in-situ gas chromatography/mass spectroscopy.[54] The decomposition of carbonates electrolytes might be attributed to the nucleophilic attack of superoxide radical, O_2^-, on these electrolytes.[56] Hence, it has been suggested that organic carbonates are not suitable as electrolytes for Li–air battery; new electrolytes that are stable during both charging and discharging process are required for the long-term operation of rechargeable Li–air battery.

McCloskey *et al.*[51] studied the discharge products of Li–air batteries in various solvents, including carbonates, dimethoxyethane (DME), and in the mixture of both as well. They have found that Li_2O_2 was formed as a discharge product in pure DME-based cells, which on charging decomposed into lithium and O_2 and the decomposition of DME occurred only at higher potentials. But on mixing the carbonates with DME, carbonate-based solvents decomposed and formed nonvolatile deposits at the cathode during discharge, which decomposed further during charging. It has been reported that Li–O_2 electrochemistry is strongly dependent on the choice of the solvent used.[51]

A mixture of nonaqueous electrolyte that contains acetonitrile and propylene carbonate was identified as a good electrolyte for Li–air battery by Zheng *et al.* Acetonitrile is extensively used as a nonaqueous solvent due to its high dielectric constant, low melting point, and low viscosity. But CH_3CN is unstable, as metallic lithium reacts with it to form methane and LiCN. Hence, CH_3CN alone could not be used as a solvent for Li–air batteries.[56] If it is mixed with either PC or γ-butyrolactone, a passivation layer can formed on the surface of the metallic lithium, which prevents further reaction between CH_3CN and Li.[56,57] Zheng *et al.* mixed CH_3CN and PC in 1:1 ratio analyzed by GC-MS and found that there was no reaction between metallic lithium and CH_3CN even after 1 week. The oxygen solubility in CH_3CN (8.1 mM) is two times higher than in PC (3.6 mM) and the viscosities of PC and CH_3CN are 2.53 and 0.341 cP, respectively. So, in a mixed electrolyte, CH_3CN contributes to the high solubility and fast diffusion of O_2, while PC stabilizes the lithium, so a proper ratio of CH_3CN and PC would be a good electrolyte for Li–air battery.

The deposition of Li_2O_2 on the cathode during discharge destroys the porous catalytic electrode slowly over time, which in turn results in poor cyclablility. A rechargeable Li–air battery with TEGDME-$LiPF_6$ and an uncatalyzed porous carbon electrode was investigated by Laoire *et al.* and found that the rechargeability of the cell was limited due to poor cycling efficiency of the lithium anode and the high impedance associated with Li_2O_2 deposited on the carbon cathode.[50] Yang *et al.* described that tris(pentafluorophenyl)borane (TPFPB) can dissolve a large amount of lithium salts such as Li_2O, Li_2O_2, LiF, and CF_3COOLi, which are insoluble in organic solvents.[58]

Even though various types of nonaqueous electrolytes including ionic liquids were studied so far, no breakthrough was observed in increasing diffusion coefficient and oxygen solubility. The effect of TPFPB as an additive was explained by Xu *et al.*[59] The discharge capacity was enhanced significantly in spite of an increase in viscosity and decrease in the conductivity and contact angle of the electrolyte. The addition of perfluorinated additives in nonaqueous electrolytes increased the oxygen solubility, which in turn increases diffusion-limited current of oxygen reduction on the GDE. There are two advantages in addition of perfluorinated additive. First, perfluorinated compounds are more effective in bringing oxygen to the interface. Second, the direction of chemical reaction between lithium and O_2 is limited, as most of the dissolved O_2 is encapsulated in the droplet of perfluorinated compound. The addition of FTBA (fluorotetrabutylammonium) to different electrolytes was investigated and found that the specific capacity and rate capability were improved significantly after the addition of FTBA.[60]

The important factor that influences the performance of Li–air system is the contact angle between the liquid electrolyte and the solid surface of the air cathodes. The contact angles of electrolytes and the air cathode determine the wetting conditions of the electrolytes. The air cathode is highly porous; therefore, low contact angle electrolytes may easily flood all the pores inside the air cathodes, thereby blocking the oxygen pathways. The high contact angle electrolytes have high polarity and cannot easily wet the air cathode surface and barely fill the pores inside the air cathode, which leads to plenty of gaps between the electrolytes and oxygen that can easily diffuse though the pores of the air cathodes. When these conditions occur, there are lots of triphase regions inside the air cathode formed, which in turn enhances the discharge capacity. There are two ways in which oxygen can flow into the triphase regions. The first way is diffusion of O_2 through the electrolyte in which the diffusion rate depends on oxygen solubility and viscosity of the electrolyte. The second way is direct diffusion of oxygen through the air gaps and diffusion rate is several orders magnitude higher than diffusion through the liquid electrolyte. As the performance of Li–air battery depends on the formation of triphasic regions formed by oxygen (gas), electrolyte (liquid), and active carbon with catalyst (solid), Xu *et al.* suggested that the amount of electrolyte added to Li–air batteries should be optimized in order to reach the maximum discharge capacity.[59]

Many studies, theoretical as well as experimental, were devoted toward the understanding of the parameters of the carbons, such as pore volume, surface area, void utilization, effective carbon mass, and so on. Nevertheless, it is very difficult to trace out which factor affects the cell performance significantly.[61–65] Zhang *et al.* examined single-wall carbon nanotube (SWNT) and carbon nanofiber (CNF) composite as air electrodes, demonstrating that the discharge capacity decreases with the thickness of the air electrode. A discharge capacity of ~2500 $mAhg^{-1}$ was achieved with a thin (~20 μm) air electrode, but it diminished to only ~400 $mAhg^{-1}$ when a thicker electrode (~220 μm) was used.[61] Zheng *et al.* studied the effect of increasing the porosity with specific capacity of the Li–air battery and observed that the specific capacity increased with increasing the porosity.[64] Tran *et al.* explained the nature of high surface area carbon materials by means of porosity and electrochemical discharge. It has been reported that micropores would be blocked by discharge product and these pores will not be accessed by oxygen or electrolyte, which makes them unavailable for electrochemical reaction.[65] They also reported a near-linear relationship between discharge capacity and average pore diameter of carbon. In this study, Tran *et al.* claimed that carbons with large pores should be selected to ensure effective three phase (oxygen, electron, and Li^+) boundaries. In order to achieve better performance, the air cathode must have mesopores as well as macropores so that more and more discharge products can be accommodated in these pores without hindering the mass transfer.

Other factors that affect the capacity and cell performance are the vaporization of the electrolyte and hydrolysis of anode. Hydrophobic ionic liquid electrolyte could be the promising candidate to solve these problems due to its hydrophobicity and negligible vapor pressure. Kuboki *et al.* used 1-ethyl-3-methylimidazolium bis(trifluoromethylsulfonyl)amide (EMITFSI) and 1-ethyl-3-methylimidazolium tetrafluoroborate ($EMIBF_4$) as the electrolytes for Li–air batteries and they achieved high discharge capacity of 1790 and 5390 $mAhg^{-1}$, respectively. The cell was tested 56 days in air and it has been found that the humidity and long discharge time did not affect the discharge behavior.[66] On the contrary, Li–air batteries, exposed to H_2O, entering into organic electrolyte from the surrounding air resulted in passivation of lithium metal.

The cations present in the supporting electrolyte show a strong influence on the reversibility of oxygen reduction and evaluation reactions, as their presence leads to different reaction mechanisms of oxygen reduction at air cathodes.[56,67]

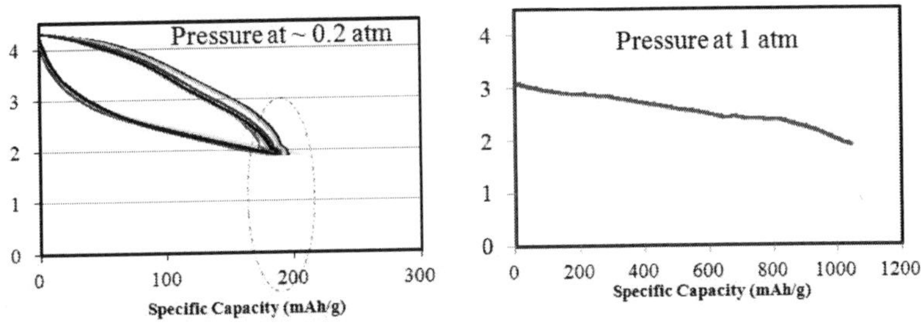

Figure 4.4 (a) Typical pseudo-capacitor (EDLC)-like behavior is observed owing to below optimal pressure. (b) At 1 atm gas pressure, the same cell works as Li–air cell as expected.

The effect of oxygen pressure on the performance of Li–air battery using nonaqueous electrolyte has been studied by many research groups. The formation of discharge product on the air cathode side affects the diffusion of oxygen and decreases the specific capacity of the battery at a high discharge rate.[68] An increase in the pressure of oxygen emerges to be helpful in preventing the blocking of pores of air cathode by the discharge products.[69]

Prabaharan *et al.*[70] studied the effect of gas pressure by employing porous carbon electrode as air cathode without involving a catalyst to elucidate effects of gas pressure in Li–air cell. It is revealed that an optimum pressure must be sustained throughout the life cycle of the battery (at least during discharge). Figures below (Fig. 4.4) exhibit gas pressure dependence on the performance of Li–air battery. In order to establish an effective triphasic region at air cathode side, an optimum gas pressure must be maintained.

In order to overcome all these problems associated with discharge product, a new Li–air rechargeable battery was developed by using aqueous/nonaqueous hybrid electrolyte. A water-stable ceramic LISICON film is used to protect the lithium metal in the aqueous electrolyte.[71,72] As H_2O and O_2 could not permeable through the LISICON film, they cannot react with the lithium metal. Moreover, the O_2 reduction product in aqueous electrolyte is soluble hydroxide, which will not choke the porous air cathode.[73–76] Zhang *et al.* demonstrated a novel rechargeable Li–air battery, which consists of Li/PEO/LTAP/HOAc-H_2O-LiOAc/Carbon. This Li–air cell could maintain a discharge–charge capacity of 250 $mAhg^{-1}$ for

15 cycles at the rate of 0.5 $mAcm^{-2}$, at 60°C and 3 atm of air. The energy density of the hybrid system was estimated on the basis of the weight of the lithium anode and HOAc and was found to be 779 $WhKg^{-1}$, which is two times higher than that of a typical Li-ion battery with graphite/$LiCoO_2$ (380 $WhKg^{-1}$).[77] Li *et al.* assembled a hybrid electrolyte Li–air battery with phosphate buffer in the catholyte. The addition of buffer to catholyte is to facilitate the reaction of O_2 with protons to form water during discharge and diffusion of Li^+ ions through the LTAP membrane to form LiH_2PO_4. While charging, water decomposes to protons and O_2 and the Li^+ ions in the catholyte get reduced and plated on the lithium metal. The discharge capacity of 221 $mAhg^{-1}$ and the energy density of ~770 $WhKg^{-1}$ were achieved at the current density of 0.5 $mAcm^{-2}$ for 20 cycles.[75] Li–air battery consists of Li/organic electrolyte/Li^+ conducting glass ceramic plate/aqueous electrolyte/Pt air cathode, which was investigated by He *et al.* It has been found that discharge voltage of this hybrid cell increased with a decrease in the concentrations of LiOH due to higher oxygen solubility in lower alkaline concentrations. The voltage efficiency was improved up to 85% with 3.32 V at the discharge and 3.9 V at the charge by the addition of $LiClO_4$ into aqueous electrolyte.[78]

Early investigation of Li–air batteries showed that the air cathode was the most important challenge for their development, although lithium anode side needs another challenging endeavor in the hybrid configuration.[75,79] Prabaharan *et al.* recently reported on their proprietary lithium metal/SE monolithic cassette as the protected lithium anode.[80]

4.5.2 *Gas Diffusion Electrode*

Air electrode plays an important role in the operation of Li–air battery. A successful air cathode must satisfy many requirements such as good electrical conductivity, high surface area, fast oxygen diffusion, stable electrode integrity, and fast ionic conductivity. In nonaqueous Li–air batteries, the air cathode is a carbon-supported porous structure, which acts as a gas transport channel involving the formation and storage of discharge product. GDE helps to accomplish an air cathode to complete the cell such as that in PEM fuel cells.[63,81] GDE in conjunction with liquid electrolyte forms a triphasic region, which is necessary to facilitate diffusion of O_2 and the adsorption of electrolyte at the surface. Porous carbon is the most widely used material in assembling air cathodes in metal-air batteries, which supports electrocatalysts, while providing venting for gas diffusion as well as offering electrical conductivity for the GDE.

Considering the air cathode, the major obstacles are the intrinsic slow reactions, high overpotential, and poor reversibility. It is obvious that porosity plays an extremely important role in GDE structure, as the pores in carbon aid the venting of O_2 gas during discharge/charge as well as accommodate the discharge products such as Li_2O_2 and Li_2O. As the pores are gradually filled with the insoluble discharge product, the reaction will end when all the spaces are used up. As a result, highly porous carbon with large pore volume is needed in order to attain high energy density. It is generally believed that the pore volume, especially the mesopore volume, must be preferred rather than the specific surface area (m^2g^{-1}) unlike supercapacitors.[67,82,83]

The design of such porous carbon as air cathode is one of the key attributes to enhance the capacity as well as better cyclability of the cell.[84]Polymer-derived activated carbons were demonstrated as potential cathodes for Li–air batteries. It was found that the carbon with the highest mesopore volume of 1.8717 cm^3g^{-1} showed the good capacity of 1852 $mAhg^{-1}$ due to an increase in storage space available for the accommodation of discharge product Li_2O_2 while the sample with lowest mesopore volume 0.8434 cm^3g^{-1} rendered low capacity of 910 $mAhg^{-1}$.[85] The data obtained from the commercial carbons with high porosity, pore volume, and surface area have been attracted for further research to improve the specific capacity of air cathode.[86–89] Mirzaeian *et al.* studied the performance in Li–air batteries prepared from carbon aerogels with controlled porosity. It was observed that carbon with a pore volume of 2.195 cm^3g^{-1} along with wide pore size of ~14 nm showed a specific capacity of 1290 $mAhg^{-1}$.[82] On the contrary, graphene and graphene-based composites have been studied so far as electrode for supercapacitors, lithium-ion batteries, and PEM fuel cells.[90,91]

Owing to exceptional properties exhibited by graphenes in niche energy storage applications, graphene has been investigated as air cathode in Li–air batteries as well.[92,93] A novel air electrode consists of an unusual hierarchical arrangement of functionalized graphene nanosheets and were found to deliver unusual capacity as high as 15,000 mAh g^{-1} in Li–air cells. Owing to the unique morphology of the graphene-based electrode in which plenty of large tunnels assist continuous oxygen flow, it does help achieve ideal triphase regions for oxygen reduction. Moreover, the discharge product Li_2O_2 prefers to nucleate and grow near functionalized lattice defect sites on graphene with functional group as a result of strong interaction with Li_2O_2.[93]

Recently, graphene nanosheets (GNSs) and its doped derivatives for instance, nitrogen doped graphene (NGNS), have been reported as an attractive air cathode material for Li–air batteries due to its structure and morphology.[94] Some works could not be reproduced though due to its extraordinary specific capacity reported. Li *et al.* reported nitrogen-doped carbon nanotubes (N-CNTs) synthesized by a floating catalyst chemical vapor deposition (FCCVD) that delivered an initial discharge capacity of 866 $mAhg^{-1}$, which is 1.5 times higher than CNTs (590 $mAhg^{-1}$) even though N-CNTs have lower surface area and pore volume. This enhancement in capacity is attributed to electrocatalytic activity of N-CNTs itself, which facilitates the cathode reaction during discharge.[94] The performance of mesocellular carbon foam prepared by nanocasting technique using mesocellular foam (MCF) silica hard template was reported by Yang *et al.* MCF-C delivered a discharge capacity of 2500 mAh g^{-1}, about 40% higher than that of super P at a current density of 0.1 $mAcm^{-2}$.[95]

As air cathode determines the power and energy density of Li–air battery, immense challenges present in air cathode itself. Such air cathode must be engineered with the following components: (i) facile substrate, (ii) electrocatalysts, (iii) pore design, (iv) electrolyte, and (v) O_2 permeable external membrane. Accordingly, it is desirable that a substrate must solely promote the formation of the discharge product (Li_2O_2). Hence, pore design is imperative, as it is preferred to be hierarchically a pore structure that permits efficient diffusion of O_2 into pores and avoids pore blockages.

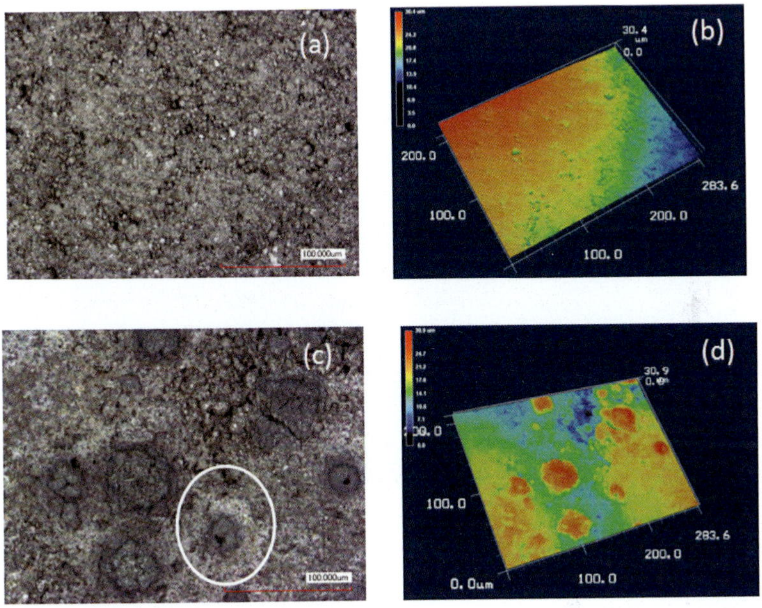

Figure 4.5 Hierarchically nanoporous carbon GDE structure scanned under laser microscopy. (a), (b) Pristine GDE and (c), (d) discharged GDE under argon atmosphere (*ex situ*).

For instance, recently Prabaharan *et al.* have demonstrated a high capacity air cathode (4000 $mAhg^{-1}$ carbon) made of hierarchically nanoporous carbon structure (mesopores/micropores) obtained from carbonaceous foams employing dual-template approach and the details of which will be reported elsewhere.[96] Figure 4.5 depicts laser microscopic (LM) images obtained on the pristine and discharged air cathode made of hierarchically porous carbon structure. Images (a) and (b) are taken prior to assemble a Li–air cell; images (c) and (d) show *ex situ* LM images on the discharged air cathode showing the patches of discharged product. Image (d) clearly illustrates the presence of discharged product deposited on the surface of carbon GDE structure.

It is well known that a more important factor in achieving high performance is the ability to make effective use of pores, not the size or total volume of pores. Keeping this in mind, Prabaharan *et al.* have synthesized a hierarchically porous carbonaceous

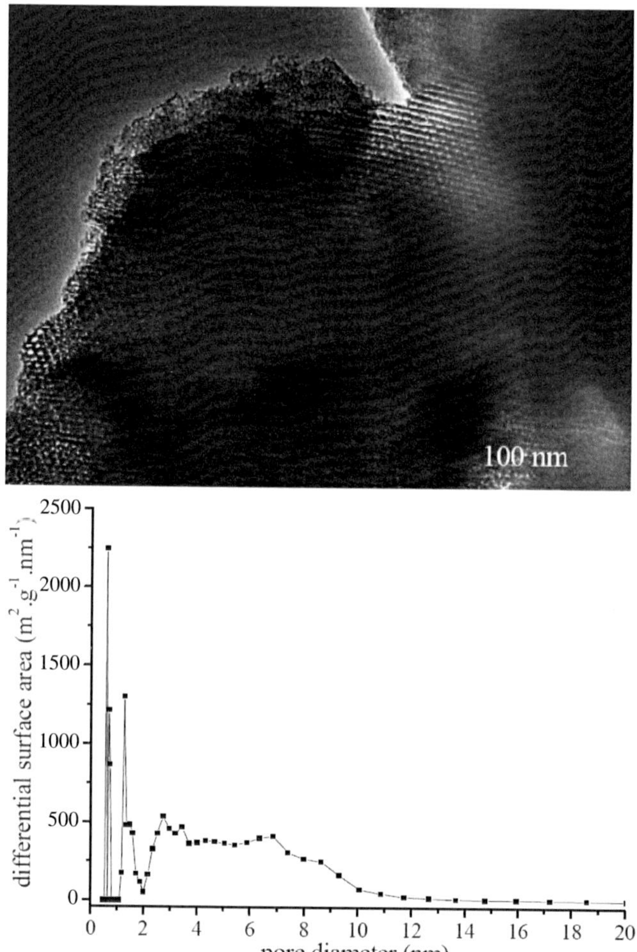

Figure 4.6 2D microstructure of hierarchically nanoporous carbon developed as an engineered carbonaceous material for air cathode. (a) HRTEM picture showing the porous structure. (b) Pore size distribution obtained through DFT calculation.

product that could accommodate a large amount of lithium oxides without hindering the transport of reactants (i.e., oxygen, electron, and Li^+) to active sites. The detailed results will be published elsewhere.

Recently, Lim *et al.* demonstrated that designing electrode morphology combined with effective catalyst decoration can enhance the rechargeability of the Li–O_2 battery over 100 cycles with full discharge and charge. The aligned carbon structure with a hierarchical micro-nano-mesh ensures facile accessibility of reaction products and provides the optimal catalytic conditions for the Pt catalyst. The new electrode is highly reversible even at extremely high current rate of 2 Ag^{-1}.[97]

4.5.3 *Role of Electrocatalysts in the Li–Air System*

The most abundant element in the Earth's crust is oxygen (O_2). The ORR is the most important reaction in life processes, namely, biological respiration, as well as in energyconverting systems such as fuel cells, metal/air batteries and so on. ORR in aqueous (acidic) solutions occurs through two pathways: the direct $4e^-$ reduction pathway from O_2 to H_2O, and the $2e^-$ electron reduction pathway from O_2 to hydrogen peroxide (H_2O_2) although in alkaline electrolyte O_2 may be directly reduced to OH^- through $4e^-$ pathway (refer Table 4.1). It is likely that in nonaqueous aprotic solvents the $1e^-$ reduction pathway from O_2 to superoxide (O_2^-) can occur.

Normally, the kinetics of ORR and OER occurring at the cathode of fuel cells, including direct methanol fuel cells (DMFCs) is very slow. In order to speed up the ORR kinetics to reach a practical usable level in a fuel cell, ORR catalyst is needed at the air cathode. Platinum (Pt)-based materials are the most practical catalysts used in PEM technology. These Pt-based catalysts are too expensive to make fuel cells commercially viable, and hence extensive research over the past several decades has been focused on development of alternative catalysts.[98–100] These alternative electrocatalysts include noble metals and alloys, carbon materials, quinone and its derivatives, transition metal macrocyclic compounds, transition metal chalcogenides, transition metal carbides and transition metal oxides. In this chapter, we focus on both noble and nonnoble electrocatalysts being used in air cathodes and the kinetics and mechanisms O_2 reduction/oxidation reaction (both ORR and OER), catalyzed by them.

Table 4.2 Maximum specific capacity and energy for Li–air cathodes compared with $LiCoO_2$ intercalation cathodes in current Li-ion batteries[100]

Specific capacities of air cathodes *vs.* $LiCoO_2$[a]	Li_2O_2	Li_2O	$LiCoO_2$
Specific capacity with respect to carbon (mAh/g carbon)	4600	6000	Not applicable
Specific capacity with respect to $C^+Li_2O_x$ or $LiCoO_2$ (mAh/g carbon)	900	1350	160
Average discharge voltage (V_{Li})	2.96	2.91	3.9
Specific energy density with respect to $C+Li_2O_x$ or $LiCoO_2$ (mWh/g cathode)	2450	3700	620

[a]Underlying assumptions for carbon black (e.g., Vulcan-XC72) based Li–air cathodes: (i) carbon packing density of 0.36 gcm^{-3} with ~15% carbon volume fraction[98,99]; (ii) 25% electrolyte volume fraction; and (iii) complete filling of the 60% void volume fraction with Li_2O_x product.

To compare cathode capacities of Li-ion with those of Li–air batteries, capacities must be normalized to the lithiated (discharged) cathode mass in both Li-ion batteries and in Li–air system. Carbon support of air cathode is also included in the latter system due to its high mass fraction. Table 4.2 shows that "five-fold higher specific capacities" and "four-fold higher specific energies" projected for Li–air cathodes compared with state-of-the-art $LiCoO_2$, whereby a considerably higher capacity could be realized by choosing a proper catalyst to facilitate the formation of Li_2O_2 or Li_2O during discharge.

4.5.3.1 Nano-electrocatalysts for efficient air cathode

Electrocatalysts play an important role in determining the electrode performance of Li–air battery. Li–air battery has the highest theoretical specific energy densities compared with any other practical electrochemical cell. Ideally, in Li–air battery, oxygen diffuses through air cathode reacts with the Li^+ that comes from the anode and forms lithium peroxide (Li_2O_2) predominantly. During the charging process, the lithium peroxide should be converted to Li^+ and returned to the anode. Air cathode plays an important role in the formation and stabilization of Li_2O_2, a major discharge product. To reuse the cell, Li_2O_2 product has to be reduced again to reverse the process. This appears to be not straight forward. And hence, the need for OER catalyst may be considered imperative. In a typical GDE structure, an optimum amount of bi-functional catalysts is loaded to kinetically control the ORR (discharge) and OER (charge) processes.

The electrocatalysts being used in Li–air battery can be roughly classified into the following four categories (see Table 4.3):

(1) Nanoporous carbon materials, including black carbons, functionalized carbon, nanostructured carbon, nanostructured diamond like carbon, graphene, N-doped graphene, hierarchical porous carbonaceous foams
(2) Precious metals and alloys (for instance, Pt, Au, Ag, Pd, and Pd-metal oxide composites)
(3) Nonprecious metals
(4) Transition metal oxides such as RuO_x, TaO_2, manganese-based oxides, cobalt oxides, NiO, binary Ni-Co, Co-Fe, Mn-Fe oxides, rare earth oxides.

Hitherto, many different kinds of electrocatalysts have been studied in the context of Li–air batteries. Each one of them is unique in terms of ORR and OER activities and obviously most of them are nanoscale products. Some showed remarkable properties such as reduced polarization and enhanced capacity. Nevertheless, in terms of cyclability, nothing is proven to be useful for long-term cycling and round trip efficiency in Li–air battery. But they provide the basis for further development. Hence, in Table 4.3, we have listed a comprehensive collection of most catalyst materials hitherto investigated to the best of our knowledge and the readers are encouraged to refer those literatures cited in Table 4.3 for further understanding. Indeed, it is out of scope of this chapter to describe their characteristics in Li–air cells and the readers are encouraged to refer cited references therein.

What is considered paramount to both ORR/OER reactions at the air cathode side is the use of bi-functional catalysts. For instance, Lu *et al.*[32] compared the effect of catalyzed (40 wt% Au/C or 40 wt% Pt/C) and pure Vulcan-XC72 carbon cathodes on the discharge and charge voltage profiles of single-cell Li–air (Li-O_2) batteries. This discharge and charge voltage profiles revealed that 40 wt% Au/C had showed the highest discharge activity (in association with ORR)

Table 4.3 Summary of various types of ORR and OER electrocatalyst predominantly used in secondary Li–air batteries and their various properties along with their carbon support at room temperature, 30°C

Name of catalyst	Average discharge potential/V_{Li}	Average charge potential/V_{Li}	Name of carbon support	[a]Discharge capacity, mAh/g carbon
Au/Pt[101]	2.7	~ 3.6	Vulcan XC72 —carbon black	~ 1000
Pt[102]	2.6	~ 4.25	Super P	470
C/Pd (40.03 Wt%)[103]	2.65	3.4	Carbon black (Norit)	855
C/PdO (36.79 Wt%)[103]	2.75	3.15	Carbon black (Norit)	556
Ru[103]	2.76	3.7	Carbon black (Norit)	577
RuO_2[103]	2.89	3.65	Carbon black (Norit)	446
$La_{0.8}Sr_{0.2}MnO_3$[102]	2.6	~ 4.25	Super P	750
Fe_2O_3[102]	2.6	~ 4.25	Super P	2700
Fe_3O_4[102]	2.6	~ 4.25	Super P	1200
CuO[102]	2.6	~ 4.25	Super P	900
Co_3O_4[102]	2.6	~ 4.25	Super P	2000
NiO[102]	2.6	~ 4.25	Super P	1600
α-MnO_2[104]	2.9–2.7	3.6	TAB	3650
α-MnO_2/Pd[104]	2.9	3.6	TAB	5450
Nanostructured NiO[105]	2.6	~ 4.25	MSP20 Maxorp	1350
$NiCo_2O_4$[8]	2.6	~ 3.3	MSP20 Maxorp	1800
$Ni_{0.8}Cu_{0.2}Co_2O_4$[8]	2.6	~ 3.1	MSP20 Maxorp	2200
$CoFe_2O_4$[102]	2.6	~ 4.25	Super P	1200

$CoMn_2O_4$ in Graphene[106]	~ 2.9	~ 3.7	AB	3000
$MnCo_2O_4$ in Graphene[107]	~ 2.95	~ 3.75	TCFP	2743
$Co(phen)_2$[108]	2.82	4.4	BP2000	4870
$Ag_2Mn_8O_{16}$[109]	2.75	3.5	SWNT/CNF	420mAh/g
Cu-Fe (1:1 atomic ratio)[110]	~ 2.5V	–	Ketjenblack EC-600 JD	1339
Au-PRO[111]	2.65	4.05	Carbon black (no specific name given)	12,500
Cu[112]	~ 2.6V	–	No carbon support used	Not specified/reported
PbRuO[111]	2.65	4.05	Carbon black (no specific name given)	10,000
BiRuO[111]	2.65	3.85	Carbon black (no specific name given)	~8000
Graphene Nano sheets[91]	2.75	3.97	Graphene	2332
N-doped Graphene[93]	2.8	2.29		11660
Mesoporous Carbon nitride (MCN) loaded with Pt[113]	2.87	3.30	AB	2500
Mesoporous Co_3O_4[55]	~ 2.85	~ 3.5	AB	2250

Note: Li-ORR reaction: $O_2 + 2Li^+ + 2e^- \rightarrow Li_2O_2$; Li-OER reaction: $Li_2O_2 \rightarrow O_2 + 2Li^+ + 2e^-$

[a]First cycle capacity only

while 40% Pt/C showed prominent OER activity during charging (decomposition reaction of Li_2O_2).

Even though it is believed that electrocatalysts are critical to improve the power density, cycling capability, and round-trip energy efficiency of Li–air batteries as in fuel cell,[67,73,74] electrocatalysts decrease the overpotential for charge process (i.e., OER).[75,114] Nevertheless, McClosekey *et al.* recently reported that, in a Li–air cell using DME as electrolyte, the investigated electrocatalysts, namely Pt, Au, MnO_2 did not help reduce the overvoltage significantly during charging.[51] Quite recently, Hun-Gi Jung *et al.* claimed that with an aide of stable electrolyte (glycol electrolyte), OER process can occur with reduced polarization voltage without any catalyst present at GDE. Interestingly, it is demonstrated that the battery can operate efficiently even at a high charge rate (as high as $3Ag^{-1}$ carbon) with limited discharge/charge polarization.[27]

Although there have been mixed opinions in the scientific scenario concerning the use of catalysts for ORR/OER reactions, the need for electrocatalysts in air electrode is not quite understood due to lack of knowledge on Li-O_2 electrochemistry in a nonaqueous system. Many recent articles questioned the true electrocatalytic effect of various catalysts[59,114] and the research outcomes can provide some insight for future studies on Li–air secondary batteries. It has been established that electrocatalysts are insensitive in nonaqueous electrolytes with large cations, such as tetrabutylammonium (TBA^+) and tetraethylammonium (TEA^+) due to the reversible nature of oxygen reduction-oxidation with one electron transfer reaction near the equivalent potential. On the contrary, in the nonaqueous electrolyte with small cations such as Li^+, the onset potential of ORR is positively shifted that makes the ORR/OER less reversible and widens the voltage gap between charge/discharge voltage.

Oxygen electrochemistry in the Li–air battery using nonaqueous electrolyte involves complicated processes especially it requires close scrutiny in understanding formation/decomposition of Li_2O_2. It is important to identify the limiting factors in the Li–O_2 reaction processes, as the latter is the prerequisite for the design and development of advanced bi-functional electrocatalysts to enable rechargeability in Li–air batteries. If all the steps involved in Li–O_2 reaction process might be electrochemically controlled,

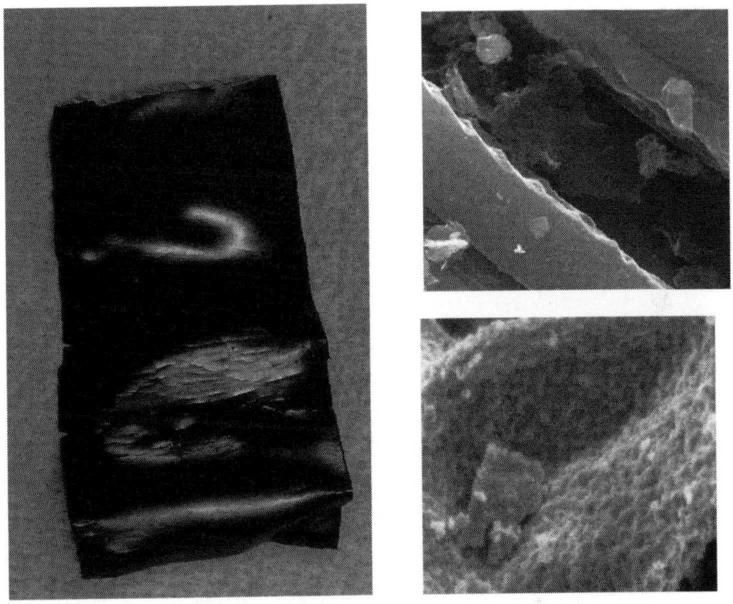

Figure 4.7 Typical GDE structure fabricated using MSP20 carbon black; nano catalysts used are NiO and $NiCo_2O_4$ in which PTFE was used as a binder.

electrocatalyst might function well. Moreover, it has been observed that some of the elementary steps in Li–O_2 reactions are non-electrochemical processes and most of the discharge products are not soluble and conductive. Arguably, use of catalyst especially for OER reaction depends upon the electrolyte system as demonstrated by Prabaharan *et al.*[80] They fabricated GDE using a nanoporous activated carbon (MAXORP MSP 20, surface area 2300 m^2g^{-1}, commercially supplied by Kansai Coke & Chemicals Co, Japan) disbursed with nanoscale catalysts such as NiO and $NiCo_2O_4$ is shown in Fig. 4.7.

It is proven that use of such electrocatalysts (NiO and $NiCo_2O_4$) helped reduce the overpotential as shown in Fig. 4.8. As can be seen, comparison of NiO and binary Ni–Co oxide catalysts revealed that the addition of these catalysts diminished polarization potential during OER and help enhance the discharge capacity (1800 mAh g^{-1} carbon) as well.

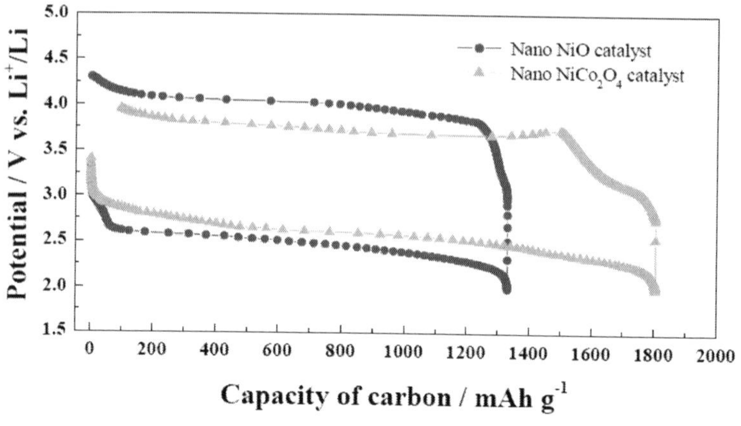

Figure 4.8 Discharge/charge performance of Air cathode showing the effect of non-precious catalysts in 1M $LiPF_6$ (EC/DMC).

4.6 Developmental Challenges

Li–air batteries differ from the conventional batteries in that the porous carbon electrode in the cell (air cathode) is used continuously to reduce oxygen accessed from the ambient environment or oxygen supplied. Consequently, the electrolyte in the cell is exposed to the atmospheric air, which is not dry enough to handle the nonaqueous electrolyte solvent soaked in the air cathode (GDE). Hence, preservation of electrolytic features of the electrolyte solvent has become one of the major concerns as far as the electrolyte is concerned. Hence, use of a highly stable electrolyte solvent is imperative to design a rechargeable Li-air cell. The major impediment while cycling Li–air is the continuous reduction of O_2 to Li_2O_2 via LiO_2. This discharge product is continuously being deposited into the pores of the carbon matrix leading to pore blocking. As a result, GDE will be curtailed from sustained O_2 flow resulting in cell deterioration and eventually the cell fails. Hence, the structure of GDE requires perfection in its design such as hierarchical porous carbons with well-defined macro/mesopore combination for handling discharge deposition with continuous flow of air. Such a design is still a due and quite a few research groups are still focused on such efforts.

Apart from electrolyte and design of GDE, another crucial challenge resides in the anode side in which lithium metal being used as a fuel for this cell reaction. Nevertheless, due to exposed atmospheric air venting at the cathode side, lithium side is found to be very much vulnerable owing to the presence of moisture in the air. Hence, care must be exercised to prevent the lithium side from being exposed to air/O_2 via GDE. This means that the cathode must not be fully wet and air/O_2 should not be allowed to diffuse through pores of GDE into the vicinity of lithium side. Although this is not a plausible situation in a cell environment due to intimate stacking, lithium side needs special care in protecting it. Such a thing can be a wish rather than being pragmatic and so protecting the lithium anode from seeing the air side is very crucial and variety of novel ways of being pursued to make it protected over air-stable/water-stable solid electrolytes.[45–47] This makes it to be a tumbling block to develop a lithium metal anode based Li–air cell as a rechargeable kind. This area needs intense research as well although exhaustive amount of work has been going in developing air cathode as well as

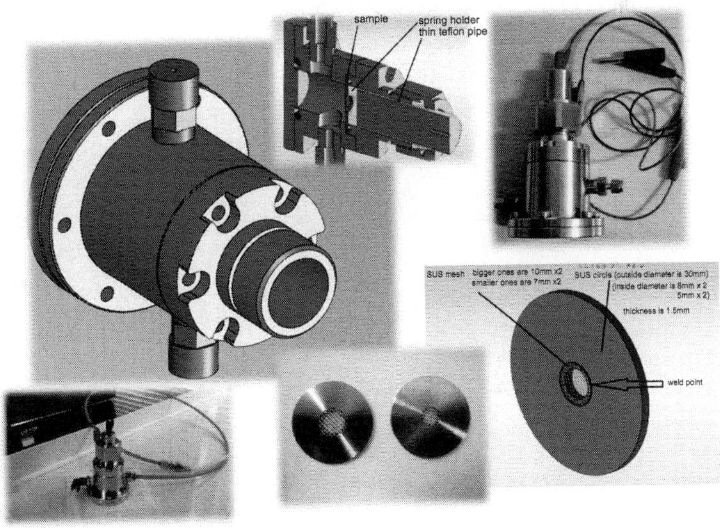

Figure 4.9 Modified Swagelok design of the SS Li–air test cell and the practical SS cell design based on our proprietary design is seen on the right side.

finding a suitable stable electrolyte. One of the remarkable progress in this direction is that a proprietary solution is being proposed by Prabaharan *et al.* to form a monolithic cassette-like structure preventing the lithium metal completely from seeing the cathode side.[115]

4.7 Our Proprietary Cell Design

As can be seen elsewhere in this chapter, we have highlighted our success in our attempt to develop a full cell version of Li–air battery using an in-house designed proprietary test cell is as shown in Fig. 4.6. It is a great challenge to make such a cell, although many researchers use coin type cells with holes at the cathode cup.

4.8 Summary

Li–air secondary battery system is classified as futuristic power source that is expected to fulfill the specific energy demand imposed by many OEM applications and electric transportation. Although the current status of development is still in its infancy, recent experimental evidences and theoretical calculations have shown that Li–air battery has the ability to provide four times higher specific energy much needed for electric vehicles, which current generation Li-ion batteries cannot meet. One of the great benefits is that oxygen non-portable, it has tremendous potential in terms of boosting its energy density at a much lower cost than its present technological competitor Li-ion batteries. Nevertheless, the current Li–air battery is limited by its poor power density, poor cyclability, and round-trip efficiency. As discharge products such as Li_2O_2 are solid and non-conductive, although smaller sized (nano-sized), Li_2O_2 are exhibiting metallic characteristics near surface. Hence, rate of discharge needs to be given special attention to ensure formation of nano-sized Li_2O_2 product. In terms of cyclability, there exists a limitation due to various issues associated with Li_2O_2 decomposition during the charge process (OER), although some nanostructured, nonprecious electrocatalysts have been shown to be expediting the decomposition process, and rate by which it occurs is slow. Overall, Li–air cell though interesting in the technological point of view, various sections of the cell need perfection before a practical and/or commercially viable battery system of this kind is being developed. Hence, considering its present pace of development though proposing R&D efforts are devoted, it is still regarded as a technology for future. Until then, Li-ion batteries will continue to enjoy its pride as a power source of choice in all electronic gadgets and electric mobility systems. This chapter was aimed at providing an overview of up-to date development and various challenges and/or issues centered on this exhilarating technology to meet out the challenges for future electric mobility.

References

1. A Galbraith, 4th International Electric Vehicle Symposium, Duesseldorf, Wet Germany, 1976.
2. G Kirishkumar, B McCloskey, A C Luntz, S Swanson, W Wilcke, *J Phys Chem Lett*, 1, 2193, 2010.
3. F Mizuno, S Nakaishi, Y Kotani, S Yokoishi, H Iba, Electrochemistry, 78, 403, 2010.
4. P G Bruce, S A Freunberger, L J Hardwick, J M Tarascon, *Nature Materials*, 11, 19, 2012 and references therein.
5. J Christenson, P Albertus, R S Sanchez-Carreera, T Lohmanna, B Kovinsky, R Liedtke, J Ahmed, A Kojic, *J Electrochem Soc*, 2012, 159, R1, 2012.
6. P Albertus, G Girishkumar, B McCloskey, R S Sanchez-Carrera, B Kozinsky, J Christensen, A C Luntz, *J Electrochem Soc*, 158, A343, 2011.
7. J Wang, Y Li, X Sun, *Nanoenergy*, 2, 443, 2013.
8. J Lu, K Amine, *Energies*, 6, 6016, 2013.
9. J S Lee, S T Kim, R Cao, N S Choi, M Liu, K T Lee, J Cho, *Adv. Energy Mater*, 1, 34, 2011.
10. W Y Li, C S Li, C Y Zhou, H Ma, J Chen, *Angew Chem Int Ed*, 45, 6009, 2006.
11. V Neburchilov, H J Wang, J J Martin, W Qu, *J Power Sources* 195, 1271, 2010; S Yang, H Knickle, *J Power Sources*, 112, 162, 2002.
12. *Encyclopedia of Electrochemical Power Sources* (Elsevier, 2009).

13. P Hamlen, T B Atwater, *Handbook of Batteries* (McGraw-Hill, 2001).
14. Q F Li, N J Bjerrum, *J Power Sources* 110, 1, 2002.
15. X Ji, L F Nazar, *J Mater Chem*, 20, 9821, 2010.
16. K M Abraham, Z Jiang, *J Electrochem Soc*, 143, 1, 1996.
17. X J Chen, A Shellikeri, Q Wu, J P Zheng, M Hendrickson, E J Plichta, *J Electrochem Soc*, 160(10), A1619, 2013.
18. M Balaish, A Kraytsberg, Y Ein-Eli, *Phys Chem*, 16, 2801, 2014.
19. R Padbury, X Zhang, *J Power Sources*, 196, 4436, 2011.
20. A Kraytsberg, Y Ein-Eli, *J Power Sources*, 196, 886, 2010
21. R Black, B Adams, L F Nazar, *Adv Energy Mater*, 2, 801, 2012.
22. Z Q Peng, S A Freunberger, L J Hardwick, Y H Chen, V Giordani, F Barde, P Novak, D Graham, J M Tarascon, P G Bruce, *Angew Chem Int Ed*, 50(18), 6351, 2011.
23. J S Hummelshoj, J Blomqvist, S Datta, T Vegge, J Rossmeisl, K S Thygesen, A C Luntz, K W Jacobson, J K Norskov, *J Chem Phys*, 132, 071101, 2010.
24. M W Chase, Jr., *J Phys Chem*, 9, 1510, 1998.
25. M D Radin, J F Rodriguez, F Tian, D J Siegel, *J Am Chem Soc*, 134, 1093, 2012.
26. J C Cannon, Y Zhang, *Ind Eng Chem Res*, 34, 3146, 1995.
27. H-G Jung, J Hassoun, J-B Park, Y-K Sun, B Scrosati, *Nat Chem*, 4, 579, 2012.
28. J Read, *J Electrochem Soc*, 149, A1190, 2002.
29. H Cheng, K Scott, *J Power Sources*, 195, 1370, 2010.
30. Y Shao, S Park, J Xiao, J-G Zhang, Y Wang, J Liu, *ACS Catal*, 2, 844, 2012.
31. C O Laoire, S Mukerjee, K M Abraham, E J Plichta, M A Hendrickson, *J Phys Chem A*, 113, 20127, 2009.
32. J Lu, Y Qin, P Du, X Luo, T Wu, Y Ren, J Wen, D J Miller, J T Miller, K Amine, *RSC Adv*, 3, 8276, 2013.
33. J Christensen, P Albertus, R S Sanchez-Carrera, T Lohmann, B Kozinsky, R Liedtke, J Ahmed, A Kojic, *J Electrochem Soc*, 159(2), R1, 2012.
34. S D Beattie, D M Manolescu, S L Blair, *J Electrochem Soc*, 156, A44, 2009.
35. P G Bruce, L J Hardwick, K M Abraham, *MRS Bull*, 36, 506, 2011.
36. F Cheng, J Liang, Z Tao, J Chen, *Adv Mater*, 23, 1695, 2011.
37. Y Sun, *Nano Energy*, 2, 801, 2013.
38. E Yeager, *J Mol Catal*, 38, 5, 1986.

39. A J Bard, L R Faulkner, in *Electrochemical Methods: Fundamentals and Applications*, New York, Wiley, 1980.
40. S Vyacheslav, S Bryantsev, J Uddin, V Giordani, W Walker, D Addison, G V Chase, *J Electrochem Soc*, 160(1), A160, 2013.
41. W Xu, K Xu, V Viswanathan, S Towne, J Hardy, J Xiao, Z Nie, D Hu, D Wang, J C Zhang, *J Power Sources*, 196, 9631, 2011.
42. A Debart, A Paterson, J Bao, P G Bruce, *Angew Chem*, 2008, 120, 4597, 2008, *Angew Chem Int Ed*, 47, 4521, 2008.
43. W Xu, J Xiao, D Wang, J Zhang, J Zhang, *Electrochem Sol State Lett*, 13(4), A48, 2010.
44. J B Bates, N J Dudney, G R Gurzalski, R A Zuhr, A Choudhury, C F Luck, J D Roberson, *J Power Sources*, 43, 103, 1993.
45. N Imanishi, T Zhang, Y Shimonishi, A Hirano, Y Takeda, O Yamamoto, 217th ECS Meeting, Abstract #748, The Electrochemical Society, 2008.
46. N Imanishi, S Hasegawa, T Zhang, A Hirano, Y Takeda, O Yamamoto, *J Power Sources* 185, 1392, 2008.
47. G Y Aleshin, D A Semenenko, A I Belova, T K Zakharchenko, D M Itkis, E A Goodilin, Y D Tretyakov, *Solid State Ionics*, 184, 62, 2011.
48. Z Peng, S A Freunberger, L J Hardwick, Y Chen, V Giordani, F Bard, Petr Novák, Duncan Graham, Jean-Marie Tarascon, Peter G Bruce, *Angew Chem Int Ed*, 50, 6351–6355, 2011.
49. J Read, *J Electrochem Soc*, 153(1), A96, 2006.
50. C Ó Laoire, S Mukerjee, E J Plichta, M Hendrickson, K M Abraham, *J Electrochem Soc*, 158(3), A302–A308, 2011.
51. B McCloskey, D Bethune, R Shelby, G Girishkumar, A Luntz, *J Phys Chem Lett*, 2, 1161, 2011.
52. S A Freunberger, Y Chen, Z Pang, J M Griffin, L J Hardwick, F Barde, P Novak, P G Bruce, *J Amer Chem Soc*, 133, 8040, 2011
53. F Mizuno, S Nakanishi, Y Kotani, S Yokoshi, H Iba, *Electrochemistry*, 78, 403, 2010.
54. W Xu, V V Viswanathan, D Wang, S A Towne, J Xiao, Z Nie, D Hu, J G Zhang, *J Power Sources*, 196, 3894, 2011.
55. A Debart, J Bao, G Armstrong, P G Bruce, *J Power Sources*, 174, 1177, 2007.
56. D Zheng, X Q Yang, D Qu, *Chem Asian J*, 6, 3306, 2011.
57. H S Lee, X Sun, X Q Yang, J McBreen, *J Electrochem Soc*, 149, A1460, 2002.
58. Y Wang, D Zheng, X Q Yang, D Qu, *Energy Environ Sci*, 4, 3697, 2011.

59. W Xu, J Xiao, D Wang, J Zhang, J Zhang, *J Electrochem Soc*, 157, A 219, 2010.
60. R R Mitchell, B M Gallant, C V Thompson, Y Shao-Horn, *Energy Environ Sci*, 4, 2952, 2011.
61. G Q Zhang, J P Zheng, R Liang, C Zhang, B Wang, M Hendrickson, E J Plichta, *J Electrochem Soc*, 157(8), A953, 2010.
62. X-H Yang, P He, Y-Y Xia, *Electrochem Commun*, 11(6), 1127–1130, 2009.
63. J Xiao, D H Wang, W Xu, D Y Wang, R E Williford, J Liu, J G Zhang, *J Electrochem Soc*, 157(4), A487, 2010.
64. J P Zheng, P Andrei, M Hendrickson, E J Plichta, *J Electrochem Soc* 158, A43, 2011.
65. C Tran, X Q Yang, D Y Qu, *J Power Sources*, 195, 2057, 2010.
66. T Kuboki, T Okuyama, T Ohsaki, N Takami, *J Power Sources*, 146, 766, 2005.
67. C J Allen, S Mukerjee, E J Plichta, M Hendrickson, K M Abraham, *J Phys Chem Lett*, 2, 2420, 2011.
68. J Read, K Mutolo, M Ervin, W Behl, J Wolfenstine, A Driedger, D Foster, *J Electrochem Soc*, 150(10), A1351, 2003.
69. X-h Yang, Y-y Xia, *J Solid State Electrochem*, 14, 109–114, 2010.
70. S R S Prabaharan, J Malathi, N Kuwata, M S Michael, J Kawamura, 13th Asian Conference on Solid State Ionics, Sendai, Japan, 17–20, July 2012.
71. L F Li, H S Lee, H Li, X Q Yang, K W Nam, W S Yoon, J McBreen, X J Huang, *J Power Sources*, 184, 517, 2008.
72. B Xie, H S Lee, H Li, X Q Yang, J McBreen, L Q Chen, *Electrochem Commun*, 10, 1195, 2008.
73. Y Wang, P He, H Zhou, *Energy Environ Sci*, 4, 4994, 2011.
74. P He, Y Wang, H Zhou, *J Power Sources*, 196, 5611, 2011.
75. L Li, X Zhao, A Manthiram, *Electrochem Comm*, 14, 78, 2012.
76. H He, W Niu, N M Asl, J Salim, R Chen, Y Kima, *Electrochim Acta* 67, 87, 2012.
77. T Zhang, N Imanishi, Y Shimonishi, A Hirano, Y Takeda, O Yamamoto, N Sammes, *Chem Commun*, 46, 1661, 2010.
78. H He, W Niu, N Mahootcheian Asl, J Salim, R Chen, Y Kima, *Electrochim Acta*, 67, 87, 2012.
79. G Y Aleshin Dmitry, A Semenenko, A I Belova, T K Zakharchenko, D M Itkis, E A Goodilin, Y D Tretyakov, *Solid State Ionics*, 18(1), 62, 2011.
80. S R S Prabaharan, J Kawamura, Y Iwai, N Kuwata, M S Michael, *Prep Pap–Am Chem Soc, Div Fuel Chem*, 57 (1), 748, 2012.

81. H Zhang, X Wang, J Zhang, Chapter 19, in *PEM Fuel Cell Electrocatalysts and Catalyst Layers: Fundamentals and Applications*, J Zhang (Ed.), 2008, Springer-Verlag, London.
82. M Mirzaeian, P J Hall, *Electrochim Acta*, 54, 7444, 2009.
83. A A Franco, K-H Xue, *ECS J Solid State Sci Technol*, 2, M3084, 2013.
84. G O Shitta-Bey, M Mirzaeian, P J Hall, *J Electrochem Soc*, 159(3) A315–A320, 2012.
85. C K Park, S B Park, S Y Lee, H Lee, H Jang, W I Cho, *Bull Korean Chem Soc*, 31(11), 3221, 2010.
86. D Wang, J Xiao, W Xu, J G Zhang, *J Electrochem Soc*, 157(7), A760, 2010.
87. Y Gao, C Wang, W Pu, Z Liu, C D P Zhang, Z Mao, *Int J Hydrogen Energy*, 1, 2012.
88. J Xiao, J Hu, D Wang, D Hu, W Xu, G L Graff, Z Nie, J Liu, J G Zhang, *J Power Sources*, 196, 5674, 2011.
89. M Pumera, *Energy Environ Sci*, 4, 668, 2011 and references therein.
90. L Qu, Y Liu, J-B Baek, L Dai, *ACS Nano*, 4(3), 1321, 2010.
91. B Sun, B Wang, D Su, L Xiao, H Ahn, G Wang, *Carbon*, 50, 727, 2012.
92. J Xiao, D Mei, X Li, Wu Xu, D Wang, G L Graff, W D Bennett, Z Nie, L V Saraf, I A Aksay, Jun Liu, Ji-G Zhang, *Nano Lett*, 11, 5071, 2011.
93. Y Li, J Wang, X Li, D Geng, M N Banis, R Li, X Sun, *Electrochem Commun*, 18, 12 (2012).
94. Y Li, J Wang, X Li, J Liu, D Geng, J Yang, R Li, X Sun, *Electrochem Commun*, 13, 668, 2011.
95. X Yang, P He, Y Xia, *Electrochem Commun*, 11, 1127, 2009.
96. S R S Prabaharan, Renal Baknov, Nicolas Burn, M S Michael, *J Power Sources*, 2014 (to be published).
97. H-D Lim, H Song, H Gwon, K-Y Park, J Kim, Y Bae, H Kim, S-K Jung, T Kim, Y H Kim, X Lepró, R O-Robles, R H Baughman, K Kang, *Energy Environ Sci*, 6, 3570, 2013.
98. W Gu, D R Baker, Y Liu, H A Gasteiger, *Handbook of Fuel Cells: Fundamentals, Technology, and Applications*, John Wiley & Sons, Chichester, Vol. 6, p. 631, 2009.
99. W Vielstich, H A Gasteiger, H Yokokawa (Eds.), p. 631, John Wiley & Sons, Chichester, 2009.
100. Yi-Chun Lu, H A Gasteiger, M C Parent, V Chiloyan, Y Shao-Horn, *Electrochem Solid-State Lett*, 13, A69, 2010.
101. Y C Lu, Z Xu, H A Gasteiger, S Chen, K H Schifferli, Y Sho-Horn, *J Am Chem Soc* 132, 12170, 2010.

102. A Debart, J Bao, G Armstrong, P G Bruce, *ECS Transactions*, 3(27), 225, 2007.

103. H Cheng, K Scott, *Appl Catal B*, 108, 140, 2011.

104. A K Thapa, K Saimen, T Ishihara, *Electrochem Solid-State Lett*, 13, A165, 2010.

105. V Giordani, S A Freunberger, P G Bruce, J M Tarascon, D Larcher, *Electrochem Solid-State Lett*, 13, A180, 2010.

106. L Wang, X Zhao, Y Lu, M Xu, D Zhang, R S Ruoff, K J Stevenson, J B Goodenough, *J Electrochem Soc*, 158(12), A1379, 2011.

107. H Wang, Y Yang, Y Liang, G Zheng, Y Li, Y Cui, H Dai, *Energy Environ Sci*, 5, 7931, 2012.

108. H Wang, X Liao, Q Jiang, X Yang, Y He, Z Ma, *Chinese Science Bulletin*, doi: 10.1007/s 11434-011-4944-7.

109. G Q Zhang, M Hendrickson, E J Plitcha, M Au, J P Zheng, *J Electrochem Soc*, 159(3), A310, 2012.

110. X Ren, S S Zhang, D T Tran, J Read, *J Mater Chem*, 21, 10118, 2011.

111. S H Oh, L F Nazar, *Adv Energy Mater*, 2, 903, 2012.

112. Y Wang, H Zhou, *Chem Commun*, 46, 6305, 2010.

113. Y Lu, Z Wen, J Jin, Y Cui, M Wu, S Sun, *J Solid State Electrochem*, 16, 1863, 2012.

114. K M Abraham, *J Power Sources*, 34, 81, 1991.

115. S R S Prabaharan, J Kawamura, N Kuwata, M S Michael, to be presented in 6th Forum on New Materials, CIMTEC 2014, 15–19 June 2014, Italy.

Chapter 5

Nano Anode Materials for Lithium-Ion Batteries

Lijun Fu,[a] Yi Shi,[a] Shu Tian,[a] Yuping Wu,[a] and Teunis van Ree[b]

[a] *New Energy and Materials Laboratory (NEML), Department of Chemistry & Shanghai Key Laboratory of Molecular Catalysis and Innovative Materials, Fudan University, Shanghai 200433, China*

[b] *Department of Chemistry, University of Venda, Thohoyandou, 0950 South Africa*

wuyp@fudan.edu.cn

Nano anode materials brought impetus for the development of electrode materials for advanced battery applications. In lithium-ion batteries (LIBs), for instance, the advanced electrode materials suffer from intrinsic electronic–ionic conductivities which limit their performance. Nevertheless, use of nanomaterials help overcome such conductivity issues as the size of the grains offer shorter diffusion length of Li^+ as well as electrons and providing preferable accommodation to the strain for lithium insertion–extraction. Nanoporous materials, on the other hand, facilitate accessibility of electrolyte and enhance the rate of performance. This chapter enunciates the various nano advantages in LIB applications focusing on various anode materials (negative electrodes), for instance, carbon nanotubes (CNTs), carbon nanofibres, Si and Sn compounds, and 3d transition metal oxides.

Nanotechnology in Advanced Electrochemical Power Sources
Edited by S. R. S. Prabaharan and M. S. Michael

ISBN 978-981-4241-43-4 (Hardcover), 978-981-4303-05-7 (eBook)
www.panstanford.com

5.1 The Development of Anode Materials for Lithium Ion Batteries

Since the birth of LIBs in the early 1990s, their development has been rapid. Nowadays they are widely used in variety of electronic mobile gadgets, power tools, and many original equipment manufacturer (OEM) gadgets. Although the manufacturing technology has majored enough, further development in terms of its performance and focus on its materials becomes mandatory. Anode material is one key component which includes carbon, alloy, oxides, phosphides, nitrides, and so on.[1] Recently, to get high rate ability, that is, to charge and discharge at high current densities, nano anode materials have attracted a great deal of attention among battery scientists.

Nanostructured materials represent a rapidly growing area in the field of LIBs, because of their substantial advantages in terms of mass transport. Nanoparticle systems typically encompasses pronounced size-effects, dimensional confinement, shorter transportation lengths for both electrons and Li^+ ions, larger electrode–electrolyte contact area and ratio of surface/volume, better accommodation of the strain of Li^+ insertion–extraction, and in some cases local anomalies.[2–5] Therefore, elaborate nanocomposites generally show higher reversible capacity,[6] higher power density, and better cycling behavior than their simpler counterparts. Of such nanomaterials, carbon nanomaterials, Si-based nanocomposites, Tin (Sn)-based nanocomposites, and nanostructured oxides have received a great deal of attention, and are considered as alternative anode materials.[1,7]

5.2 Nanocarbon Anode Materials

Hitherto, graphitic carbons are still the dominant anode materials for LIBs. This indicates the importance of carbonaceous anode materials in LIBs. Here, we mainly introduce CNTs and carbon nanofibers as anode candidates for LIBs.

5.2.1 *Carbon Nanotubes*

Carbon nanotubes have been used in various applications since the primary detection in 1991, including its potential as anode materials of LIBs.[8] CNTs are the hollow tubes which are rolled up from 2D grapheme sheets with diameters in the nanometer range and lengths usually on the micro scale. Based on the number of tube walls, CNTs can be classified into single-walled carbon nanotubes (SWNTs) and multi-walled carbon nanotubes (MWNTs). SWNTs are usually 1 nm to 2 nm in diameter and tens of micrometers in length. The specific area can be as large as 2630 m^2/g. MWNTs are made up of coaxial cylindrical carbon layers with an interlayer distance of 0.34–0.37 nm, which is a little larger than that of graphite (0.335 nm).[9] As result, CNTs combine the structural features of both graphite and fullerene, which is very attractive for intercalation process.

The surface and pore structures of CNTs determine their storage performance of Li ion. There are three kinds of intrinsic pore structure for CNTs. The first pore is the hollow tube cavity, the CNTs are often capped with semi-sphere shaped fullerene molecules. For the Li ion insertion into the interior tube, it is necessary to take a decapping treatment. However, since the tube is quite long with a narrow diameter, even the CNTs are open; it is still not easy for the Li ion to access. The second pore structure is the interstitial pores (for SWNTs), shown in Fig. 5.1.[9] The SWNTs are easy to aggregate into bundles with a diameter of several tens of nanometers and form a triangular lattice structure through van der Waals forces. The interstitial pores are related with the spaces surrounded by the adjacent in-bundle SWNTs. These interstitial pores provide a pathway for Li ions to access the central portion of the bundle from radial direction instead of the axial direction. The third pore structure is the interlayer space between the coaxial cylindrical carbon layers (for MWNTs). This is another accommodation site for Li ions. Besides the intrinsic pores, there are also aggregated pores in CNTs between aligned MWNTs or SWNT bundles, or even formed by randomly packed MWNTs or SWNT bundles. They also contribute to the storage performance.

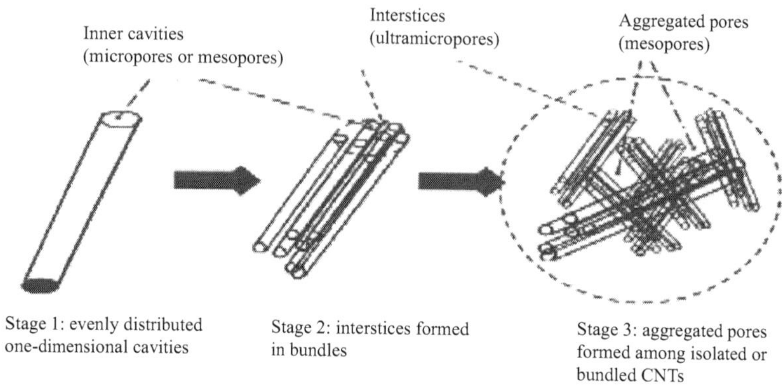

Figure 5.1 Pore formation model of SWNTs.[9]

Several problems hinder the application of CNTs as an anode material for LIBs. One problem for nanotubes is that the irreversible capacity is quite large and side reactions which probably occur on the surface of tubes with large specific surface area.[10] Another problem is the microtexture irregularities, like oxygen defect in the nanotubes, C–O–H, play a significant role in Li adsorption through $-C-O^{-}Li^{+}$, this attraction is so strong that the extraction of Li ions is less favorable, resulting in the hysteresis between insertion and de-insertion of Li[11] and low coulombic efficiency. Besides, the environment of the inner surface of the empty nanotube is similar to the basal plane of graphite, which is difficult for Li storage. The result why the Li storage of the inner space of the empty SWCNT is negligible.[12] *Ab initio* and density function theory (DFT) calculation also show that the insertion of Li ions through the side wall or through the capped zone of closed CNTs is energetically unfavorable unless structural defects exist.[13]

At first, we will expound on SWNTs. Purified SWNTs show a reversible saturation composition of $Li_{1.7}C_6$,[14] simulation results show that possibility of Li intercalation in SWNT bundles can reach a high saturation density of Li_3C_6,[15] higher than the LiC_6 ideal value for graphite and reported MWNTs.[11] During the Li intercalation process, the Li fills the SWNT bundle continuously and randomly, thus there are no staging phenomena, which are opposed to that of graphite.[16]

A lot of efforts were made to improve the performance by increasing the first coulombic efficiency as well as the reversible capacity. Ball-milling[10] and chemical etching[17] are adapted to open tube ends or form defects in the tube sidewalls, therefore, Li can intercalate inside the nanotube interior space. Especially after ball-milling, the reversible saturation composition is $Li_{2.7}C_6$ (~1000 mAh/g), and the first coulombic efficiency is improved as well. After chemical etching, the average length of nanotubes could be reduced from 10 μm to 4 μm and reversible Li storage capacity could be increased from LiC_6 to LiC_3.

The encapsulations of organic molecules, like 9,10-dichloro-anthracene, β-carotene, and coronene was found to be effective to increase the reversible storage capacity. Especially for the SWNT with coronene, the reversible capacity is 736 mAh/g, which is about 2.5 times greater than that of original empty tube. This is due to the steric hindrance of the organic molecules in the tube. The electrolyte molecules and solvated Li ions cannot enter the tube. The desolvation of the solvated Li ions occur at the entrance of the tubes, while the organic molecules offer accommodation of Li ions. However, the cycling performance and coulombic efficiency still need to be improved for practical application.[12]

Whereas, theoretical computations mainly focus on doping heteroatom to improve the storage performance of SWNTs. In the case of BC_3,[18] it is electron deficient. B-atoms have the ability to accept electron density from the adsorbed Li atoms, the energy barrier for Li adsorption could be decreased and enhance the Li storage capability. However, nitrogen-doping forms an electron-rich structure and cannot effectively accept electrons from the adsorbed Li atoms, as a result, hinders Li adsorption in SWNTs.[19] Recent DFT computations show that, CN_x with pyridine-like structures are electron deficient and decreases the energy barrier for Li penetrating pyridine-like defects and presents potential to enhance Li storage.[20]

As to MWNTs, they were investigated as an anode material for reversible Li-ion intercalation as early as 1998.[21] Lots of methods were adopted for preparation of MWNTs, and different MWNTs show various electrochemical performance.[16] Compared to SWNTs, the irreversible capacity of MWNTs is quite large. A typical example

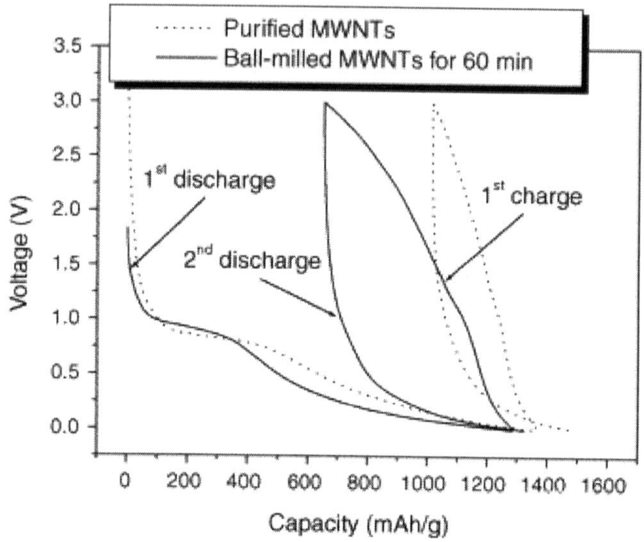

Figure 5.2 Charge–discharge curves of Li insertion–extraction from purified MWNTs and MWNTs ball-milled for 60 min. Reprinted from Ref. 22, Copyright 2006, with permission from Elsevier.

is that the chemical vapor deposition (CVD) prepared MWNTs could adsorb a large amount of Li, 925 mAh/g during the first cycle, while the reversible capacity is 505 mAh/g,[11] and the hysteresis of voltage between Li insertion and extraction is huge (Fig. 5.2).[22] Also, for a perfect structure with capped ends, Li ions cannot be inserted between the grapheme layers and into the inner core through the basal plane.[22]

Microtexture is a key factor to determine the Li storage performance of MWNTs, such as microtextural irregularities in the tube walls and graphitization. By heat-treating, the microtextural irregularities in the tube walls decrease and the nanotubes present better textural organization and graphitization, leading to not only decreased Li insertion but also extraction, thus both reversible and irreversible capacity decreases. It is consistent with the following report. Slightly graphitized MWNTs exhibit a high specific charge capacity of 640 mAh/g, in contrast, the well-graphitized MWNTs shows a lower capacity of 282 mAh/g during the first cycle.[23]

Chemical etching was adopted to improve the Li storage performance of MWNTs. After etching for 10 h, the reversible

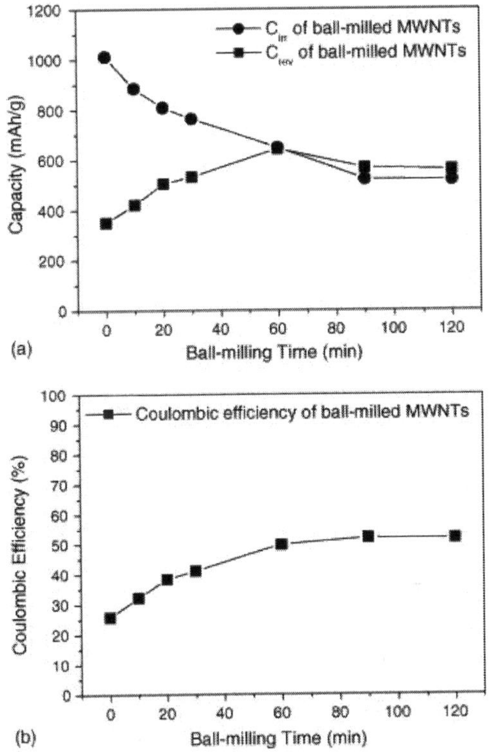

Figure 5.3 (a) Change in reversible capacity and irreversible capacity; (b) Coulombic efficiency of ball-milled MWNTs as function of ball-milling time. Reprinted from Ref. 22, Copyright 2006, with permission from Elsevier.

capacity increases from $Li_{0.9}C_6$ (351 mAh/g) to $Li_{1.8}C_6$ (681 mAh/g). The lateral defects, opened ends, edges of grapheme layers, and surface functional groups containing hydrogen and oxygen were formed on surface of the etched MWNTs with a reduced length after chemical etching. These structural and chemical modifications improve the insertion of Li ions into the etched MWNTs. However, the irreversible capacity also increases with the chemical etching, due to the larger surface area after chemical etching.[24]

Ball-milling is an effective method to increase reversible capacity of MWNTs from $Li_{0.9}C_6$ (351 mAh/g) to $Li_{1.7}C_6$ (641 mAh/g) (Fig. 5.2), in the meantime decrease the undesirable irreversible capacity (Fig. 5.3a) and increase the efficiency from 25 to 50% (Fig. 5.3b). And the capacity retention ability is better compared with the original MWNTs. The mechanism of the improvement is still not clear yet. But decreasing the voltage hysteresis after ball-milling should be one possibility (Fig. 5.2).[22] Cutting conventional micrometer-long CNT to short segments by solid state reaction achieved the same effect.[25]

5.2.2 *Carbon Nanofibers*

With the increasing research interest in CNTs, carbon nanofibers were also investigated as anode material for LIBs.

Various methods were adopted to fabricate carbon nanofibers with different performance for Li-ion insertion and extraction. High crystalline carbon nanofibers prepared by catalytic CVD presents high capacity (431 mAh/g).[26] Relatively, crystalline carbon nanofibers obtained by applying a high DC voltage to a mist of 2-Propanol at low temperature (673 K) exhibits a large reversible capacity of approximately 340 mAh/g.[27] Carbon nanofiber web was fabricated by the combination of electrospinning of polymer poly-acrylonitrile (PAN) and thermal treatment in an inert atmosphere. This conductive web exhibits high surface area, presents high reversible capacity, approximately 450 mAh/g at a current of 30 mA/g, and shows good rate performance, 350 mAh/g at a current of 100 mA/g.[28] However, same as CNTs, the carbon nanofibers exhibit large irreversible capacity during the first cycles, leading to poor capacity retention performance.

Nuclear magnetic resonance (NMR) measurement[29] reveals that the mechanism of reversible Li storage in carbon naofibers takes place by surface storage (Li with pure ionic character) and intercalation into the carbon matrix (lower ionic character Li). The properties of both carbon matrix and surface are important to the electrochemical performance of carbon nanofibers. For the carbon matrix, as mentioned above, the high crystalline carbon nanofibers presents high capacity.[26] For the surface properties, since the surface area of carbon nanofibers is large and the electrode suffers large irreversible capacity due to the side reaction on the surface. Besides, the carbon nanofibers always capture water or other molecules such as O_2 from the atmosphere which are harmful to the electrochemical performance.[29]

Surface modification was used to address this problem by impregnating lithium oxalate ($Li_2C_2O_4$) to carbon nanofibers, a stable passivating surface film can be formed, which reduces the side reaction of electrolyte on the surface of carbon nanofibers. Therefore, the capacity retention ability is better compared with the bared carbon nanofibers. However, for both of the bare carbon nanofibers and impregnated oxalates, the reversible capacity is around 200 mAh/g, which is still lower compared with the standard graphite electrode. With thermal and hydrothermal treatments to remove these harmful impurities, the reversible capacity increased, especially after hydrothermal treatment at 150°C for 2.5 h and then drying under vacuum, the reversible capacity of carbon nanofibers maintains between 302 and 285 mAh/g in the first 50 cycles. After chemical etching in refluxing acid, the reversible capacity increases over 400 mAh/g, and the extra capacity is due to the Li that exhibits pure ionic character.[29]

5.3 Nano Si-Based Anode Materials

Besides carbon materials, the most promising candidate for anode material is silicon (Si). It has a theoretical capacity as high as $Li_{4.4}Si$ (4200 mAh/g).[1,30] In addition to its outstanding capacity, Si is the second most abundant element on earth. Therefore, Si attracts high attention for researchers.

The main problem of Si is its volume expansion during Li insertion–extraction process, that is, 400% volume expansion of the Si lattice occurs after insertion. This volume expansion causes poor electronic contact between the Si particles and finally, large irreversible capacity and continuous capacity fading. When increasing the pressure of the cell or the amount of the conductive additives, the electrochemical performance could be improved. Amorphous Si shows better electrochemical performance than crystalline Si,[31] due to the homogeneous expansion of volume of amorphous Si.

Reducing the Si particle size to nanoscale is a most widely studied method to solve this problem.[32] When the particle size is nanoscale, the volume expansion is reduced. The discharge capacity is 2775 mAh/g and reversible capacity is 2097 mAh/g, after 10 cycles, the reversible capacity remains 1729 mAh/g. However, even the volume change could be reduced, the capacity fading still remains for nano-Si.

Fabricating composites of Si and conductive materials is quite effective to improve the capacity retention ability of nano-Si. The conductive materials include Li inactive materials and Li active materials.[33]

TiN is inactive Li. It has high electronic conductivity and high mechanical strength to withstand the volume change of Si on cycling. TiN–Si nanocomposites prepared by ball-milling for 12 h exhibit a stable capacity of ~300 mAh/g,[34] indicate a promising future of this method.

Graphite is a commercial available anode material for LIBs right now. The electrochemical performance of the Si could be improved by combing the features of both Si and graphite. By pelletizing a mixture of nano-Si/graphite/petroleum pitch powders, followed by pyrolysis at 1000°C under argon (Ar) atmosphere, a nano-Si/graphite/carbon composite of spherical shape is synthesized. The composites show a high reversible capacity of ~700 mAh/g and a high initial coulombic efficiency of 86%.[35]

Metal is high conductive material. Preparing nanocomposites of metal and Si is another way to improve electrochemical performance of Si. With sol–gel method, a layer of cobalt oxide (Co_3O_4)–Co was coated on the surface of Si. The modified Si obtains high reversible capacity of 2650 mAh/g with high first coulombic efficiency of 88%. The cycling performance is good after 12 cycles and the capacity could remain as 80%. It is due to the cobalt metal on the coating layer, which provides good electron conductivity.[36] Thin film technology was adopted to prepare metal and Si composites. Si–Sn nanocomposites were prepared *via* vacuum deposition on a metal foil substrate. The as-prepared film presents higher conductivity than a pure Si film. After 500 cycles, the 350 nm thick film retains a capacity of 1400 mAh/g under 10C rate current.[37] To modify the morphology of current collector is another approach to improve Li storage ability of Si. Three dimensional Copper current collector was prepared by multi-step electrochemical deposition. Si film deposited on the 3D current collector showed higher conductivity and better cycling performance than that on 2D current collector.[38]

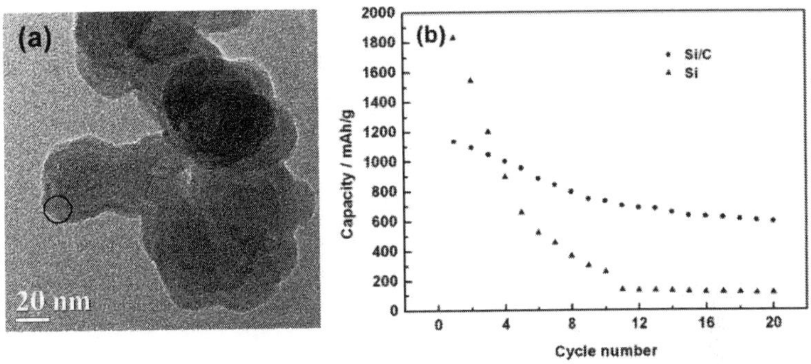

Figure 5.4 (a) Transmission electron micrograph (TEM) of core-shell Si–C nanocomposite; (b) The cycling behavior of the core-shell Si–C nanocomposite. Reprinted from Ref. 39, © *Pure and Applied Chemistry* 2006.

Another effective method is to achieve core-shell structured nanocomposites. The shells can be C and SiO_x. In the case of Carbon-coated nano-Si composite, the preparation is as follows:[39] as anode material for LIBs. At first, PAN-coated Si nanoparticles are formed by emulsion polymerization and the PAN-coated precursor is heat-treated under Ar to generate a Si–C core-shell nanocomposite (Fig. 5.4a). The conductive carbon shell envelops the Si nanoparticles and suppresses the aggregation of the Si nanoparticles during electrochemical cycling, subsequently ameliorating the capacity retention of this composite anode material (Fig. 5.4b).

The Si–SiO core-shell nanocomposite is prepared by a sol-gel method in combination with a subsequent heat-treatment process.[40] At first, molten Si with high purity was irradiated by laser and gasified under an Ar atmosphere, and the condensed Si nanoparticles were collected on a water-cooled wall. Then, the obtained Si nanoparticles were dipped into a stirred tetraethylorthosilane (TEOS) sol, followed by the processes of pH adjustment, aging, drying, and heat treatment. Finally, a core-shell structured Si–SiO nanocomposite was synthesized. Because the Si-hydroxyl groups in the TEOS sol possess high activity, the silica gel can be easily adsorbed onto the surface of the Si nanoparticles. The as-synthesized Si–SiO composites yield a core-shell structure. Besides, the crystalline structure of the obtained SiO shell is highly

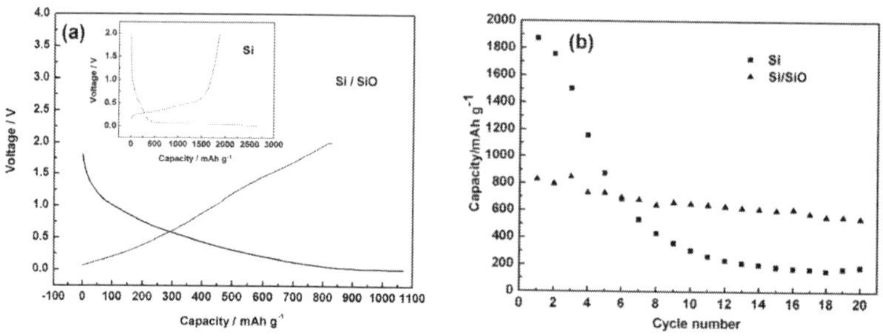

Figure 5.5 (a) Charge–discharge profiles in the first cycle; (b) Cycling behaviors of the Si–SiO nanocomposite and the virginal Si nanoparticles. Reprinted from Ref. 40, Copyright 2007, with permission from Elsevier.

compatible with that of nano-Si, with a face-centered cubic structure, which can be seen from the X-ray diffraction (XRD) patterns. After the coating, the core-shell Si–SiO nanocomposite displays better reversibility of Li insertion–extraction and higher coulombic efficiency than the virginal Si nanoparticles (Fig. 5.5a). Besides, good capacity retention for the Si–SiO core-shell nanocomposite was obtained (Fig. 5.5b). The SiO shell acts as a barrier to prevent the aggregation of the Si nanoparticle and serves as a buffer as well to alleviate the volume expansion during the lithiation process. Consequently, the capacity retention of the core Si nanoparticles is apparently improved. As to the decrease of the reversible capacity of Si–SiO nanocomposite compared with the virginal Si nanoparticles, this can be explained mainly by the fact that the reversible capacity of the SiO shell is lower than that of the Si nanocomposite.

This research as well as the core-shell Si–C research indicates that the core-shell structure offers a promising method to improve the electrochemical performance of nano-Si as electrode material for LIBs.

Mesoporous Si–C core-shell nanowires (NWs) were prepared by using a SBA-15 template. As-prepared NWs exhibit excellent first reversible capacity of 3136 mAh/g and a coulombic efficiency of 86% at a rate of 0.2C (600 mA/g) between 1.5 and 0 V. After 80 cycles, the capacity remains at about 2700 mAh/g at 0.2C. The rate performance is good and the capacity at 2C (6000 mA/g) is about 2500 mAh/g.[41]

5.4 Nano Sn-Based Anode Materials

5.4.1 *Metallic Sn*

Tin exhibits a high theoretical Li storage ability of $Li_{22}Sn_4$, that is, 994 mAh/g, thus it has been investigated as another promising candidate as anode material for LIBs. However, similar with Si, it suffers huge volume expansion during the formation process of Li_xSn, resulting in poor capacity retention ability. Decrease of the particle size of Sn metal is popular approach for improving electrochemical performance. It is easily expected that the passage from bulk to nanostructure results in (1) a higher interfacial area and shorter length for Li-ion transportation, leading to higher charge–discharge rates and (2) reduction of volume change accompanying with Li insertion–extraction. However, problem of the capacity fading still remains and requests further modification. The efforts made mainly focus on the concept involves the intimate mixture of Sn and another material, where the added material plays a role of buffer for the volume expansion, including graphite, carbon, and other metals.

Graphite could be used as a soft matrix which buffers volume expansion of Sn, at the same time, graphite could also offer accommodation for Li. Sn nanoparticles with particle size of 10 or 3.5 nm were prepared from 1,10-Phenanthroline/$SnCl_4$ precursor. By adding graphite at the beginning of the synthesis procedure, Sn nanoparticles could load on the graphite homogeneously. The 10.3%, 3.5 nm Sn–graphite nanocomposite displays a high reversible capacity of 415 mAh/g, after 60 cycles, the capacity retains 91.3% of that in the first cycle.[42]

Carbon could also be a structural buffer and Li active materials for improvement of Sn nanoparticles Li storage performance. Amorphous carbon could be coated on the Sn nanoparticles by using hydrothermal method at 180°C with glucose as carbon precursor. The amorphous C–Sn nanocomposites enhance dimensional stability during cycling and present a higher reversible capacity of 681 mAh/g than bare Sn and 98% retention after 50 cycles (Fig. 5.6a), although the amorphous carbon contributes 68 mAh/g irreversible capacity for the first time (inserted plot in Fig. 5.6b).[43] When adopting allytriphenyltin (ATP-Sn) as precursor, through solid-

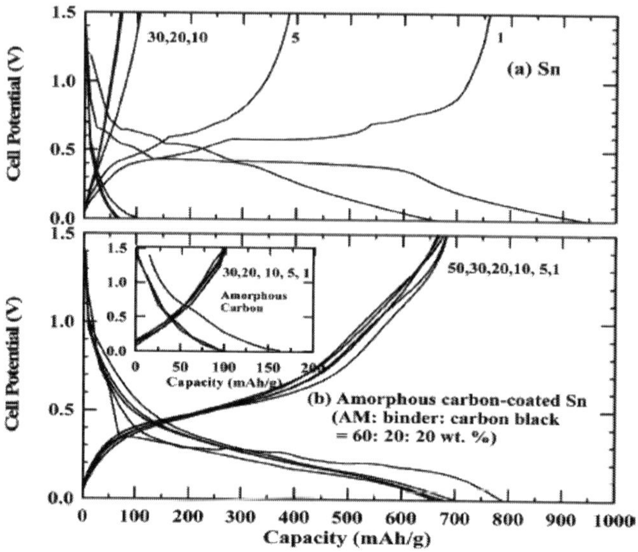

Figure 5.6 (a) Voltage profiles of tin nanoparticles; (b) Amorphous C–Sn nanocomposites in a coin-type half cell (Li metal as a counter electrode) between 0 and 1.5 V at 0.1C rate (1C = 800 mAh/g). Reprinted with permission from Ref. 43. Copyright 2005 American Chemical Society.

state pyrolysis, hollow Sn nanopartricles encapsulated with carbon were obtained. These C–Sn nanocomposites with hollow structure exhibit high reversible capacity of 550 mAh/g and excellent rate performance. The hollow morphology facilities Li-ion diffusion and limits the volume expansion of Sn, resulting into improved Li storage performance.[44] Another organometallic tin precursor was infiltrated into an organic resorcinol-formaldehyde gel, followed by calcination under Ar, Sn–C nanocomposites were obtained with a homogeneous Sn nanoparticles dispersion in carbon matrix (Fig. 5.7). They present high reversible capacity of 500 mAh/g at a current of 0.8C (about 0.2 A $cm^{-2}g^{-1}$), which stays in constant for 100 cycles. When increasing the current to 5C, the capacity recovers 40% of that at 0.8C.[45] CNTs can also be used as a carbon buffer for Sn–C nanocomposites. However, the high irreversible capacity should be reduced.[46]

Another strategy to enhance the cycling performance of nano-Sn as anode material for Li-ion battery is preparing M_xSn_y alloys, where

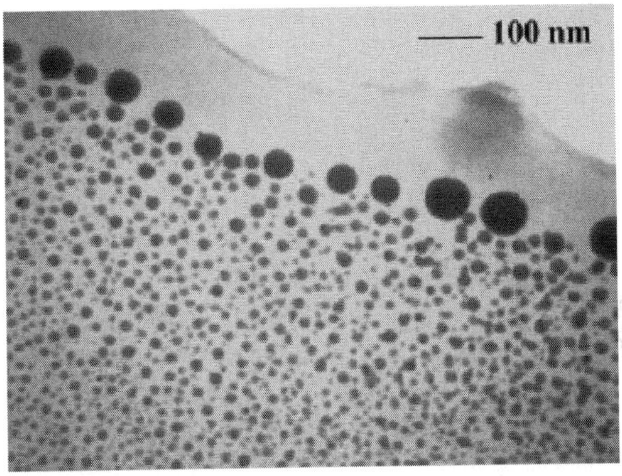

Figure 5.7 TEM image of a Sn–C composite material. Reprinted from Ref. 45, with permission. Copyright © 2007 WILEY-VCH Verlag GmbH & Co. KGaA, Weinheim.

M is electrochemically inactive transition metal (M = Fe, Ni, Cu, and Co). M also provides a matrix that buffers volume changes occurring with the lithiation–delithiation processes, therefore the mechanical integrity between Sn nanoparticles and with current collector can be maintained. Both preparation of alloy nanopowders and thin films are popular strategies.

Amorphous Sn–Co alloy presents good electrochemical performance. In the amorphous alloy Sn_xCo_{1-x}, there is little amount of Co_3Sn_2 existing besides the amorphous phase when $0.43 < x < 0.6$. When $x = 0.4$, the reversible capacity of alloy is 650 mAh/g. However, the capacity decreases a lot during cycling due to the aggregation of Sn to form Sn-rich region. The capacity fading becomes serious with the decrease of Co amount in the alloy. Adding carbon into the alloy to form amorphous $(Sn_{0.55}Co_{0.45})_{1-y}C_y$ composites enhanced the capacity retention ability.[47]

Nickel (Ni) could alloy with Sn in the form of Ni_3Sn and Ni_3Sn_2. The Li storage performance varies from the alloy prepared by different approaches. Amorphous Ni–Sn alloy prepared by reduction at room temperature shows nano-scale particle size and high discharge capacity (above 1000 mAh/g), but following with large irreversible capacity loss in the following cycles. Crystalline Ni_3Sn_2

reduced by H_2 at high temperature presents micro-scale size and very stable reversible capacity of more 300 mAh/g over 40 cycles.[48] Modification of copper current collector is another way for improvement of the Ni–Sn alloy performance. By applying anodic alumina oxide (AAO) template, Cu nanorods were electrodeposited on the surface of copper current collector, where the Ni–Sn alloy was electrodeposited. With this structure, the Ni–Sn alloy presents a high reversible capacity of 500 mAh/g, without capacity fading for 200 cycles.[49] By using polystytrene (PS) spheres as a template, Sn–Ni alloy was electrodeposited into the interstitial spaces of the PS template. Three dimensional Sn–Ni alloy was obtained by dissolving PS into tetrahydrofuran (THF) solvent. The reversible capacity of first cycle is 692 mAh/g, 62% capacity retains at the 75th cycle.[50]

Copper is another popular metal to alloy with Sn for improvement. However, Cu–Sn alloy shows poor cycling performance.[51] When coating with a carbon layer, the capacity is stabilized with a capacity of ~400 mAh/g for 50 cycles.[52] Three dimensional foam structured Cu–Sn alloy was fabricated *via* an electrochemical deposition process with a hydrogen over-voltage. Hydrogen bubble decomposed from the solvent becomes a template during the electrode position process, and the metal growth toward the gas bubble is prohibited leading to formation of 3D foam structured alloy. This porous structured alloy delivers a reversible capacity of about 400 mAh/g up to 30 cycles. When increasing the current rate to 10 mA/cm^2, the capacity remains more than 50% of the capacity at 0.5 mA/cm^2.[53]

5.4.2 *SnO_2*

SnO_2 is regarded as a potential anode material for its high theoretical specific capacity (781 mAh/g). One of the problems that impede the application of SnO_2 is that the large volume expansion (~300%) during the charge–discharge process,[54] resulting in pulverization leading to poor electrical contact and hence, the limited cycling life of electrode.

Fabrication of low dimensional SnO_2, that is, nanorods/wires and nanotubes, is one such approach to circumvent this problem. SnO_2 NWs can be prepared on a Si substrate *via* thermal evaporation method combined with self-catalyzed procedure, the as-prepared

SnO_2 NWs still suffer from capacity fading, but the capacity retention ability is better compared with SnO_2 nanopowders.[55] Traditional 1D SnO_2 nanostructures still need to be mixed with binder and carbon black and further pressed onto current collector when fabricating as an electrode. This process introduces additional undesirable interfaces in the electrode, resulting in more side reaction happened on these interfaces. By adopting a hydrothermal method, SnO_2 nanorods could be synthesized on the metal substrates (Ni or Fe–Co–Ni foil) directly in a large scale. Nanorods with a diameter of 60 nm and a length of 670 nm deliver a reversible capacity of 580 mAh/g after 100 cycles at 0.1C with a slight capacity fading in the first 40 cycles. When increasing the rate to 5C, the capacity retains at 350 mAh/g demonstrating a good rate performance as well.[56]

The concept to prepare SnO_2 particles with nanoporous or hollow structure[57,58] is effective to solve the problem of volume expansion of SnO_2 during charge–discharge process. These two structures provide empty space, which could partially accommodate the large volume change. Mesoporous SnO_2 was synthesized with a hard template of KIT-6 mesoporous silica. The as-prepared mesoporus SnO_2 exhibits high surface area (160 m^3/g) and reversible capacity as high as 800 mAh/g at 0.2C (1°C = 400 mA/g). The capacity retention is 773 mAh/g up to 50 cycles. At 10C rate, the reversible capacity is 98% of that at 0.2C, presenting a good rate performance.[59]

Another direction is to improve the electron conductivity of SnO_2 by preparing SnO_2–C composites in most occasions. Various approaches were used for preparation of SnO_2–C composites, such as hydrolysis method,[60] hydrothermal method[61] and oxidation of SnO_2 precursor in a supercritical CO_2–methanol mixture containing CNTs.[62]

As mentioned earlier, CNTs exhibit unique one dimensional tubular structure and high conductivity, thus it is a promising nanomaterial to improve the electron conductivity of SnO_2. An ordered tetragonal single-crystal SnO_2 NW arrays with *in situ* formation of CNTs coating was achieved by using AAO membrane as a template *via* a sol–gel route.[63] A typical SEM micrograph of the AAO membrane is shown in Fig. 5.8. Amorphous carbon nanotube (a-CNT)-coated single-crystal SnO_2 NWs with a mean diameter of approximately 80 nm were synthesized by annealing the anodic

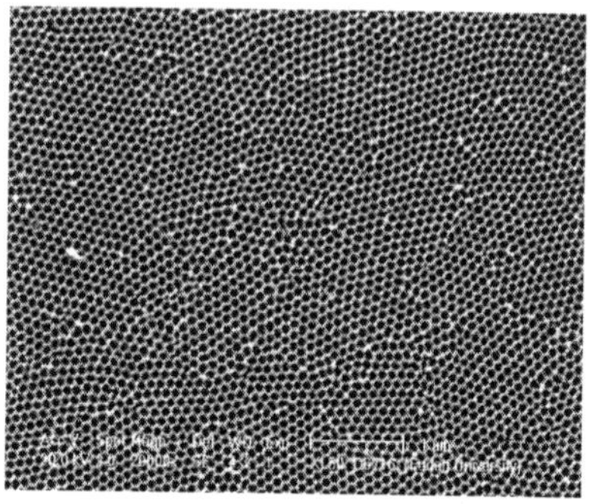

Figure 5.8 Scanning electron micrograph (SEM) image of anodic alumina oxide (AAO) membrane. Reprinted with permission from Ref. 63. Copyright 2008 American Chemical Society.

Figure 5.9 SEM micrographs of SnO_2 NWs after the AAO template was (a) partially and (b) totally removed by immersing in 0.5 and 1 M NaOH solution for 2 h and 4 h, respectively; (c) TEM of single SnO_2 wire; the inset is the selected area electron diffraction (SAED) pattern of the single-crystal SnO_2 NW in (c); (d) HRTEM micrograph of the area marked in (c) with a square. The black arrow indicates the *in situ* formed thin layer. Reprinted with permission from Ref. 63. Copyright 2008 American Chemical Society.

alumina membrane in air filled with a SnO_2 sol made from $SnCl_2 \cdot 2H_2O$ and citric acid.

Uniform SnO_2 NWs were observed after the alumina was partially (Fig. 5.9a) and totally (Fig. 5.9b) removed by the alkali solution. The typical high resolution TEM micrograph (Fig. 5.11d) of the single SnO_2 NW (Fig. 5.9c) shows the single-crystal structure. The interplanar spacing of the marked (110) facets was measured to be 0.332 nm. Analysis of the SAED pattern reveals that the tetragonal SnO_2 crystal structure is consistent with the rutile SnO_2.

A carbon layer with an average thickness of about 7 nm was observed on the surface of the SnO_2 NWs, indicated with a black

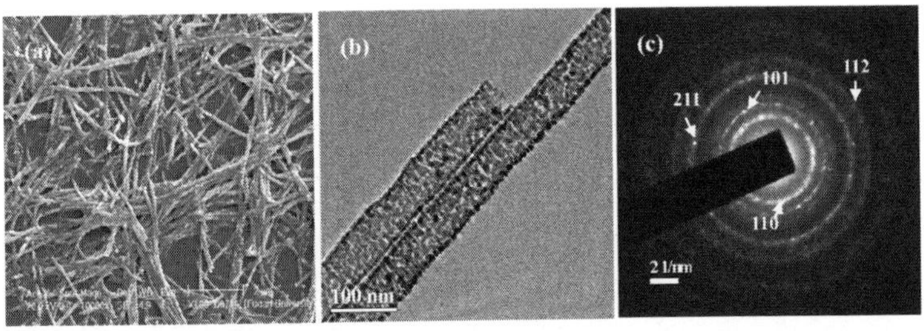

Figure 5.10 (a) SEM; (b) TEM of polycrystalline SnO_2 NWs after the AAO template was totally removed by immersing in 1 M NaOH solution for 4 h; (c) the corresponding SAED pattern of the polycrystalline SnO_2 NWs. Reprinted from Ref. 64, © *Pure and Applied Chemistry* 2008.

arrow (Fig. 5.9c,d). The outer carbon layer can be removed after further annealing the template-free SnO_2 NWs in air at 450°C for 1 h, while the color transformation from black to pure white can be obviously observed.

When the concentration of the SnO_2 sol is decreased, carbon nanotube-coated polycrystalline SnO_2 NWs can be synthesized.[64] As seen in Fig. 5.10a,b, the NWs are composed of many nanoparticles which are situated in the walls of the CNTs with a mean diameter of 70 nm (according to the reproduced AAO nanopores). It seems that the crystallization process was weakened since the interconnection between the nuclei was lessened when the SnO_2 precursor was applied. As a result the scattered nanoparticles were unable to grow into one single crystal in the calcination process. The SEAD pattern in Fig. 5.10c confirms the polycrystalline SnO_2 microstructure.

The galvanostatic charging and discharging curves in Fig. 5.11 are in agreement with the normal slopes of powder SnO_2, and there is no great difference between single-crystal and polycrystalline SnO_2 NWs. The polarization of the single-crystal NWs generally seems to be larger than that of the polycrystalline NWs during the charging and discharging processes and the reversible capacity of the single-crystal NWs is a little less than that of the polycrystalline NWs. It is clear that the interstices in the polycrystalline SnO_2 NWs allow liquid electrolyte to penetrate into the tube so that the inside SnO_2 material is easily activated, while the activation process of the single-crystal SnO_2 NWs is not fast because the tubes are relatively full.

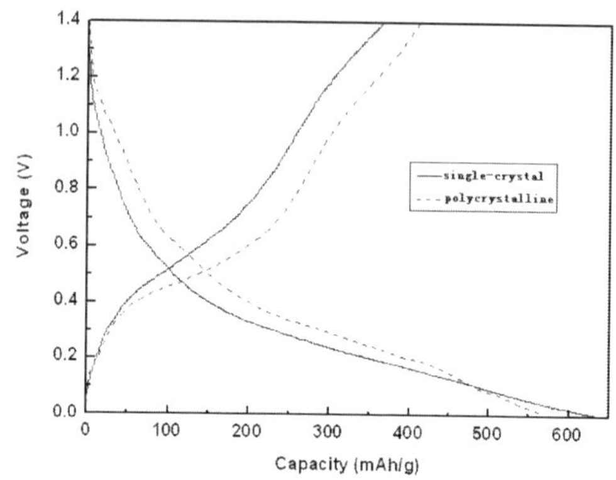

Figure 5.11 Galvanostatic charge and discharge curves of the single-crystal NWs and polycrystalline NWs in the 2nd cycle. Reprinted from Ref. 64, © *Pure and Applied Chemistry* 2008.

Nanowire arrays of a-CNT-coated single crystal and polycrystalline SnO_2 NWs can be easily fabricated by this method. They show great promise for application as anodes in LIBs for use in nanodevice power supply. In addition, other NW arrays of a-CNTs coated oxides can also be prepared by this process.

5.5 Titanium Oxide-Based Nano Anode Materials

Titanium oxide-based anode materials have been found to be one of the good candidates as Li-ion host. TiO_2 has a volumetric capacity of 1307 mAh/cm^3 if when one Li is inserted, that is, $LiTiO_2$. This will be higher than that of graphite (837 mAh/cm^3), along with its low cost and non-harm features.[65] It is of great interest to be investigated as a candidate for Li-ion storage. Various polymorphs of TiO_2 are available.[66,67] Here, we mainly focus on the Li storage performance of anatase and rutile TiO_2, TiO_2(B), and $Li_4Ti_5O_{12}$.

5.5.1 *Anatase TiO_2*

Anatase TiO_2 is generally considered to be the most electroactive Li insertion host among all TiO_2 polymorphs. Many efforts were made to investigate anatase TiO_2 for Li storage. The Li insertion and extraction occur at the potential around 1.7 V and 2.0 V *vs.* Li^+/Li. The voltage is higher compared with that of other anode materials mentioned. Nano anatase TiO_2 performs much better compared with the micro-scale particles, due to the shorter pathway of Li ions into TiO_2 bulk The nano TiO_2 shows a reversible capacity of 170 mAh/g, while for the micro-TiO_2, the first reversible capacity is 30 mAh/g.[68] For further improvement of the capacity and rate performance of TiO_2, the concept of preparing nanoporous TiO_2 was widely used. Porous structure provides high surface areas readily accessible to electrolyte, as a result, reduces the transport lengths for both electronic and ionic transport. The hierarchically mesoporous TiO_2 sub-micron spheres with high porosity were obtained after removing the $CdSO_4$ component from the composite by washing with acid solution. The reported nano-sized building blocks result in excellent Li insertion–extraction performance.[69]

Recently we fabricate 3D mesoporous microparticles of anatase titania by adopting polystyrene (PS) microparticles as a template followed by calcination in atmospheric air to remove the PS template.[70] The preparation process of the 3D mesoporous titania is schematically shown in Scheme 5.1. As shown in Fig. 5.12, the cross-section of the prepared 3D mesoporous TiO_2 has a face centered

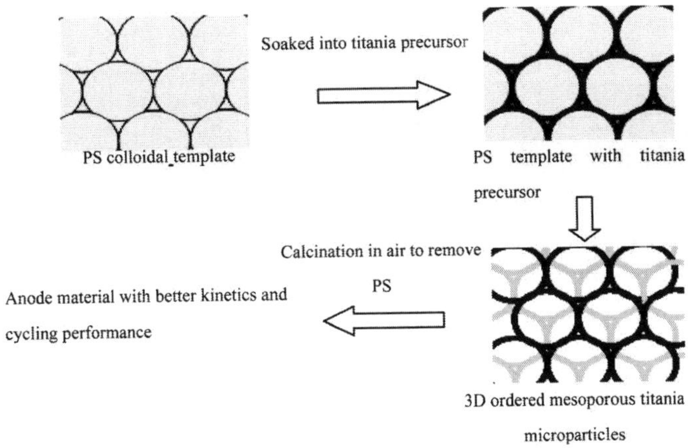

Scheme 5.1 Preparation process of the 3DOM titania microparticles formed by adopting colloidal polystyrene crystals as a template. Reprinted from Ref. 71, Copyright 2006, with permission from Elsevier.

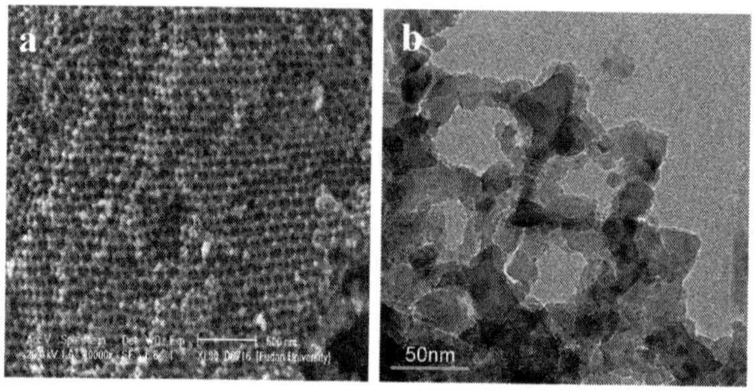

Figure 5.12 (a) SEM and (b) TEM of the synthesized 3DOM titania microparticles. Reprinted from Ref. 71, Copyright 2006, with permission from Elsevier.

cubic (fcc) structure, with a distance between centers of adjacent pores of about 120 nm, as a result of the replication of the PS template structure.

Although the porous structure facilitates Li-ion diffusion, the cycling performance of the 3D is not very good, which is due the poor electronic conductivity of the 3D structure. After coating the 3D TiO_2 with a carbon layer by adopting sucrose as carbon precursor, TiO_2–C nanoporous composites were synthesized which provides an electron conductive network. This network in combination with the nanoporous structure benefits the transportation for both Li ions and electrons during charge and discharge process. From the charge–discharge curves in Fig. 5.13a, the nanoporous TiO_2 presents a reversible capacity of 81.7 mAh/g and 47.6 mAh/g at 0.125 C and 0.5 C, while after carbon coating, the capacity is 84.0 mAh/g and 70.8 mAh/g at 0.125 C and 0.5 C, differently (Fig. 5.13b). Furthermore, the coulombic efficiency at high rate increases after carbon coating.[72] Three-dimensional porous RuO_2–TiO_2 nanocomp-sites were prepared to enhance the electron conductivity. RuO_2 is electron conductive, this 3D porous structure together with electron conductive network allow both Li^+ and e^- fast transportation and resulting in good cycling and rate performance.[69]

A core–shell structured nano TiO_2–C can be prepared as Scheme 5.2. The initial TiO_2 is anatase, PAN was coated on the surface of nano-TiO_2 particles with emulsion polymerization, after calcination at 800°C, nano TiO_2–C composites were obtained

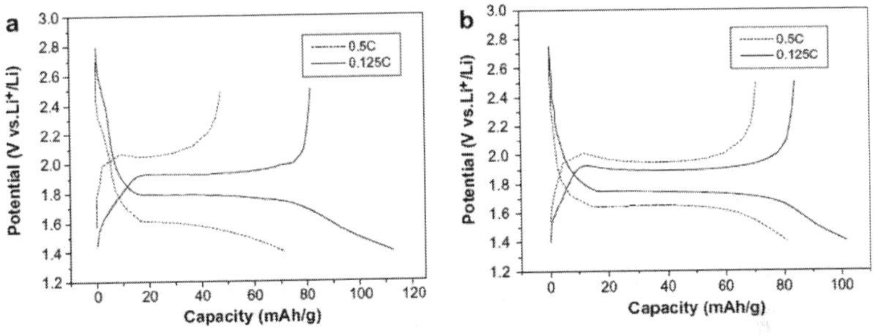

Figure 5.13 The discharge and charge curves in the first cycle of the nanoporous TiO_2 and TiO_2–C electrode in 1 M solution of DEC/EC/DMC (1:1:1, *v*/*v*/*v*) at different discharge and charge rates: (a) nanoporous TiO_2 and (b) nanoporous TiO_2–C electrode. Reprinted from Ref. 72, Copyright 2009, with permission from Elsevier.

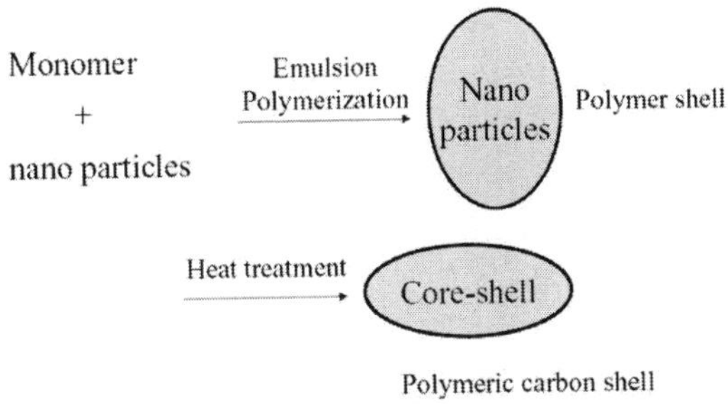

Scheme 5.2 Transmission electron micrograph of TiO_2–C core-shell nanoparticles.[71]

(Fig. 5.14). From Fig. 5.15, it is clear that the nano TiO_2–C composites is electroactive toward Li insertion and exhibits a reversible capacity of 122 mAh/g for the first cycle and remains at 96.7% after 10 cycles (Fig. 5.15b). Nano TiO_2–C composites present even better performance than that of anatase TiO_2 (Fig. 5.15a), this is due to the presence of conductive carbon shell, which facilitates the electron transfer and reduces the aggregation of nanoparticles.[71] In addition, the carbon shell ensures good cycling performance, which could not be simply solved by other kinds of nano TiO_2.

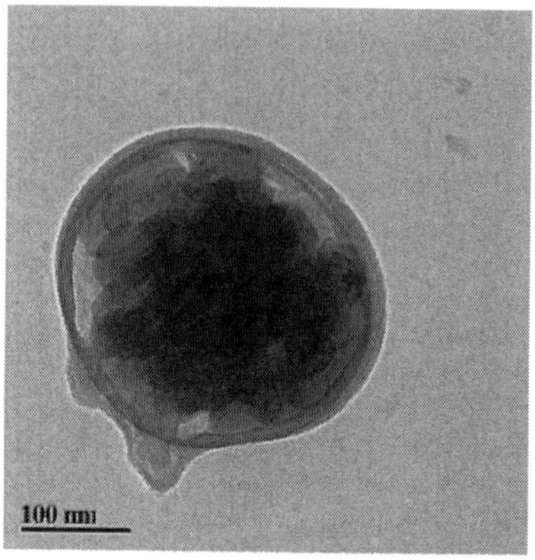

Figure 5.14 TEM of TiO_2–C core-shell nanoparticles. Reprinted from Ref. 71, Copyright 2006, with permission from Elsevier.

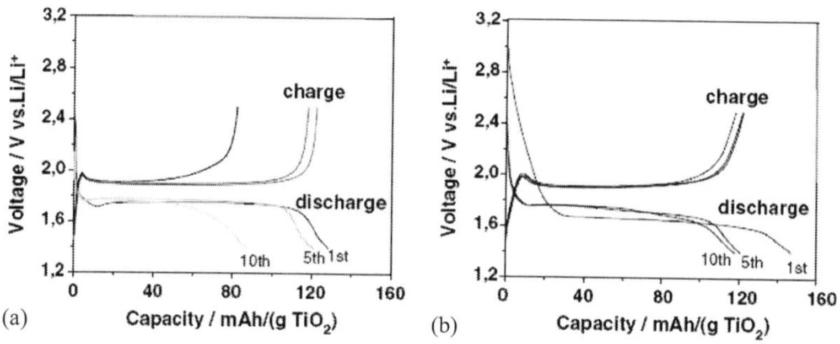

Figure 5.15 Discharge and charge profiles of (a) the anatase nano-TiO_2 particles and (b) rutile TiO_2–carbon core-shell nanocomposite at a constant current density of 0.1 mA/cm^2 in the voltage range 1.4–2.5 V. Reprinted from Ref. 71, Copyright 2006, with permission from Elsevier.

5.5.2 *Rutile TiO_2*

Rutile TiO_2 is reported to be negligible for Li insertion, which is caused by the sluggish Li diffusion along *ab*-plane. When reducing the particle size to nano-scale, Rutile TiO_2 shows high Li insertion activity. 0.8 mol Li can be inserted into nanorutile TiO_2, only 0.1–0.25 mol of Li can be inserted into micro-scaled rutile. Nanorutile TiO_2 also shows excellent retention ability on cycling and a high rate performance with a high capacity of 100 mAh/g at 10 C current rate.

The good Li storage of nanorutile TiO_2 might because Li stores at the surface or amorphous surface layer of nanorutile TiO_2.[73,74]

5.5.3 *TiO_2(B)*

TiO_2(B) was first synthsized in 1980 by ion exchange of K^+ for H^+ in $K_2Ti_4O_9$ to form a hydrated hydrogen titanate followed by calcination at 500°C to form to TiO_2(B).[75] Like rutile and anatase, TiO_2(B) is composed of corrugate sheets of edge- and corner-sharing TiO_2 octahedra, with the sheets being linked together by bridging oxygen atoms to form a 3D network However, the structure of TiO_2(B) shows an better capability as an intercalation host due to its significant voids and perovskite-like pathways formed by octahedra arrangement.[76] As comparison, the performance of anatase nanoparticles, TiO_2(B) NWs and bulk TiO_2(B) is shown in Fig. 5.16 (the particle size of anatase is equivalent to the diameter of width of TiO_2(B) NWs). It is obvious that both TiO_2(B) NWs and bulk TiO_2(B) show superior intercalation capacity and cycling performance to those of anatase nanoparticles.[77]

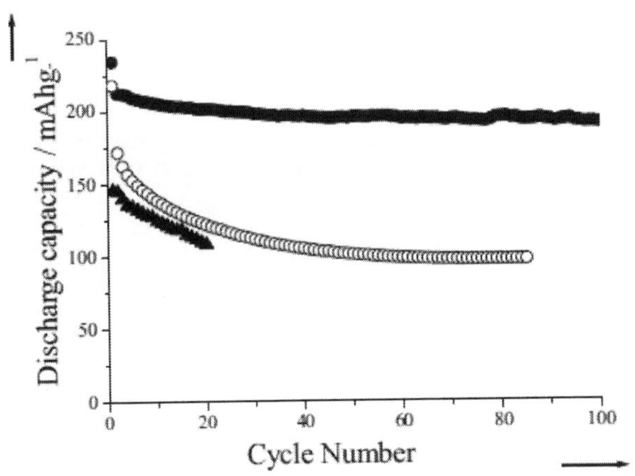

Figure 5.16 Comparison of intercalation capacity *vs.* cycle number for TiO_2(B) NWs (□), TiO_2(B) bulk (○) and anatase nanoparticles (▲) at the rate of 50 mA/g between voltage limits of 1 and 2.5 V. Reprinted from Ref. 77, with permission. Copyright © 2005 WILEY-VCH Verlag GmbH & Co. KGaA, Weinheim.

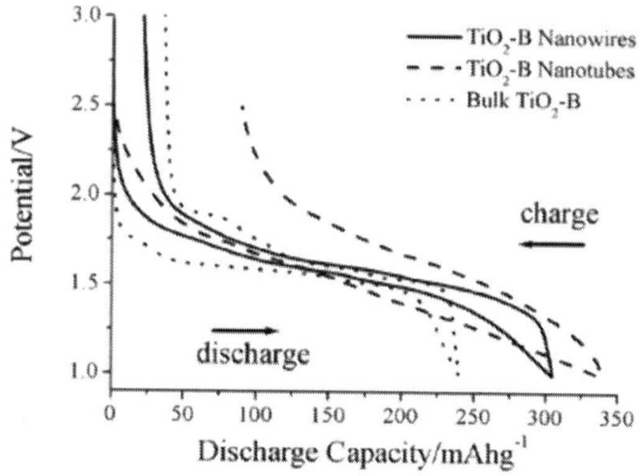

Figure 5.17 First charge–discharge curve for TiO_2(B) NWs, nanotubes and bulk TiO_2(B) at 10 mA/g between 1 and 3 V *vs.* Li/Li^+. Reproduced from Ref. 78 by permission of the The Electrochemical Society.

TiO_2(B) NWs attract more research interest since they show superior electrochemical performance as a Li^+ insertion host than their bulk counterpart (Fig. 5.16). TiO_2(B) NWs were prepared *via* a simple hydrothermal reaction. They could accommodate 0.91 Li to form $Li_{0.91}TiO_2$ (305 mAh/g) at 10 mA/g and show excellent rate performance, the intercalation capacities are around 200, 155, 125, 100 and 85 mAh/g at the rate of 200, 500, 1000, 2000 and 3000 mA/g. XRD patterns for TiO_2(B) NWs at various states of charge shows no obvious expansion of lattice after Li-ion insertion.[77]

By adjusting hydrothermal reaction parameter, TiO_2(B) nanotubes could be synthesized. The intercalation capacity could be up to 338 mAh/g ($Li_{1.01}TiO_2$), which is greater than the TiO_2(B) NWs and bulk TiO_2(B) (Fig. 5.17). Due to the small dimensions and thin wall of the nanotube, the charge and discharge curve is smooth for TiO_2(B) nanotube. TiO_2(B) nanotubes present excellent rate capability, and the intercalation capacity reduces only 10% when increasing the current rate from 10 mA/g to 200 mA/g, which is comparable to that of NWs.[78]

For all these three TiO_2(B) electrode, an apparent irreversible capacity exists during the first cycle. Among them, the TiO_2(B) nanotube shows the greatest irreversible capacity. Since it is believed that there is no surface layer formation above 1 V, the irreversible capacity may be due to the poor electronic conductivity of TiO_2(B).[78]

When applying TiO_2(B) NWs to fabricate TiO_2(B)–GPE–$LiFePO_4$ battery, a capacity of 225 mAh/g is delivered at C/5 with good cycle life, which is larger than the similar cell containing $Li_4Ti_5O_{12}$ as anode. The rate performance is excellent for TiO_2(B)–GPE–$LiFePO_4$ battery, 80% of the capacity at 5/C could be retained when increasing the rate to 5C. The operating voltage for the battery approaches 2 V. The cell voltage is reduced compared with a cell containing a graphite anode The disadvantage of lower voltage is counterbalanced by advantages such as superior intrinsic safety, inherent overcharge protection and avoidance of Li plating. When a cathode, $LiNi_{0.5}Mn_{1.5}O_4$, with higher potential is used to fabricate TiO_2(B)–GPE–$LiNi_{.5}Mn_{1.5}O_4$ battery, a relatively high operating voltage of 3.1 V can be achieved, at the meantime, the advantage mentioned above is maintained. It delivers a capacity of 225 mAh/g with excellent capacity retention at C/5 between 2 and 3.5 V.[79]

5.5.4 *Nano $Li_4Ti_5O_{12}$*

$Li_4Ti_5O_{12}$ attracts high research interests as anode materials for LIBs, due to its good reversibility and structure stabilization (zero-strain insertion material) during Li intercalation and de-intercalation process. Like TiO_2, it has a mid-discharge voltage close to 1.55 V *vs.* Li^+/Li, which could use as an anode combined with high-voltage cathodes for LIBs.[80] The Li insertion potential is above the reduction potential of common electrolyte, as a result, there will be no decomposition of electrolyte. Besides, the battery security can be improved since there is no Li metal deposition at this potential.[81]

$Li_4Ti_5O_{12}$ was traditionally synthesized by a high temperature solid-state reaction of TiO_2 and Li_2CO_3 or LiOH, which resulting in microcrystalline products.[82] To develop $Li_4Ti_5O_{12}$ materials with high rate performance, a short transport length for Li^+ and electrons a larger surface area for easy electrolyte accessibility are required. The use of nanocrystalline $Li_4Ti_5O_{12}$ has shown to be effective.

The ball-milling of the conventional microsized $Li_4Ti_5O_{12}$ was first adopted to prepare nanocrystalline $Li_4Ti_5O_{12}$.[83] It is reported that 700 nm $Li_4Ti_5O_{12}$ particles were obtained after treating the conventional microsized $Li_4Ti_5O_{12}$ with vibrating ball-milling. The nanoparticles present a reversible capacity of more than 160 mAh/g at 1C rate, and without capacity fading up to 100 cycles. The capacity retains 86% at the rate of 10C, indicating an excellent rate capacity.[84]

Sol–gel method is frequently used for prepare nano $Li_4Ti_5O_{12}$.[85,86] In the case of the sol–gel method, it is necessary to heat $Li_4Ti_5O_{12}$ at high temperature to obtain well-crystallized $Li_4Ti_5O_{12}$, that is, 800°C or even higher to prepare a pure $Li_4Ti_5O_{12}$, resulting in an undesirable particle growth and aggregation of nanoparticles Nanocrystalline $Li_4Ti_5O_{12}$ with particle size of less than 10 nm was obtained *via* a polyol process. Ethylene glycol solution of titanium tetra isopropoxide ($Ti(O\text{-}i Pr)_4$) and Li_2O_2 were refluxed at 197°C for 12 h, followed by drying and calcinating at different temperature. When calcinating the precursor at 500°C, $Li_4Ti_5O_{12}$ with particle size of 10 nm were produced, it shows an initial charge capacity of 280 mAh/g, after a capacity fading at the first 10 cycles, it maintains a good cycling performance with the capacity of 230 mAh/g up to 60 cycles.[87] By using P123 as surfactant, a welldispersed nano-sized $Li_4Ti_5O_{12}$ was yield, which shows higher reversible capacity of more than 150 mAh/g at a current density of 0.1 A/g. When increasing the current to 1 A/g, the capacity retains 88.6%. While for the $Li_4Ti_5O_{12}$ prepared without P123, only 30% remains at the current of 1 A/g.[81]

Besides nanoparticles, various $Li_4Ti_5O_{12}$ materials with different architectures were fabricated to facilitate Li-ion transfer. Three-dimensionally ordered macroporous (3DOM) $Li_4Ti_5O_{12}$ was synthesized by soaking metal organic aqueous precursor into a poly(methylmethacrylate) colloidal crystal templates and followed by calcinating at 675°C. This 3DOM material presents enhanced rate performance.[88] Flower-like $Li_4Ti_5O_{12}$ consisting of nanosheets was synthesized *via* hydrothermal process in glycol solution and following calcination at 500°C. The as-prepared $Li_4Ti_5O_{12}$ shows a capacity of 217.6 mAh/g at 0.2C with very stable capacity retention ability. When increasing the current rate to 8C, the initial capacity if 165.8 mAh/g and stays around 152 mAh/g after 100 cycles.[89] Hollow spherical $Li_4Ti_5O_{12}$ was prepared through the macro-emulsion method in a two-phase solution consisting of *n*-heptane

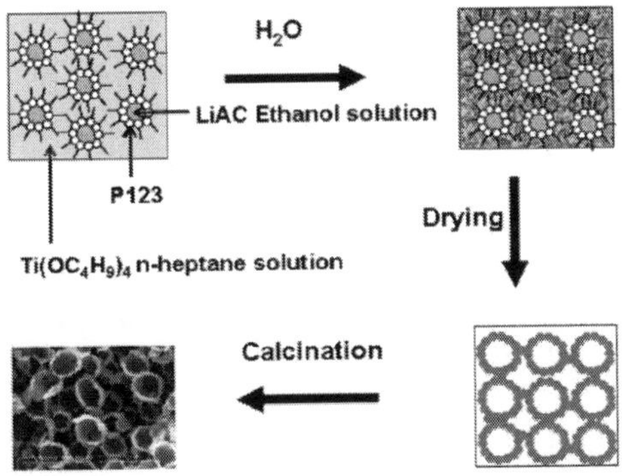

Scheme 5.3 Schematic of hollow spherical $Li_4Ti_5O_{12}$ fabrication. Reproduced from Ref. 90 by permission of the The Electrochemical Society.

and ethanol, with P123 triblock copolymer as the emulsifier. The preparation process is shown in Scheme 5.3. The hollow spherical is composed with $Li_4Ti_5O_{12}$ nanoparticles. The as-prepared hollow spherical $Li_4Ti_5O_{12}$ presents a capacity of 140 mAh/g and remains at 135 mAh/g after 500 cycles. It exhibits good rate performance with a capacity of 162, 150, 140, 137, 128, 115, and 95 mAh/g at the rate of 0.5, 1, 2, 2.5, 5, 10 and 20C (Fig. 5.16).[90]

$Li_4Ti_5O_{12}$ is reported to be an insulator.[91] Thus another direction of development of $Li_4Ti_5O_{12}$ is to improve the electron conductivity. Metallic Ag and C were used to fabricate Ag–$Li_4Ti_5O_{12}$[92] and C–$Li_4Ti_5O_{12}$ composites.[93] CNT–$Li_4Ti_5O_{12}$ composites were prepared *via* the sol–gel method, in which CNTs were added into the precursor, followed by calcination. The prepared composites exhibit better rate capacity and capacity retention than the pure $Li_4Ti_5O_{12}$.[94] Carboncoated $Li_4Ti_5O_{12}$ can be prepared *via* solid-state reaction by adding sucrose to the TiO_2 (anatase), Li_2CO_3 precursor, followed by calcination at 750°C for 12 h and then 850°C for 24 h in the air. The asprepared $Li_4Ti_5O_{12}$–C composites display a reversible capacity of 160 mAh/g at 0.1C and good cycling performance.[95]

Beside carbon and metal, TiN is another good candidate to improve the electronic property of $Li_4Ti_5O_{12}$. A conductive thin layer of TiN was formed on the surface of $Li_4Ti_5O_{12}$ *via* nitrogen thermal treatment using ammonia gas as the nitrogen source. The thin TiN layer improves the electron conductivity and enhances the electrochemical performance of $Li_4Ti_5O_{12}$ at high current density

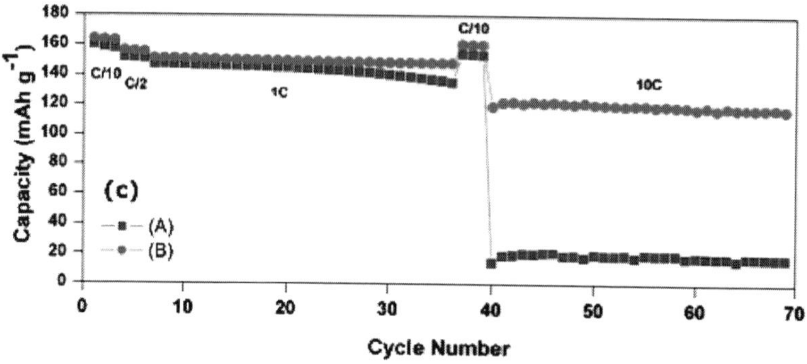

Figure 5.18 Rate and cycling performance of (A) pristine and (B) 10 min-nitridated $Li_4Ti_5O_{12}$. Reprinted with permission from Ref. 96. Copyright 2008 American Chemical Society.

(Fig. 5.18). A 10 min-nitridated $Li_4Ti_5O_{12}$ displays a capacity above 160 mAh/g at 0.1C, it retains at 120 mAh/g at 10C, presenting much better rate performance than the pristine $Li_4Ti_5O_{12}$.[96]

5.6 Other Oxides-Based Nano Anode Materials

Some more oxides were also investigated as anode material for LIBs, most of them are transition metal oxides, such as, cobalt oxide, nickel oxide, chromium oxide (Cr_2O_3) and molybdenum oxide. The basis for electrode reaction is not Li intercalation but reversible conversion reaction:[97]

$$M_xO_y + 2yLi^+ + 2ye^- \leftrightarrow yLi_2O + xM \quad (5.1)$$

where M is transition metal. They show higher capacity than graphite, but the voltage for Li insertion and extraction is relatively high. They caught high research interest to meet the continuous demanding of electronic devices with lower operating voltage.[98]

5.6.1 *Co-Based Oxides*

Among the transition metal oxides, cobalt oxides show highest reversible capacity (750 mAh/g and 1000 mAh/g for CoO and Co_3O_4, respectively).[97]

In most occasions, cobalt oxides suffer severe capacity fading upon cycling, which hinders their real application One of the reasons is the volume change during the Li insertion and extraction, which may result in aggregation of particles and loss of electrical contact, and finally poor cycling performance.[99] One general accepted strategy is to synthesize nanosized materials with tailored structure such as 1D nanostructured and mesoporous materials.

Template assistance approach is quite a popular method to prepare 1D and mesoporous structure. When using a template of AAO membrane and $Co(NO_3)_2{\cdot}6H_2O$ as precursor, Co_3O_4 nanotubes were synthesized. The initial discharge capacity is 850 mAh/g at the current rate of 50 mA/g, which is superior to that of nanorods and nanoparticles, that is, 830 mAh/g and 815 mAh/g, respectively. The capacity retention ability is not satisfied.[100] A sonochemistry method is another approach for synthesis of Co_3O_4 nanotubes by adopting CNTs as a template and $Co_4(CO)_{12}$ as precursor. $Co_4(CO)_{12}$ and CNTs (treated with nitric acid) were mixed in hexane and sonoicated 1 h at room temperature, followed by calcination at 500°C in oxygen for 3 h. During the sonication, $Co_4(CO)_{12}$ can be readily decomposed to Co and CO, and Co atom could be easily oxidized into CoO_x due to oxygen atmosphere in the solution. The CoO_x with positive charge can compactly deposit on the surface of CNTs with negative charge due to the acid treatment. Finally the porous Co_3O_4 nanotubes with a diameter of about 30 nm can be obtained by calcination at 500°C in oxygen. The porous Co_3O_4 nanotubes exhibit a high second discharge capacity of 1200 mAh/g at a current density of 50 mA/g and good capacity retention ability up to 20 cycles.[101] Mesoporous and NW Co_3O_4 were synthesized by adopting SBA-15 (1D pore structure), KIT-6 (3D pore structure) as hard templates and $Co(NO_3)_2{\cdot}6H_2O$ as precursor. Both of them deliver a charge capacity of approximately 900 mAh/g in the voltage range of 0.005–3.0 V at the rate of 100 mA/g. The capacities for both nanomaterials exceed the theoretical value for conversion reaction and increase in the first 10–15 cycles, this is due to the increasing formation of a polymeric surface layer on the high surface area material. The mesoporous Co_3O_4 show a capacity of approximately 800 mAh/g after 60 cycles, while NWs suffer a sever capacity decrease upon cycles.[102] However, the method to prepare

1D materials by adopting AAO, SBA-15 or KIT-6, the mass product is impracticable, which is the bottleneck for its practical application.

Viruses can assemble on multiple length scales, and they could also be used as template for designing self-assembling battery electrode. Co_3O_4 NWs electrode were synthesized at room temperature with this method. E4 virus clone and tetraglutamate (EEEE–) fused to Nterminus of each copy of the major coat p8 protein with 100% expression was incubated in aqueous cobalt chloride solution (1 mM) for 30 min to promote cobalt ion binding. After reduction with sodium tetrahydridoborate ($NaBH_4$) and spontaneous oxidation in water, the monodisperse, crystalline Co_3O_4 NWs were prepared. When fabricate a half cell with Li metal foil as negative electrode, the Co_3O_4 NWs delivered a reversible capacity ranging from 600 mAh/g to 750 mAh/g. When a hybrid clone called AuE4 virus, engineered p8 proteins containing the gold-binding peptide motif randomly packaged onto the virus progeny, was incubated in a 5 nm gold colloid suspension, resulting in 1D arrays of Au nanoparticles bound to the gold-binding peptides distributed among p8 proteins. After removal of excess unbound gold nanoparticles *via* centrifugation, Co_3O_4 was nucleated and grown by EEEE– functionality, and the hybrid nanostructures of 5 nm Au nanoparticles dispersed in Co_3O_4 NWs were obtained. The hybrid Au–Co_3O_4 NWs delivers a capacity of ranging from 1000 to 1200 mAh/g, indicating that a small amount of Au markedly improves the electrochemical performance of Co_3O_4 NWs. Since the negatively charged M13 viruses can form ordered, 2D liquid crystalline layers on top of electrostatically cohesive films of linear poly(ethylene imine) (LPEI)/poly(acrylic acid) (PAA). This is driven by competitive electrostatic interactions, the interdiffusion of poly-electrolyte and the anisotropic shape of M13 viruses. By using this technique to order E4 viruses and subsequently grow Co_3O_4 on the virus coat proteins, 2D organized ensembles of NWs were achieved. The polymer electrolyte could act as a solid electrolyte because the relatively high ionic conductivity of LPEI and PAA pairs. When evaporating Cu to the Co_3O_4 NWs/polymer layer as current collector, and adopting Li foil as negative electrode, the Co_3O_4 NWs/polymer presents a reversible capacity of approximately 800 mAh/g at 1.12C and approximately 500 mAh/g at 5.19C.[103] Co_3O_4 NWs and nanotubes could also be prepared without template assistance.

A self-supported topotactic transformation approach was used to synthesize Co_3O_4 nanotubes. Using cobalt nitride as a precursor in ammonia solution, at the early stage of preparation, needle-like β-$Co(OH)_2$ nanorods with growth direction along [001] was formed due to the highly anisotropic hexagonal crystal structure. In air mediation, β-$Co(OH)_2$ nanorods began to experience a rock efflorescence, forming small platelets on the surface of the nanoneedles. When treated the nanoneedles with air in 100°C, the β-$Co(OH)_2$ platelets were oxidized to Co_3O_4. At the same time, the solid materials in the core region of β-$Co(OH)_2$ nanoneedles were continuously evacuated through dissolution and reoxidation and deposited on the prior formed Co_3O_4 wall, and the Co_3O_4 nanotubes could be obtained. The as-prepared Co_3O_4 nanotubes present ultrahigh discharge capacity, >2000 mAh/g, in the first cycle at the current rate of 50 mA/g, and the first reversible capacity is 950 mAh/g. It shows good cycling performance in the first 30 cycles.[99]

The Co_3O_4 NW arrays were synthesized *via* an ammonia evaporation-induced method. Typically, 10 mmol $Co(NO_3)_2$ and 5 mmol NH_4NO_3 were dissolved in 15 mL 30 wt% ammonia solution and 35 ml H_2O. After stirring for half an hour in air, the pink color turned into black due to the oxidation of Co(II) Ti Co(III). After heating the solution at 90°C, a clean Ti foil was introduced and further heated for 12 h at the same temperature for NW growth. The NW arrays with a mesoporous structure were obtained after heating at 250°C for 2 h to remove the trapped and adsorbed water. The obtained Co_3O_4 NW arrays on the Ti foil deliver a discharge capacity of 1124 mAh/g and reversible capacity of 859 mAh/g and also retain good cycle ability in the first 20 cycles. When increasing the current rate to 8, 20 and 50C, the capacity retains 85%, 69% and 50% of that at 1C[104] Nano Co_3O_4 arrays were also synthesized by adopting hydrothermal method with Co–$(CH_3COO)_2{\cdot}4H_2O$ as precursor. The asprepared 1D arrays of Co_3O_4 present a discharge capacity of approximately 1150 mAh/g at 0.2°C. The reversible capacity after 11 cycles is 841 mAh/g.[102]

Besides synthesis of nanosized materials with particular structure, preparation of active/inactive composites is another strategy to solve the problem of volume expansion upon cycling.

Co_3O_4–C composites were prepared to prevent the aggregation of the particles and loss of the electrical contact. Co_3O_4–C composite produced by spray pyrolysis technique presents a high reversible capacity of 800 mAh/g and good cycling performance for the first 50 cycles.[103]

Layered cobaltite ($Ca_3Co_4O_9$) presents a misfit structure builtup with alternating single CdI_2-type $[CoO_2]_\infty$ layers and triple [CaO–CoO–CaO] rock salt type layers. The proposed structural model for involving a supercell with $b \approx 8b_1 \approx 13b_2$ is shown in Scheme 5.4. This layered structure enables electrons behave in an anisotropic

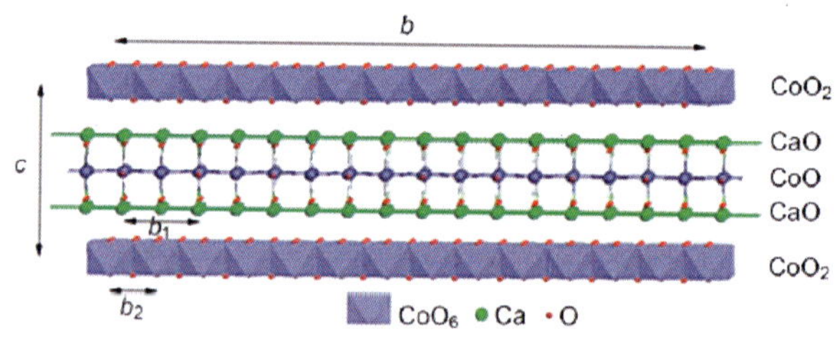

Scheme 5.4 Proposed structural model for $Ca_3Co_4O_9$ involving a supercell with $b \approx 8b_1 \approx 13b_2$. Reprinted from Ref. 105, with permission. Copyright © 2007 WILEY-VCH Verlag GmbH & Co. KGaA, Weinheim.

structural environment and results in high electrical capacity. It is observed that after the reaction with Li in the first cycle, $Ca_3Co_4O_9$ is decomposed into metallic Co, Li_2O and CaO. During the charge process, metallic Co is oxidized to α-CoO and finally regenerates to Co_3O_4, which reacts with Li in the following cycles. CaO cannot react with Li due to its strong bond and acts as an inactive material which could alleviate the volume change, mechanical stress and aggregation of Co nanograins during cycling. $Ca_3Co_4O_9$ presents a reversible capacity of 761 mAh/g in the first cycle and retains at $\approx$500 mAh/g after 50 cycles, at the current of 0.2C (1C is defined as 12 Li ions per hour and per formula unit of $Ca_3Co_4O_9$). The capacities at 1, 5 and 10C are 660, 320 and 190 mAh/g, respectively.[105]

Another inherent drawback is the large irreversible capacity during the first cycle. A positive electrode material with excess Li could serve as a Li reservoir and compensate for the large irreversible capacity. When chemically lithiated spinel $Li_{1.3}Mn_2O_4$ is adopted as cathode and a $CoO/Li_{1.3}Mn_2O_4$ Li-ion cell with an M_r equal to 6.2 was assembled (where M_r is the optimum mass ratio between CoO and $Li_{1.3}Mn_2O_4$), a capacity of 110 mAh/g of the overlithiated spinel is obtained, while a comparable $CoO/LiMn_2O_4$ Li-ion cell exhibits a capacity of 90 mAh/g, indicating the anode materials with large irreversible capacity could be incorporated into future generations of LIBs.[98]

5.6.2 *Ni-Based Oxides*

NiO has a theoretical capacity of 718 mAh/g, when it is used as anode material for LIBs, the cycling performance of NiO is poorer than other transition metal oxides, such as CoO; however, it is less expensive than CoO.[97]

Spherical clusters of NiO nanoshafts were prepared by chemical precipitation of nickel sulfate followed by thermal decomposition at 280°C in air. The as-prepared NiO nanoshafts present a high specific surface area (113 m^2/g), they deliver a reversible capacity of approximately 900 mAh/g in the first cycle, but the cycling performance is quite poor.[106]

NiO nanowalls were synthesized directly on the nickel foil by the plasma assisted oxidation method. The height of the nanowall is over 100 nm, the thickness is less than 40 nm. The first discharge capacity at the current rate of 448 mA/g is 1253 mAh/g and the reversible capacity is 794 mAh/g. When increasing the current density to 895 and 1343 mA/g, the reversible capacities were 833 and 506 mAh/g, respectively. It shows good capacity retention capability at different current rates.[107]

NiO is a semiconductor and after first Li insertion, the pristine particles are disintegrated and resulting in poor electrical contact, which gives rise to poor cycling performance. One effective way to solve this problem is to compose the pristine NiO with a conductive material, like carbon, metal and conductive polymer.

Net-structured NiO was prepared *via* microwave radiation method with $Ni(CH_3COO)_2$ as precursor, followed by calcination at 500°C. The as-prepared NiO was dispersed into glucose solution and subsequently carbonized under hydrothermal condition at 180°C, and the net-structured NiO–C nanocomposites were obtained. TEM images show that the NiO network is homogenously filled with amorphous carbon. The electrochemical performance is improved after carbon coating, after 40 cycles, the reversible capacity of NiO–C nanocomposites is 429 mAh/g, while only 178 mAh/g for NiO electrode.[108]

Nanocomposites of NiO–Ni were prepared by calcining a mixture of $Ni_2(OH)_2CO_3$ and ethanol in a pre-heated tube furnace at 700°C for 45 min in air. From TEM image, it is shown that nickel nanoparticles (<10 nm) are dispersed in the NiO matrix (~100 nm). In the voltage range 3.0–0.02 V (*vs.* Li/Li^+) under a specific current density of 100 mA/g, NiO–Ni nanocomposites deliver a higher discharge capacity, 1152.4 mAh/g, than that of pure NiO (997.4 mAh/g), which might due to the large quantity of lattice defects in NiO–Ni composite, which was quenched from 700°C to room temperature in air. The initial coulombic efficiency of NiO–Ni nanocomposites is improved for NiO–Ni composites, that is, 71.2%, which is better than that of the pure NiO electrode (64.9%). The NiO–Ni nanocomposites also show better capacity retention ability than pure NiO The capacity of NiO–Ni composites retains at approximately 700 mAh/g up to 50 cycles, while only 400 mAh/g for pure NiO electrode.[109]

After nickel foam supported porous NiO film was prepared *via* chemical deposition with $NiSO_4$ as precursor, the polyaniline (PANI) was deposited on the porous NiO film with electrodeposition technique. The NiO film is constructed with interconnected nanoflakes with a thickness of 20 nm. After deposition of PANI, the film maintains net-like structure. Cyclic voltammerty results reveal that after depositing PANI, the polarization is reduced compared with NiO electrode. SEM images show that after 20 cycles, NiO film is disintergrated and particles aggregate severely, while for NiO/PANI film maintains the original morphology. The capacity of NiO/PANI electrode after 50 cycles is 520 mAh/g, higher than that of NiO electrode, 440 mAh/g.[110]

5.6.3 Cr_2O_3

There are drawbacks for transition metal oxides mentioned above. They limits their practical application is the average working voltage which is too high as an anode. This value is too high and decreases the voltage and energy density of corresponding Li-ion batteries. Working voltage (U) is mainly determined by the electromotive force (emf, E) and partly influenced by the overpontential η ($U = E \pm \eta$). Li *et al.* (2004) found that Cr_2O_3 shows relative high electrochemical activity and low emf,[4] the average charging voltage for Cr_2O_3 is ~1.2 V, much lower than most of the other transition metal compounds (>1.8 V), which is a significant advantage for as Cr_2O_3 an anode material for LIBs.[111]

According to the Nernst equation calculation, emf for Cr_2O_3 is 1.085 V *vs.* Li/Li^+. The average discharge voltage is lower than the emf and the average charge voltage is high than that, which means the polarization of Cr_2O_3 is quite high. Meanwhile, phase transformation reaction causes large volume variation, giving rise to poor electronic contact of active particles on current collector. As a result, the pure Cr_2O_3 electrode shows quite poor capacity retention ability. Preparation of the core-shell C–Cr_2O_3 composites is a good way to enhance the cycling performance, from the *ex situ* SEM observation, it is found that the carbon layer could avoid Cr_2O_3 electrode cracking after first cycle.[111]

Tying-up the current collectors with electrode materials is another way to gain better electrochemical performance for Cr_2O_3. A Cr_2O_3 thin film was obtained by ablating a pure polycrystalline chromium target under a controlled-oxygen pressure onto titanium-polished disk. It is found that the electrochemical performance is different and independent of the film thickness. The capacity of the film with a thickness below 175 nm is sustainable over hundreds cycles. However, rapid capacity decay is observed for thicker films. A progressive surface flaking exists upon cycling, resulting in electronic disconnection. This is more pronounced for thicker films. For thin films, electronic path could be maintained due to tunneling between metallic Cr particles. It could be easily expected that capacity fading should be more catastrophic for Cr_2O_3 microsized powers.[112]

To enlarge the surface area of Cr_2O_3 electrode is viable to improve electrode performance as mentioned before, since it facilities the transport of ions and electrons transfer.

An electrochemically active current collector was designed from the metallurgical aspect. Stainless steel disk was heat-treated under nitrogen with 10% hydrogen The thermal treatment promotes a migration of the metallic elements with enrichment of Cr at the expense of Fe and Ni. The surface composition shows drastic difference between initial and treated samples in Fe, Cr, and Ni, going from 72%, 18%, and 10% for the fresh sample to 53, 37, and 10% for the treated sample. The Cr atoms react with the trace of oxygen in reducing gas to form chromite phase, while the other metals would be reduced by H_2/O_2 ppm combination. SEM results indicate that, after treatment, the smooth surface of stainless steel becomes porous and poorly packed. The films with a thickness of 100, 290, and 500 nm present capacities of 0.08, 0.13 and 0.2 mAh/cm^2 upon 30th cycle, while the non-treated sample shows no electrochemical capacity. The oxide layer formed at 700°C can be cycled as much as 800 times between 0.02 and 2.5 V. When the stainless steel disks were firstly corroded, followed by the thermal treatment, the porosity of the film increases, and so is the capacity, indicating a significant influence of electrode surface area.[112]

By adopting KIT-6 silica as template, $Cr(NO_3)_3 \cdot 9H_2O$ as precursor, the mesoporous Cr_2O_3 was prepared. The as-synthesized Cr_2O_3 electrode presents higher capacity as well as higher irreversible capacity than its bulk counterpart. From the potential of the charge and discharge plateaus, a strong decrease in polarization for the mesoporous electrode compared to the bulk for the first 40 cycles is observed. However both the mesoporous and bulk electrodes show severe capacity decay for the first 15 cycles. Observation with TEM shows that, after discharging, the electrolyte decomposition products (EDP) not only covers the particles, but also fills up the pores. After recharged mesoporous Cr_2O_3 electrode, the EPD could be preserved, while for bulk Cr_2O_3 electrode, the EPD on the surface vanishes after recharging. The most likely reason is due to the mesoporous texture, where the redissolution reaction of EPD is slowed down by the limited ratio of electrolyte/EPD. The preserved EPD upon cycling could minimize the volume expansion and contractions; however at the same time, increase the irreversible capacity for first cycle.[113]

5.6.4 *Nano MoO_2*

In 1978, Molybdenum dioxide (MoO_2) was reported as a possible candidate as anode materials.[114] Following researches show that MoO_2 exhibits higher capacity than commercial carbonaceous mate-

Figure 5.19 (a,b) SEM; (c,d) TEM images of the as-prepared tremella-like MoO_2 at different magnifications. The SAED pattern (inset of c) was obtained on the nanosheet (marked by the frame) shown in (c), indicating the typical MoO_2 polycrystalline structure. The interplanar spacing of 1.1 nm of this novellayered structure is identified in (c). Reprinted from Ref. 116, Copyright 2008, with permission from Elsevier.

rials as an anode material for LIBs. Morphology plays an important role in Li insertion–extraction activity and cycling stability, in the case of fissile MoO_2 prepared by rheological phase reaction, it shows an initial reversible capacity of 484 mAh/g with 98% capacity retained after 20 cycles. While for spherical MoO_2 prepared with hydrothermal method, it presents higher initial reversible capacity of 769 mAh/g with a rapid capacity fading in the following cycles.[115]

Recently, tremella-like molybdenum dioxide was synthesized consisting of nanosheets by a hydrothermal method involving the reduction of MoO_3 to MoO_2 by an ethylenediamine (EDA) reducer and Fe_2O_3 assisting agent.[116] The prepared product presents a tremella-like structure consisting of ultrathin nanosheets with a

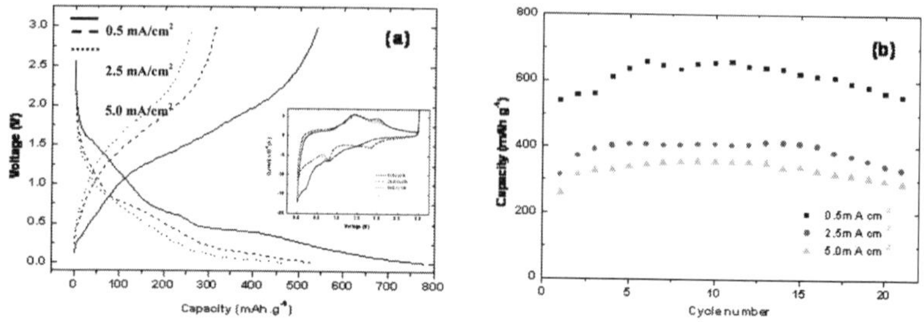

Figure 5.20 (a) The discharge and charge curves for the initial cycle; (b) Cycling performance of the tremella-like MoO_2 in the range of 0.01–3 V *vs.* Li metal at the current density of 0.5, 2.5 and 5.0 mA/cm^2, respectively. The inset in (a) shows the cycling voltammograms for the as-prepared MoO_2 in the range of 0–3 V *vs.* Li metal at the scan rate of 0.05 mV/s. Reprinted from Ref. 116, Copyright 2008, with permission from Elsevier.

thickness less than 10 nm (Fig. 5.19a). The SAED pattern (inset of Fig. 5.19a) clearly confirms the hexagonal MoO_2 polycrystalline structure. Furthermore, the TEM image at high magnification (Fig. 5.19d) exhibits a novel layered structure with the interplanar spacing of 1.1 nm from the side view of the nanosheets, which does not correspond to any interplanar crystal spacing of the typical MoO_2 crystal. This structure is probably due to the intercalation of excessive EDA molecules in the reaction system. As a result, the material shows high reversible capacity and good cycling performance even at a current density as high as 5.0 mA/cm^2 (Fig. 5.20ab). The electrodes deliver high reversible capacities of 538.0, 314.6, and 259.4 mAh/g at current densities of 0.5, 2.5 and 5.0 mA/cm^2, respectively. In subsequent cycles, the reversible capacities increases up to 650, 400, and 350 mAh/g. This increase is probably due to the ultra-thin nanosheet layers becoming saturated with the electrolyte upon discharge and charge after several cycles, and additionally an activation process may be obtained. In Fig. 5.20a, apart from the irreversible cathodic discharge capacity which may relate to the reduction of solution species to form a passivating surface film on the anode (see the small peak at about 0.6 V), there are two redox couples at 1.8/2.0 V and 0.75/1.46 V, respectively, corresponding to the phase transitions in the partially lithiated

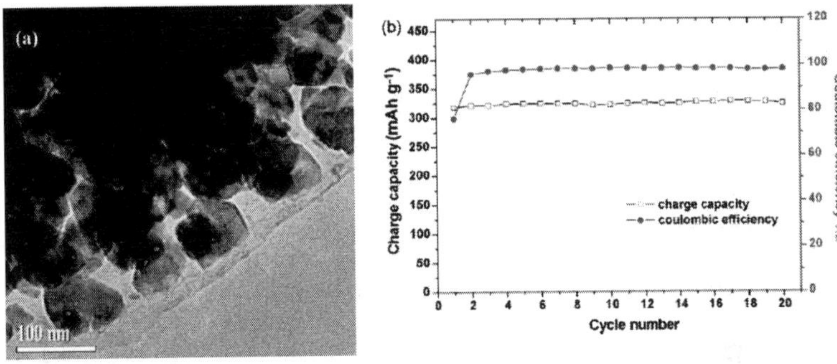

Figure 5.21 (a) TEM images of the as-obtained MoO_2; (b) Cycling behavior of the as-obtained MoO_2 tested in the range 0.01–3.00 V *vs.* Li/Li^+ at the current density of 5.0 mA/cm^2. Reprinted from Ref. 119, Copyright 2008, with permission from Elsevier.

Li_xMoO_2. This novel MoO_2 material shows potential application as anode material for LIBs.

The morphology control of MoO_2 involves the high temperature CVD[117] and H_2 reduction.[118] Recently a method was developed to prepare MoO_2 through reduction of MoO_3 with ethanol vapor at 400°C.[119] After ethanol vapor treatment, the mircro-sized starting material becomes nanoparticles (approximately 100 nm), and a thin layer of carbon was formed on the surface of MoO_2 nanoparticles, which is due to the pyrolysis of ethanol vapor (Fig. 5.21a). The MoO_2–C nanocomposites display an initial capacity of 318 mAh/g with capacity retention of 100% after 20 cycles in the potential range 0.01–3 V (Fig. 5.21b).

5.6.5 *Fe-Based Oxides*

Iron oxides are also considered to be a candidate for anode material for LIBs because of their low cost, low toxicity and high level of safety.[120] Amounts of researches were performed on α-Fe_2O_3. There are three steps involving in the lithiation process of α-Fe_2O_3, written as follows:[121]

$$Fe_2O_3 + 2Li^+ + 2e^- \rightarrow Li_2(Fe_2O_3) \quad (5.2)$$

$$Li_2(Fe_2O_3) + 4Li^+ + 4e^- \rightarrow 2Fe + 3Li_2O \quad (5.3)$$

$$2Fe + 2Li_2O \leftrightarrow 2Fe^{II}O + 4Li^+ + 4e^- \quad (5.4)$$

For the α-Fe_2O_3 nanoparticles (20 nm), 0.5 mol Li can be reversibly intercalated in the potential range of 1.5–4.0 V *vs.* Li/Li^+ when decreasing the intercalation cutoff voltage to 0.9 V Li/Li^+, 2.0 mol Li could be intercalated (Eq. 5.2), but they cannot be extracted by electrochemical charging without destroying the crystal structure. When further discharge with Li to 0.005 V *vs.* Li/Li^+, 8.5 mol Li could react with Fe_2O_3, accompanying a destruction of crystal structure and formation of iron metal Li_2O and polymer layer on iron metal coming from the decomposition of electrolyte (Eq. 5.3).[122] When charging the electrode to 3.0 V, conversion reaction of nanoscaled Fe with Li_2O results in the release of Li ions and electrons (Eq. 5.4).[97]

It is reported that the particle size plays a drastic role in the lithiation process of α-Fe_2O_3 Nanoparticles display a better performance compared with the micrometric ones.[122] Various morphology displays different Li intercalation performance. By adopting AAO template, α-Fe_2O_3 nanotubes were produced. The initial discharge capacity (Li insertion) is 1415 mAh/g in the potential range of 0.005–3.0 V with capacity fading in the following cycles at 100 mA/g[123] α-Fe_2O_3 nanoflakes were prepared on the Fe film with a thermal heating method, and they show a stable capacity of 680 mAh/g up to 80 cycles when cycling at 65 mA/g in the potential range of 0.005–3.0 V.[121] α-Fe_2O_3 nanorods were prepared *via* a hydrothermal method. They exhibit a capacity of 763 mAh/g after 30 cycles.[124]

Fe_3O_4 is another promising candidate as anode materials for LIBs. It is one of the few transition metal oxides with high electronic conductivity. Similar to other anode materials mentioned above, Fe_3O_4 particles with different nano-architecture were prepared for different Li storage performance. Fe_3O_4 based Cu nano-architectured electrodes were prepared with a two-step design consisting of the electrochemically assisted template growth of Cu nanorods onto a current collector followed by electrochemical plating of Fe_3O_4. The asprepared electrode with a deposition time of 150 s shows high reversible capacity of 800 mAh/g and excellent capacity retention It displays good rate performance, the capacity at 8C is 80% of that at C/32 though the hysteretic profile is still maintained.[125] The carboncoated Fe_3O_4 nanospindles were prepared by partial reduction of monodispersed hematite nanospindles obtained by hydrothermal process. The Fe_3O_4–C nanospindles show high reversible capacity of 745 mAh/g and 600 mAh/g at 0.2C and 0.5C respectively, with a sustained cycling performance of capacity 530 mAh/g up to 80 cycles.[126]

5.7 Future Aspects

Nano anode materials brought significant improvement for the development of LIBs in terms of shorter diffusion length of Li^+ as well as electrons and providing preferable accommodation to the strain for Li insertion–extraction. Nanoporous materials facilitate accessibility of electrolyte and enhance the rate performance. However, the more side reaction caused by the increased surface area should be taken into consideration for practical application. The strategy to combine advantages of different materials is a promising way for further development.

Acknowledgment

Financial support from the National Basic Research Program of China (973 Program No. 2007CB209702), Shanghai Committee of Science and Technology (09QH1400400), and the South African National Research Foundation (UID No. 67295) is gratefully acknowledged.

References

1. Y. P. Wu, X. B. Dai, J. Q. Ma, Y. J. Chen, *Lithium Ion Batteries: Practice and Applications*, Chemical Industry Press, Beijing (2004).
2. A. S. Arico, P. G. Bruce, B. Scrosati, J. M. Tarascon, W. V. Schalkwijk, *Nat. Mater.* 4 (2005) 366.
3. J. Maier, *Nat. Mater.* 4 (2005) 805.

4. Y. F. Zhukovskii, P. Balaya, E. A. Kotomin J. Maier, *Phys. Rev. Lett.* 96 (2006) 58302.
5. H. Li, P. Balaya, J. Maier, *J. Electrochem. Soc.* 151 (2004) A1878.
6. N. Li, C. R. Martin, *J. Electrochem. Soc.* 148 (2001) A164.
7. L. J. Fu, H. Liu, Y. P. Wu, E. Rahm, R. Holze, H. Q. Wu, *Prog. Mater. Sci.* 50 (2005) 881.
8. T. Tsumura, A. Katanosaka, I. Soumaa, T. Ono, Y. Aihara, J. Kuratomi M. Inagaki, *Solid State Ionics* 135 (2000) 209.
9. C. Liu, H. M. Cheng, *J. Phys. D: Appl. Phys.* 38 (2005) R231.
10. B. Gao, C. Bower, J. D. Lorentzen, L. Fleming, A. Kleinhammes, X. P. Tang, L. E. McNeil, Y. Wu, O. Zhou, *Chem. Phys. Lett.* 327 (2000) 69.
11. E. Frackowiak, S. Gautier, H. Gaucher, S. Bonnamy, F. Beguin *Carbon* 37 (1999) 61.
12. S. Kawasaki, Y. Iwai, M. Hirose, *Carbon* 47 (2009) 1081.
13. T. Kar, J. Pattanayak, S. Scheiner *J. Phys. Chem. A* 105 (2001) 10397.
14. B. Gao, A. Kleinhammes, X. P. Tang, C. Bower, L. Fleming, Y. Wu, O. Zhou, *Chem. Phys. Lett.* 37 (1999) 61.
15. J. J. Zhao, A. Buldum, J. Han, J. P. Lu, *Phys. Rev. Lett.* 85 (2009) 1706.
16. Y. P. Wu, E. Rahm, R. Holze, *J. Power Sources* 114 (2003) 228.
17. H. Shimoda, B. Gao, X. P. Tang, A. Kleinhammes, L. Fleming, Y. Wu, O. Zhou, *Phys. Rev. Lett.* 88 (2002) 015502.
18. Z. Zhou, J. Zhao, X. Gao, Z. Chen, J. Yan, P. Schleyer, M. Morinaga, *Chem. Mater.* 17 (2005) 992.
19. Z. Zhou, X. P. Gao, J. Yan, D. Y. Song, M. Morinaga, *Carbon* 42(2004) 2677.
20. Y. Li, B. Tan, Y. Wu, *Nano Lett.* 8 (2008) 265.
21. G. Che, B. B. Lakshmi, E. R. Fisher C. Martin, *Nature* 393 (1998) 346.
22. J. Y. Eom, D. Y. Kim, H. S. Kwon, *J. Power Sources* 157(2006) 507.
23. G. T. Wu, C. S. Wang, X. B. Zhang, H. S. Yang, Z. F. Qi, P. M. He, W. Z. Li, *J. Electrochem. Soc.* 146 (1999) 1696.
24. J. Y. Eom, H. S. Kwon, J. Liu, O. Zhou, *Carbon* 42 (2004) 2589.
25. X. X. Wang, J. N. Wang, H. Chang, Y. F. Zhang, *Adv. Func. Mater.* 17 (2007) 3613.
26. S. H. Yoon, C. W. Park, H. Yang, Y. Korai, I. Mochida, R. T. K. Baker, N. M. Rodriguez, *Carbon* 42 (2004) 21.
27. T. Doi, A. Fukuda, Y. Iriyama, T. Abe, Z. Ogumi, K. Nakagawa, T. Ando, *Electrochem. Comm.* 7 (2005) 10.

28. C. Kim, K. S. Yang, M. Kojima, K. Yoshida, Y. J. Kim, Y. A. Kim, M. Endo, *Adv. Func. Mater.* 16 (2006) 2393.
29. G. F. Ortiz, R. Alcántara, P. Lavela, J. L. Tirado, *J. Electrochem. Soc.* 152 (2005) A1797.
30. J. O. Besenhard (Ed.), *Handbook of Battery Materials*, Wiley-VCH, Weinheim, 1999, pp. 363–380.
31. S. Ohara, J. Suzuki, K. Sekine, T. Takamura, *J. Power Sources* 119–121 (2003) 591.
32. H. Li, X. Huang, L. Chen, Z. Wu, Y. Liang, *Electrochem. Solid State Lett.* 2 (1999) 547.
33. U. Kasavajjula, C. Wang, A. J. Appleby, *J. Power Sources* 163 (2007) 1003.
34. I. Kim, P. N. Kumta, G. E. Blomgren, *Electrochem. Solid State Lett.* 3 (2000) 493.
35. J. H. Lee, W. J. Kim, J. Y. Kim, S. H. Lim, S. M. Lee, *J. Power Sources* 176 (2008) 353.
36. Y. M. Kang, S. M. Lee, M. S. Sung, G. J, Jeong, J. S. Kim, S. S. Kim, *Electrochim Acta* 52 (2006) 450.
37. M. Suzuki, J. Suzuki, K. Sekine, T. Takamura, *J. Power Sources* 146 (2005) 452.
38. C. Jiang, M. Ichihara, I. Honma, H. Zhou, *Electrochem. Acta* 52 (2007) 6470.
39. T. Zhang, L. J. Fu, J. Gao, L. C. Yang, Y. P. Wu, H. Q. Wu, *Pure Appl. Chem.* 78 (2006) 1889.
40. T. Zhang, J. Gao, H. P. Zhang, L. C. Yang, Y. P. Wu, H. Q. Wu, *Electrochem. Commun.* 9 (2007) 886.
41. H. Kim, J. Cho, *Nano Lett.* 8 (2008) 3688.
42. Y. Wang, J. Y. Lee, C. Deivaraj, *J. Electrochem. Soc.* 151 (2004) A1804.
43. M. Noh, Y. Kwon, H. Lee, J. Cho, Y. Kim, M. G. Kim, *Chem. Mater.* 17 (2005) 1926.
44. G. Cui, Y. S. Hu, L. Zhi, D. Wu, I. Lieberwirth, J. Maier, K. Mullen, *Small* 3 (2007) 2066.
45. G. Derien, J. Hassoun, S. Panero B. Scrosati, *Adv. Mater.* 19 (2007) 2336.
46. T. P. Kumar, R. Ramesh Y. Y. Lin, G. T. K. Fey, *Electrochem. Commun.* 6 (2004) 520.
47. J. R. Dahn, R. E. Mar, A. Abouzeid, *J. Electrochem. Soc* 153(2) (2006) A361.
48. Q. F. Dong, C. Z. Wu, M. G. Jin, Z. C. Huang, M. S. Zheng, J. K. You, Z. G. Lin, *Solid State Ionics* 167 (2004) 49.

49. J. Hassoun, S. Panero, P. Simon, P. L. Taberna, B. Scrosati, *Adv Mater.* 19 (2007) 1632.
50. F. Ke, L. Huang, H. Jiang, H. Wei, F. Yang, S. Sun, *Electrochem. Commun.* 9 (2007) 229.
51. K. D. Kepler, J. T. Vaughey M. M. Thackeray, *J. Power Sources* 81–82 (1999) 383.
52. K. Wang, X. M. He, L. Wang, J. G. Ren, C. Y. Jiang, C. R. Wan, *J. Electrochem. Soc.* 153 (2006) A1859.
53. H. C. Shin, M. Liu, *Adv. Func. Mater.* 15 (2005) 582.
54. T. Brousse, R. Tetoux, U. Herterich, D. M. Schleich, *J. Electrochem. Soc.* 145 (1998) 1.
55. M. S. Park, G. X. Wang, Y. M. Kang, D. Wexler, S. X. Dou, H. K. Liu, *Angew. Chem. Int. Ed.* 46 (2007) 750.
56. H. Liu, G. Wang, J. Park, J. Wang, H. Liu, C. Zhang, *Electrochem. Commun.* 54 (2009) 1733.
57. S. Han, B. Jang, T. Kim, S. Oh, M. T. Hyeon, *Adv. Func. Mater.* 15 (2005) 1845.
58. X. W. Lou, Y. Wang, C. Yuan, J. Y. Lee, L. A. Archer, *Adv. Mater.* 18 (2006) 2325.
59. H. Kim, J. Cho, *J. Mater. Chem.* 18 (2008) 771.
60. G. Chen Z. Wang, D Xiao, *Chem. Mater.* 20 (2008) 6951.
61. Z. Wen, Q. Wang, Q. Zhang, J. Li, *Adv. Func. Mater.* 17 (2007) 2772.
62. G. An, N. Na, X. Zhang, Z. Miao, S. Miao, K. Ding, Z. Liu, *Nanotechnology* 18 (2007) 435707.
63. N. H. Zhao, G. J. Wang, Y. Huang, B. Wang, B. D. Yao, Y. P. Wu, *Chem. Mater.* 20 (2008) 2612.
64. N. H. Zhao, L. J. Fu, L. C. Yang, T. Zhang, G. J. Wang, Y. H. Deng, Y. P. Wu, T. van Ree, *Pure Appl. Chem.* 80 (2008) 2283.
65. C. Natarajan, K. Setoguchi, G. Nogami, *Electrochem. Acta* 43 (1998) 3371.
66. Q. Wang, Z. Wen, J. Li, *Adv. Func. Mater.* 16 (2006) 2141.
67. P. G. Bruce B. Scrosati, J. Tarascon *Angew Chem. Int. Ed.* 47 (2008) 2930.
68. V. Subramanian, A. Karki, K. I. Gnanasekar, F. P Eddy, B. Rambabu, *J. Power Sources* 159 (2006) 186.
69. Y. G. Guo, Y. S. Hu, J. Maier, *Chem. Commun.* 26 (2006) 2783.
70. L. J. Fu, T. Zhang, Q. Cao, H. P. Zhang, Y. P. Wu, *Electrochem. Commun.* 9 (2007) 2140.

71. L. J. Fu, H. Liu, H. P. Zhang, C. Li, T. Zhang, Y. P. Wu, R. Holze, H. Q. Wu, *Electrochem. Commun.* 8 (2006) 1.
72. L. J. Fu, L. C. Yang, Y. Shi, B. Wang, Y. P. Wu, *Micro. Meso. Mater.* 117 (2009) 515.
73. Y. G. Guo, Y. S. Hu, W. Sigle, J. Maier, *Adv. Mater.* 19 (2007) 2087.
74. Y. S. Hu, L. Kienle, Y. G. Guo, J. Maier, *Adv. Mater.* 18 (2006) 1421.
75. R. Marchand, L. Brohan, M. Tournoux, *Mater. Res. Bull.* 15 (1980) 1129.
76. M. Tournoux, R. Marchand, L. Brohan, *Prog. Solid State Chem.* 17 (1986) 33.
77. A. R. Armstrong, G. Armstrong, J. Canales, R. Garcia, P. G. Bruce, *Adv. Mater.* 17 (2005) 862.
78. G. Armstrong, A. R. Armstrong, J. Canales, P. G. Bruce, *Electrochem. Solid State Lett.* 9 (2006) A139.
79. G. Armstrong, A. R. Armstrong, P. G. Bruce, P. Reale, B. Scrosati *Adv. Mater.* 18 (2006) 2597.
80. A. Guerifi, S. Sévigny, M. Lagacé, P. Hovington, K. Kinoshita, K. Zaghib, *J. Power Sources* 119 (2003) 88.
81. T. Jiang, S. Zhang, X. Qiu, W. Zhu, L. Chen, *Electrochem. Commun.* 9 (2007) 930.
82. T. Ohzuku, A. Ueda, N. Yamamoto, *J. Electrochem. Soc.* 142 (1995) 1431.
83. K. Zaghib, M. Simoneau, M. Armand, M. Gauthier, *J. Power Sources* 81–82 (1999) 300.
84. K. Nakahara, R. Nakajima T. Matsushima, H. Majima, *J. Power Sources* 117 (2003) 131.
85. M. Venkateswarlu, C. H. Chen, J. S. Do, C. W. Lin, T. C. Chou B. J. Hwang, *J. Power Sources* 146 (2005) 204.
86. S. Bach, J. P. Pereira-Ramos, N. Baffier, *J. Power sources* 81-82 (1999) 273.
87. D. H. Kim, Y. S. Ahn, J. Kim *Electrochem. Commun.* 7 (2005) 1340.
88. E. M. Sorensen, S. J. Barry, H. K. Jung, J. R. Rondinelli, J. T. Vaughey, K. R. Poeppelmeier, *Chem. Mater.* 18 (2006) 482.
89. Y. F. Tang, L. Yang, Z. Qiu J. S. Huang *Electrochem. Commun.* 10 (2008) 1513.
90. J. Huang, Z. Jiang, *Electrochem. Solid State Lett.* 11 (2008) A116.
91. C. Y. Ouyang, Z. Y. Zhong, M. S. Lei, *Electrochem. Commun.* 9 (2007) 1107.
92. S. Huang, Z. Wen, X. Zhu, Z. Gu, *Electrochem. Commun.* 6 (2004) 1093.

93. L. Cheng, X. L. Li, H. J. Liu, H. M. Xiong, P. W. Zhang, Y. Y. Xia, *J. Electrochem. Soc.* 154 (2007) A692.
94. J. Huang, Z. Jiang, *Electrochem. Acta* 53 (2008) 7756.
95. G. J. Wang, J. Gao, L. J. Fu, N. H. Zhao, Y. P. Wu, T. Takamura, *J. Power Sources* 174 (2007) 1109.
96. K. S. Park, A. Benayad, D. J. Kang, S. G. Doo, *J. Am. Chem. Soc.* 130 (2008) 14930.
97. P. Poizot, S. Laruelle, S. Grugeon, L. Dupont, J. M. Tarascon, *Nature* 407 (2000) 496.
98. F. Badway, I. Plitz, S. Grugeon, S. Laruelle, M. Dollé, A. S, Gozdz, J.-M. Tarascon, *Electrochem. Solid State Lett.* 5 (2002) A115.
99. X. W. Lou, D. Deng, J. Y. Lee, J. Feng, L. A. Archer, *Adv. Mater.* 20 (2008) 258.
100. W. Y. Li, L. N. Xu, J. Chen *Adv. Func. Mater.* 15 (2005) 851.
101. N. Du, H. Zhang, B. Chen, J. Wu, X. Ma, Z. Liu, Y. Zhang, D. Yang, X. Huang, J. Tu, *Adv. Mater.* 19 (2007) 4505.
102. K. M. Shaju, F. Jiao, A. Débart, P. G. Bruce, *Phys. Chem. Chem. Phys.* 9 (2007) 1837.
103. K. T. Nam, D. W. Kim, P. J. Yoo, C. Y. Chiang, N. Meethong, P. T. Hammond, Y. M. Chiang, A. M. Belcher, *Science* 312 (2006) 885.
104. Y. F. Li, Z. Zhou, L. B. Wang *J. Chem. Phys.* 129 (2008) 104703.
105. D. W. Kim, Y. D. Ko, J. G. Park, B. K. Kim, *Angew. Chem. Int. Ed.* 46 (2007) 6654.
106. L. Yuan, Z. P. Guo, K. Konstantinov, P. Munroe, H. K. Liu, *Electrochem. Solid State Lett.* 9 (2006) A524.
107. B. Varghese, M. V. Reddy, Y. Zhu, S. L. Chang, T. C. Hoong, G. V. Subba Rao, B. V. R. Chowdari, A. T. S. Wee, C. T. Lim, C. H. Sow, *Chem. Mater.* 20 (2008) 3360.
108. X. H. Huang, J. P. Tu, C. Q. Zhang, J. Y. Xiang, *Electrochem. Commun.* 9 (2007) 1180.
109. X. H. Huang, J. P. Tu, B. Zhang, C. Q. Zhang, Y. Li, Y. F. Yuan, H. M. Wu, *J. Power Sources* 161 (2006) 541.
110. X. H. Huang, J. P. Tu, X. H. Xia, X. L. Wang, J. Y. Xiang, *Electrochem. Commun.* 10 (2008) 1288.
111. J. Hu, H. Li, X. Huang, *Electrochem. Solid State Lett.* 8 (2005) A66.
112. S. Grugeon, S. Laruelle, L. Dupont, F. Chevallier, P. L. Taberna, P. Simon, L. Gireaud, S. Lascaud, E. Vidal, B. Yrieix, J.-M. Tarascon, *Chem. Mater.* 17 (2005) 5041.

113. L. Dupont, S. Laruelle, S. Grugeon, C. Dickinson, W. Zhou, J.-M. Tarascon, *J. Power Sources* 175 (2008) 502.

114. J. J. Auborn, Y. L. Barberio, *J. Electrochem. Soc.* 134 (1987) 638.

115. Y. Liang, S. Yang, Z. Yi, J. Sun, Y. Zhou, *Mater. Chem. Phys.* 93 (2005) 395.

116. L. C. Yang, Q. S. Gao, Y. H. Zhang, Y. Tang, Y. P. Wu, *Electrochem. Commun.* 10 (2008) 118.

117. S. V. Pol, V. G. Pol, V. G. Kessler, G. A. Seisenbaeva M. Sung, S. Asai, A. Gedanken, *J. Phys. Chem. B* 108 (2004) 6322.

118. B. C. Satishkumar, A. Govindaraj, M. Nath, C. N.R. Rao, *J. Mater. Chem.*10 (2000) 2115.

119. L. C. Yang, Q. S. Gao, Y. Tang, Y. P. Wu, R. Holze, *J. Power Sources* 179 (2008) 357.

120. P. G. Bruce, *Solid State Sci.* 7 (2005) 1456.

121. M. V. Reddy, T. Yu, C. H. Sow, Z. X. Shen, C. T. Lim, G. V. Subba Rao, B. V. R. Chowdari, *Adv. Func. Mater.* 17 (2007) 2792.

122. D. Larcher, C. Masquelier, D. Bonnin, Y. Chabre, V. Masson, J. B. Leriche, J. M. Tarascon, *J. Electrochem. Soc.* 150 (2003) A133.

123. J. Chen, L. Xu, W. Li, X. Gou, *Adv. Mater.* 17 (2005) 582.

124. J. Liu, Y. Li, X. Huang, R. Ding, Y. Hu, J. Jiang, L. Liao, *J. Mater. Chem.* 19 (2009) 1859.

125. P. L. Taberna, S. Mitra, P. Poizot, P. Simon, J. M. Tarascon, *Nat. Mater.* 5 (2006) 568.

126. W. M. Zhang, X. L. Wu, J. S. Hu, Y. G. Guo, L. J. Wan, *Adv. Func. Mater.* 18 (2008) 3941.

Chapter 6

Interface between Transition Metal Oxides–Based Electrodes and Lithium Salts Electrolytes: A Physicochemical Approach

Sylvain Franger
Physico-Chimie de l'Etat Solide, Institut de Chimie Moléculaire et des Matériaux d'Orsay, UMR CNRS 8182, Université Paris XI, 91405 Orsay cedex, France
sylvain.franger@u-psud.fr

A physical model is proposed for the interpretation of peculiar ac-impedance spectroscopy diagrams obtained with transition metal oxides based electrodes in the presence of lithium salts ($LiClO_4$, $LiBF_4$, $LiAsF_6$, or $LiPF_6$) dissolved in organic solvents (PC, EC, DMC, etc.). This model is based on the principle of an adsorption of the electroactive species onto the surface of the electrodes. It can explain most of the numerous distorted plots reported in the literature and its reliability is also discussed by performing other complementary experiments.

Nanotechnology in Advanced Electrochemical Power Sources
Edited by S. R. S. Prabaharan and M. S. Michael

ISBN 978-981-4241-43-4 (Hardcover), 978-981-4303-05-7 (eBook)
www.panstanford.com

6.1 Introduction

The kinetics of heterogeneous surface reaction mechanisms play a crucial role both in the understanding of fundamental phenomena (surface reactivity, co-deposition, first stages of oxidation) and in the synthesis and use of low-dimension systems (clusters and catalytic systems, thin films for magnetism and microelectronics, electrochemistry). Thus, the performances of secondary batteries depend on the kinetics of reactions that take place at both electrodes during cycling. Now if the expansion of the lithium-based batteries (lithium-metal and lithium-ion) seems assured in portable electronics, thanks to their large weight energy density, it is still strongly slowed down in other fields, such as the electrical vehicle (EV), where better cycling performances are required. Improving the capacity retention upon cycling of the electrode material and consequently replying to what is currently one of the major objectives of intercalation compounds electrochemistry for the lithium-based batteries require to better understand the aging and degradation mechanisms of these materials during cycling. These fundamental interfacial mechanisms are still not completely understood, their parameters are not well known, and therefore, not brought under control yet. Modeling the kinetics at the electrodes is necessary to determine these parameters.

Among all the transition metal oxides studied as possible cathodic materials for lithium batteries, the 4 V ternary oxides such as $LiCoO_2$[1] and $LiNiO_2$[2] have been subject of numerous previous studies. In the field of lower voltages materials, $LiMnO_2$, V_2O_5, MoO_3, or TiO_2 are the most popular. All of them have been tested in alternating current (ac)-electrochemical impedance spectroscopy (EIS) mostly to determine the kinetic of the lithium diffusion into the host lattice through the calculation of the chemical diffusion coefficient D_{Li}, which can be extract from the Warburg region (semi-infinite diffusion) that appears in the Nyquist representation, $Z_{Im} = f(Z_R)$. EIS is based on the analogies that exist between electrochemical reactions in the cell and electrical circuits.[3,4] For insertion electrodes, the Randles equivalent circuit is commonly admitted to fit the data. But many experiments performed on $LiCoO_2$, $LiNiO_2$, or $LiMnO_2$ have shown deviation from this model[5–8] when others, such as V_2O_5, MoO_3, or TiO_2, can be simply treated with it.[9–14] For the former compounds, we can observe two semicircles and a huge evolution of the diagrams in function of the potential of the electrode. Very few explanations have been proposed for the understanding of such behavior. The most popular is the formation of a film on the surface of the electrode.[8,15] In this model, the high frequency semicircle corresponds to the film itself and the medium-low frequency semicircle refers to the charge transfer. But the major problem with this theory is the lack of influence with time. The model supposes that the film is formed immediately after introduction into the electrolyte solution and does not grow during experiment, which seems unlikely.

The nanostructured nature of some active material crystallites, as well as the electronic structure of the transition metal of the bulk material, would be studied here in order to shed more light on the behavior of the electrode/electrolyte interface and its evolution in operating conditions, and finally, relevant EIS models will be proposed.

6.2 Experimental Section

Most of the electrochemical results/data used for this study are extracted from the existing literature.

For "homemade" experiments, the experimental conditions were as following:

- The electrodes were prepared by mixing the active material (80% wt) with graphite (7.5% wt), acetylene black (7.5% wt), and PVdF as binder agent (5% wt) in N-methyl-pyrolidone and then casting the slurry on an aluminum collector (12 mm thickness).
- EIS measurements were performed with a VMP3 multichannel potentiostat/FRA (BioLogic). The amplitude of the polarization signal was 10 mV peak-to-peak and the equilibrium condition was supposed to be reached when potential drift was $dE/dt < 0.1$ mV.h^{-1}.

The fit of the ac-impedance diagrams was made with the Zplot software (Scribner Association).

6.3 Physicochemical Approach Proposed

6.3.1 *Background*

Adsorption on oxide surfaces follows significantly different patterns from that on pure metals. In the first place, the relatively ionic nature of the solids leads to a predominance of acid/base, or donor/acceptor, interactions.[16] Cations sites are Lewis acids, and may interact with donor molecules, such as organic solvents, through a combination of electrostatics (ion–dipole attraction) and orbital overlap. Oxide ions similarly act as basic sites and can interact with acceptors such as Li^+. As a matter of fact, recent investigations have shown that alkali cations on metal oxide surfaces interact very strongly with O^{2-} ions on the surface and cover the surface uniformly in the first monolayer.[16,17] It has been clearly demonstrated that lithium ions, for instance, could adsorb very easily on δ-MnO_2[18] and on ternary oxides, such as $Li_{0.45}MnO_{2.1}$ (and react at elevated temperature to form more lithiated compounds).[19–21] But several spectroscopic studies have also shown the presence of organic degradation products at the interface electrode/electrolyte[8,22] so that it is more convenient to assume that the lithium ions are certainly solvated with organic molecules from the electrolyte[23] and that it is presumably these coordinated entities that adsorb onto the surface of the transition metal oxides through a donor/acceptor interaction with partial decomposition of the organic structure at high potential *vs.* Li as it has already been mentioned in literature.[24,25] This hypothesis is also consistent with the fact that only d^n oxides give birth to this adsorption phenomenon ($LiCoO_2$, $LiNiO_2$, $LiMnO_2$) and not d^0 oxides (TiO_2, $Li_4Ti_5O_{12}$, V_2O_5, or MoO_3).

Actually, the d orbitals of surface cations can be divided roughly into "t_{2g}-like" and "e_g-like" ones (Fig. 6.1). Clearly the surface e_g orbital is oriented so as to achieve strong σ-type overlap with the

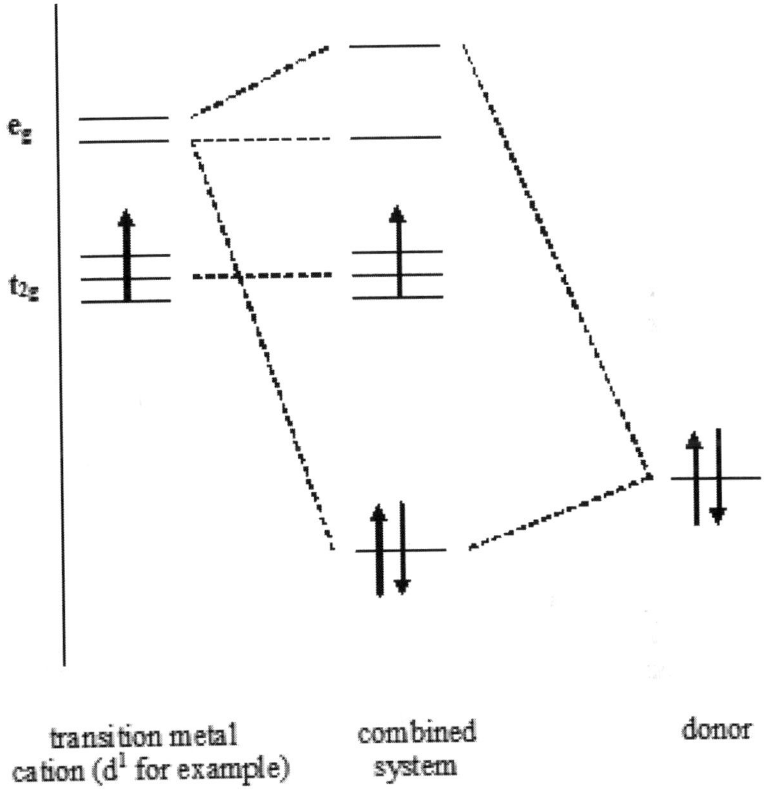

Figure 6.1 Example of a combination between nd[1] TM-oxide and donor orbital.

occupied orbital of a donor adsorbate, whereas the t_{2g} orbitals will have weaker overlap. Orbital interaction will, therefore, raise the energy of the e_g orbital, but may hardly affect t_{2g}. With transition metal cations having relatively few d electrons (and that includes d^0 oxides, such as V_2O_5), only the t_{2g} orbitals are occupied and relatively very little interaction should be seen between these and adsorbates.

These facts prompted us to propose the following electrochemical processes for the behavior of transition metal oxide–based electrodes:

$$\begin{cases} \underset{C_{\text{el}}}{[\text{Li}^+]_{\text{org}}} + \underset{\Gamma(1-\theta)}{\text{surface}} \overset{k}{\longleftrightarrow} \underset{\Gamma\theta}{\left([\text{Li}^+]_{\text{org}}, \text{surface}\right)} & (5.1) \\ \underset{\Gamma\theta}{\left([\text{Li}^+]_{\text{org}}, \text{surface}\right)} + e^- \underset{k_a}{\overset{k_c}{\leftrightarrow}} \underset{C_0\Gamma}{(\text{Li}^+, e^-)_{\text{oxide}}} + \underset{(1-\theta)}{\text{surface}} & (5.2) \\ \underset{C_0}{(\text{Li}^+, e^-)_{\text{oxide}}} \overset{D_{\text{Li}}}{\longrightarrow} \underset{C}{(\text{Li}^+, e^-)_{\text{oxide}}} & (5.3) \end{cases}$$

where C_{el}, C_0, and C are, respectively, the concentration of solvated lithium ions in the electrolyte solution, the concentration of lithium ions in the cathodic material near the interface electrode/electrolyte, and the concentration of lithium ions in the cathodic material at equilibrium; Γ is the surfacic concentration of host sites, on the electrode, for adsorbates; and θ is the rate of sites occupied.

6.3.2 *Kinetic Equations*

From these electrochemical equations that correspond to elementary processes, we can now deduce the following expressions.

The faradic current is defined by:

$$I_{\text{F}} = -FS\Gamma\,[k_c\theta - k_a(1-\theta)C_0] \tag{6.4}$$

with

$$k_c = k^0 \exp\left[\frac{-\alpha nF}{RT}(E - E^0)\right]$$

$$k_a = k^0 \exp\left[\frac{(1-\alpha)nF}{RT}(E - E^0)\right]$$

The occupied sites with adsorbed species are described by:

$$\frac{d\theta}{dt} = kC_{\text{el}}(1-\theta) - k_c\theta + k_a(1-\theta)C_0 = \frac{I_{\text{F}}}{FS\Gamma} + kC_{\text{el}}(1-\theta) \tag{6.5}$$

And the diffusion phenomenon satisfies the Fick's laws:

$$-D_{\text{Li}}\frac{\partial C(x)}{\partial x} = \Phi = -k_c\Gamma\theta + k_a\Gamma(1-\theta)C_0 \tag{6.6}$$

$$\frac{\partial C(x)}{dt} = D_{\text{Li}}\frac{\partial^2 C(x)}{\partial x^2} \tag{6.7}$$

with these limits conditions:
$x = 0$ corresponds to $C\,(0) = C_0$ and $x = \delta$ (diffusion layer thickness) corresponds to $C\,(\delta) = C$.

The linearization of the above expressions and the use of the Laplace transformation leads to:

$$\text{From (5.5): } \overline{\Delta\theta}\, j\omega = \frac{\overline{\Delta I_F}}{FS\Gamma} + k(1-\theta)\overline{\Delta C_{el}} - kC_{el}\overline{\Delta\theta} \tag{6.8}$$

The electrolyte solution is enough concentrated (generally 1 mol.L^{-1}) to assume that C_{el} is constant with time, so we can simplify Eq. 5.8:

$$C_{el} = c_{st} \Rightarrow \overline{\Delta C_{el}} = 0 \Rightarrow \overline{\Delta I_F} = FS\Gamma(j\omega + kC_{el})\overline{\Delta\theta} \tag{6.9}$$

From Eqs. 6.6 and 6.7: $\overline{\Delta C} = -T\overline{\Delta\Phi}$ with $T = \frac{th\left(\delta\sqrt{\frac{j\omega}{DLi}}\right)}{\sqrt{j\omega DLi}}$ so we can write:

$$\overline{\Delta C} = -\frac{T}{[1 + Tk_a\Gamma(1-\theta)]}\{(-k_c\Gamma - k_a\Gamma C_0)\overline{\Delta\theta} + [-b_c k_c\Gamma\theta + b_a k_a\Gamma(1-\theta)C_0]\overline{\Delta E}\} \tag{6.10}$$

But we know that

$$\overline{\Delta I_F} = \left(\frac{\partial I_F}{\partial E}\right)_{\theta,C}\overline{\Delta E} + \left(\frac{\partial I_F}{\partial \theta}\right)_{E,C}\overline{\Delta\theta} + \left(\frac{\partial I_F}{\partial C}\right)_{\theta,E}\overline{\Delta C} \tag{6.11}$$

with

$$\frac{\overline{\Delta E}}{\overline{\Delta I_F}} = Z_F;$$

and

$$\left(\frac{\partial I_F}{\partial E}\right)_{\theta,C} = \frac{1}{R_{ct}}$$

where R_{ct} symbolize the charge transfer resistance.

So, finally, we can determine the faradic impedance:

$$Z_F = \frac{1 + \frac{k_c + k_a C_0}{j\omega + kC_{el}} + \frac{Tk_a(1-\theta)\Gamma(-k_c - k_a C_0)}{[1+Tk_a\Gamma(1-\theta)](j\omega + kC_{el})}}{\frac{1}{R_{ct}}\left[\frac{1}{1+Tk_a\Gamma(1-\theta)}\right]}$$
$$= R_{ct} + R_{ct}Tk_a\Gamma(1-\theta) + \frac{R_{ct}(k_c + k_a C_0)}{j\omega + kC_{el}} \tag{6.12}$$

The faradic impedance is composed of three impedances in series:

$$Z_F = R_{ct} + Z_w + Z_1$$

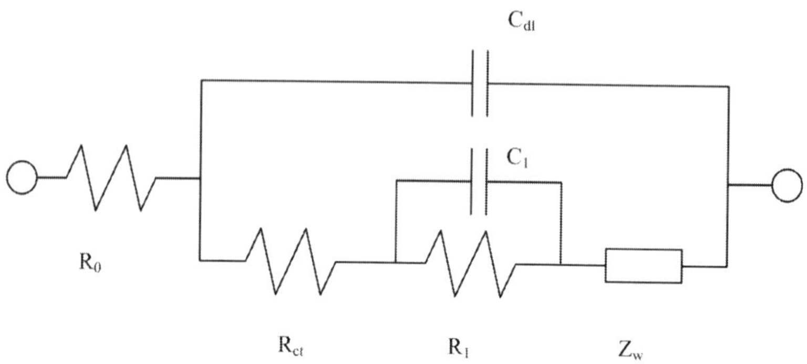

Figure 6.2 Electrical equivalent circuit of the proposed model (adsorption followed by electro-insertion).

where Z_W is the Warburg diffusion impedance, here $Z_W = R_{ct}Tk_a\Gamma(1-\theta)$.

Z_1 is more complex but is actually composed of two other impedances in parallel:

$$\frac{1}{Z_1} = \frac{j\omega}{R_{ct}(k_c + k_a C_0)} + \frac{kC_{el}}{R_{ct}(k_c + k_a C_0)} = j\omega C_1 + \frac{1}{R_1}$$

where

$$R_1 = \frac{R_{ct}(k_c + k_a C_0)}{kC_{el}} \text{ and } C_1 = \frac{1}{R_{ct}(k_c + k_a C_0)}.$$

Finally, we can build an electrical equivalent circuit for this model of adsorption (Fig. 6.2).

6.4 Confrontation/Discussion

With this model, and the simulation of the impedance plots that correspond, we can see that several trends are pointed out. First of all, the charge transfer resistance R_{ct} (which refers to the high frequencies semi-circle in Nyquist plots) is almost constant with the potential E of the electrode whereas R_1 (which corresponds to the second semi circle on Nyquist plots) is decreasing exponentially with E to reach a minimum at $E = E^\circ$ (E is the standard potential of the redox couple $M^{(n+1)+}/M^{n+}$) and is increasing again exponentially when $E > E^\circ$ (Fig. 6.3).

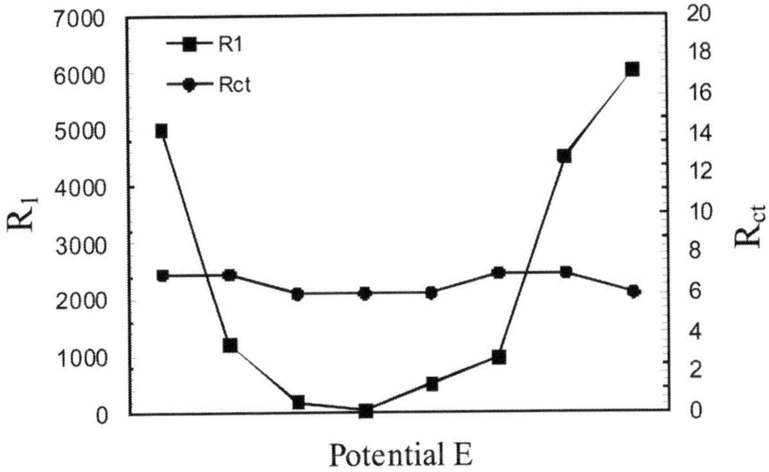

Figure 6.3 Evolution of the variations of R_1 and R_{ct} with the electrode potential (example of $LiNiO_2$).

This explains the stability of the high frequencies semi-circle and the huge variations with the medium-low frequencies semi-circle observed on experimental ac-impedance diagrams of $LiCoO_2$, $LiNiO_2$ (E = 3.7 V *vs.* Li) (Fig. 6.4) and $LiMnO_2$.[6–8]

The Warburg region (that means the straight line at 45° in the medium-low frequencies) is only observed when R_1 is not too high, that is, for $E \sim E^{\circ}$. When the potential is greater, in absolute value, than the standard potential, this domain, relative to the semi infinite diffusion, is hidden by the $R_1//C_1$ adsorption-dipole.

In order to confirm this model of adsorption on surface of d^n transition metal oxide-based electrodes, we have performed cyclic voltammetry experiments on the compounds $LiCoO_2$, $LiNiO_2$ (d^n oxides), and $Li_4Ti_5O_{12}$ (d^0 oxide) (Fig. 6.5).

Electrodes have been tested in function of the scan rate v, between 0.2 mV.s^{-1} and 0.002 mV.s^{-1}. The anodic (I_{pa}) and cathodic (I_{pc}) peaks are then reported in function of v and in function of $v^{1/2}$ (Table 5.1).

It appears clearly (Fig. 6.6a,c) that the anodic peak current for all compounds is a linear function of the square scan rate, $I_{pa} = f(v^{1/2})$, which proves that the oxidation reaction is controlled by the diffusion process of Li^+ ions into the host lattice and an average

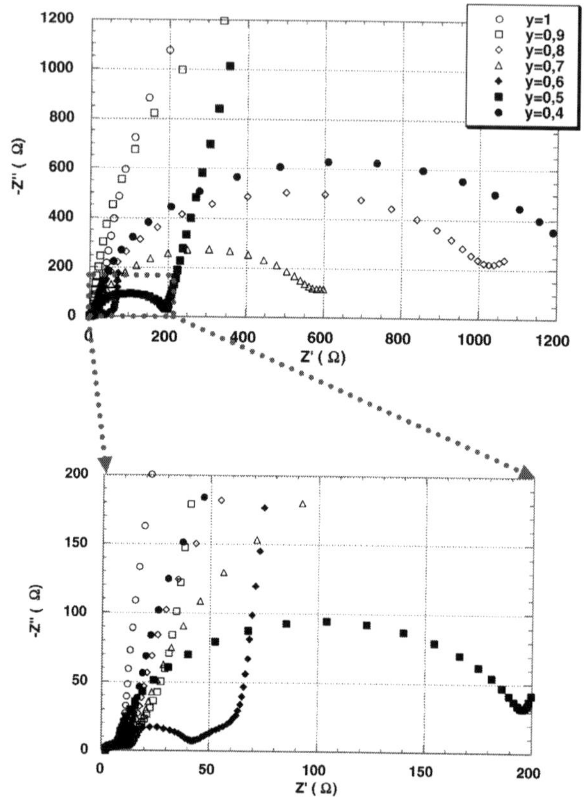

Figure 6.4 Experimental EIS data observed with Li_yNiO_2. Reprinted from Ref. 7, Copyright 2001, with permission from Elsevier, and adapted from Ref. 28.

value for the chemical diffusion coefficient of Li^+ ions in $LiCoO_2$, $LiNiO_2$, and $Li_4Ti_5O_{12}$ can be calculated by using the following relation:

$$I_{pa} = 2,69.10^5 \times S \times D_{Li}^{1/2} \times v^{1/2} \times C \quad (6.13)$$

where S is the electroactive surface area of the electrode (cm^2), C is the variation of the lithium content ($cm^3.mol^{-1}$), and v is the scan rate ($V.s^{-1}$). A molar volume of 19 $cm^3.mol^{-1}$, 20.5 $cm^3.mol^{-1}$, and 35.2 $cm^3.mol^{-1}$ was used for $LiCoO_2$, $LiNiO_2$, and $Li_4Ti_5O_{12}$, respectively.

Table 6.1 Anodic and cathodic peaks in function of the scan rate for three different materials

Intensities (mA)							
Scan rate ($mV.s^{-1}$)		$LiCoO_2$[25,26]		$LiNiO_2$[8,28]		$Li_4Ti_5O_{12}$	
v	$v^{1/2}$	I_{pa}	I_{pc}	I_{pa}	I_{pc}	I_{pa}	I_{pc}
0.0028	0.053	-	-	-	-	0.4389	0.3595
0.0086	0.093	-	-	-	-	0.8581	0.6066
0.010	0.100	0.0275	0.0232	-	-	-	-
0.0172	0.131	-	-	-	-	1.2070	0.8188
0.020	0.141	0.0475	0.0289	0.0313	0.0279	-	-
0.035	0.187	-	-	-	-	1.6965	1.0980
0.050	0.224	0.0750	0.0407	0.0719	0.0438	-	-
0.0697	0.264	-	-	-	-	2.2705	1.5301
0.100	0.316	0.1102	0.0580	0.1188	0.0625	-	-
0.200	0.447	-	-	0.1875	0.1	-	-

Thus, a mean value for D_{Li} can be determined from the slopes of the corresponding curves and we obtained (with S arbitrary taken as the geometric surface area of the electrode): $D_{\mathrm{Li}} \sim 10^{-9}$ $cm^2.s^{-1}$ for $LiCoO_2$, $D_{\mathrm{Li}} \sim 10^{-10}$ $cm^2.s^{-1}$ for $LiNiO_2$ and $D_{\mathrm{Li}} \sim 10^{-10}$ $cm^2.s^{-1}$ for $Li_4Ti_5O_{12}$. These results are well consistent with those described in the previous studies.[7,8,26–28]

On the opposite, for the reduction step, a linear relation I_{pc} *vs.* v is found for both d^n oxides ($LiCoO_2$ and $LiNiO_2$) (Fig. 6.6b), which demonstrates that the insertion process is not controlled here by the diffusion process of lithium ions from the electrolyte as in the case of d^0 transition metal oxides like V_2O_5, MoO_3, or TiO_2[9–14] and, for instance, like our d^0 reference compound, $Li_4Ti_5O_{12}$, where I_{pc} follows, just like I_{pa}, a linear function of the square scan rate (Fig. 6.6c): $I_{pc} = f(v^{1/2})$.

This shows unambiguously that the electroactive species, for $LiNiO_2$ and $LiCoO_2$, are already present at the electrode and confirms the hypothesis of the adsorption on such compounds.

So that, in accordance with the theoretical model proposed by E. Laviron[29–31] for electroactive species adsorbed on a solid electrode surface, we can write the corresponding equation:

$$I_{pc} = \frac{F^2}{4RT} \times S \times v \times \Gamma_{eq} \tag{6.14}$$

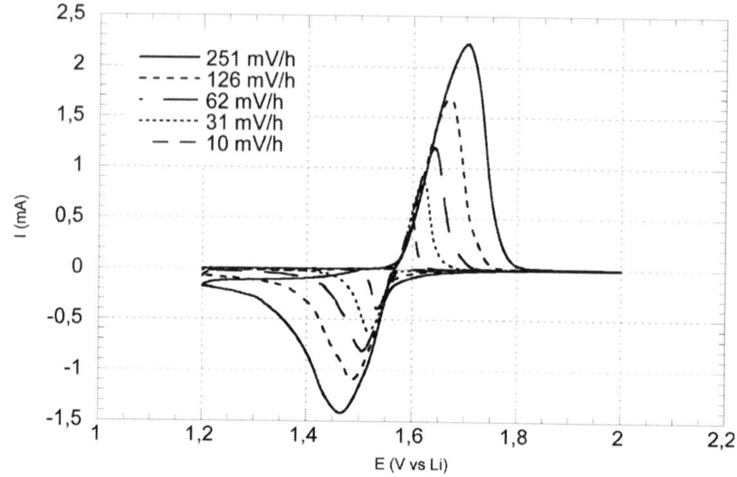

Figure 6.5 Cyclic voltametric curves relative to $Li_4Ti_5O_{12}$ at different scan rates (homemade experiments).

where Γ_{eq} is the quantity of adsorbed species onto the surface at equilibrium (mol.cm^{-2}).

A mean value of the surfacic concentration of adsorbed species on both electrodes can also be determined from the slopes of the relative curves and we obtained (with S taken here as BET surface area): $\Gamma_{eq} = 10^{-9}$ mol.cm^{-2} for both $LiCoO_2$ and $LiNiO_2$.

6.4.1 *Limitation of this Model at High Voltages (>4.3 V vs. Li) Especially in Presence of Propylene Carbonate*

The adsorption model is reliable between OCP potential of the electrode to 4.3 V *vs.* Li. Beyond this upper limit, another preponderant phenomenon takes place, as already shown in some former studies[21,24,25] the electro-oxidation of the solvent (the propylene carbonate in particular), which leads to the polymerization of the degradation products and the formation of a film. This layer covers the surface of the electrode and in ac-impedance spectroscopy, we can see a typical diagram for this phenomenon.

Another electrical equivalent circuit must then been taken to fit the data and the exploitation of the high frequencies semicircle, which refers now to the film, in front of the transition metal oxide

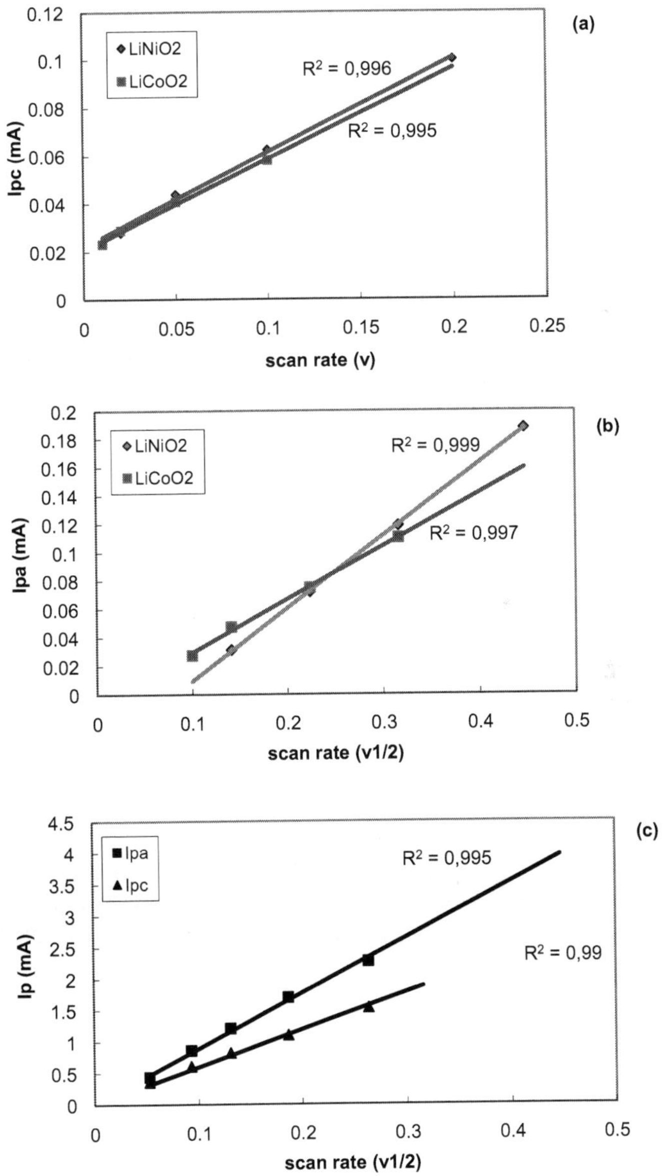

Figure 6.6 Laws followed by the intensities in function of the scan rate for (a) I_{pa} concerning $LiNiO_2$ and $LiCoO_2$; (b) I_{pc} concerning $LiNiO_2$ and $LiCoO_2$; and (c) I_{pa} and I_{pc} for $Li_4Ti_5O_{12}$.

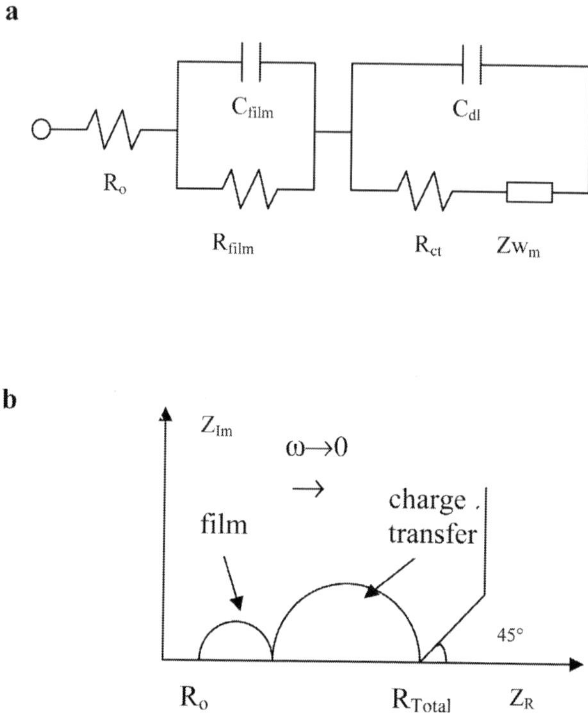

Figure 6.7 (a) representation of the electrical equivalent circuit in the case of the film formation; (b) simulation of the corresponding impedance diagram in the Nyquist plan.

surface, allows to estimate the relative thickness of this polymer layer[21,32,33] (Fig. 6.7).

We have the following relations:

$$R_{\text{film}} = \frac{d}{\sigma \times S} \qquad C_{\text{film}} = \frac{\varepsilon \times \varepsilon_0 \times S}{d}$$

where d is the thickness of the film, S is the surface, σ is the conductivity, and ε the permittivity.

Two examples are given for $LiCoO_2$ and $LiNiO_2$ (Fig. 6.8) with the assumption that the film is principally made of polypropylene oxide (the electrolyte used in that case was $LiClO_4$ 1 mol.L^{-1} dissolved in PC) charged with ions from the electrolyte. In these conditions, we can estimate, in respect to values proposed in previous studies[33–37], its conductivity σ ($10^{-8} < \sigma < 10^{-7}$ S.cm^{-1}) and its permittivity

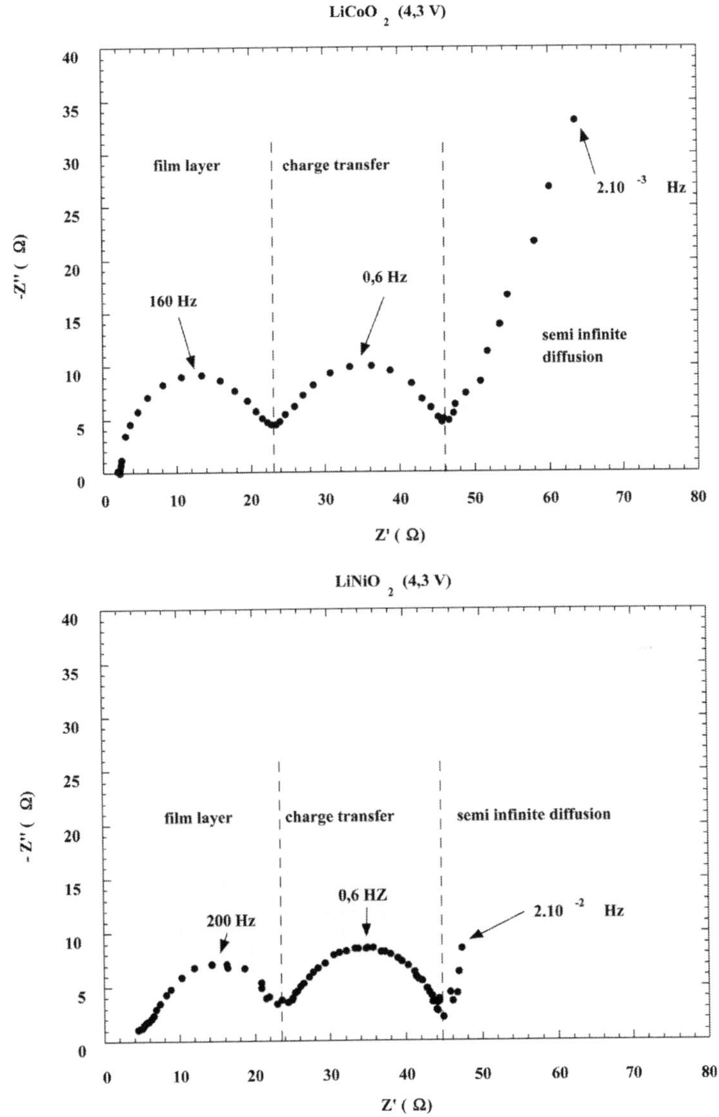

Figure 6.8 EIS diagrams of $LiCoO_2$ (up) and $LiNiO_2$ (down) polarized at high potential (>4.3 V /Li) in $LiClO_4$ 1M/PC, during 10 min, at room temperature (homemade experiments).

Table 6.2 Evaluation of the thickness of the film formed onto the surface of the electrodes

Data extracted from the ac-impedance diagrams (Fig 6.8)	$LiCoO_2$	$LiNiO_2$
R_{film}	22 Ω	18 Ω
Thickness (max.)	~ 170Å	~ 140Å

(dielectric constant) ε ($10 < \varepsilon < 20$), and conclude the calculation. The results are shown in Table 5.2.

6.4.2 *Limitation of this Model with the Specific Surface of the Particles: the Case of the Nanostructured Materials*

Most of the compounds that have been considered as electroactive materials for secondary lithium battery exhibits relatively slow kinetics of lithium insertion (as already mentioned in the previous paragraphs, $D < 10^{-9}$ cm^2.s^{-1}). This is principally due to the low intrinsic ionic conduction (few vacant sites that can ensure the mobility of the lithium ions in the host crystal lattice). To overcome this difficulty, many authors have already suggested reducing the crystallites size. Actually, for such electrochemical systems, the lithium diffusion pathway is less detrimental in a nanograin than in a micrometer-sized particle (Fig. 6.9).

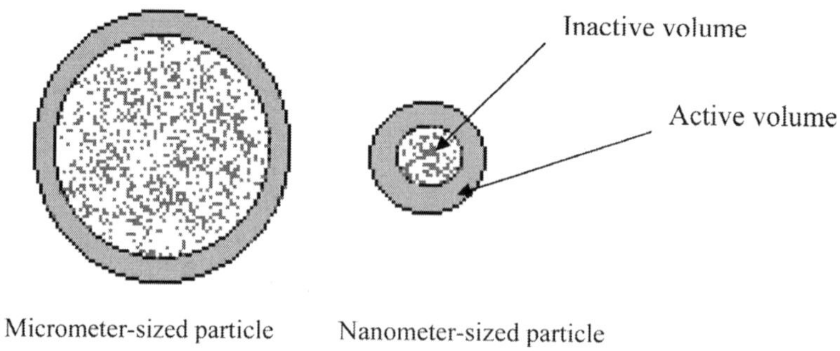

Figure 6.9 Comparison of the active volume between a micrometer and a nonometer-sized particles.

Indeed, when it is limiting, the diffusion controls the flow of the charge carrier (current density) that a battery can deliver. If one takes, for a reduction process, the hypothesis of a simple mono-directional phenomenon, one can write this flow as:

$$j = \frac{z \cdot F}{S} \cdot \frac{\partial x}{\partial t} \tag{6.15}$$

With z: lithium ion charge (1 for Li^+)

F: Faraday

S: electroactive surface (lithium ions front)

x: quantity of inserted Li^+ ions

From first Fick's law, one can also write the flow as:

$$j = -z.F.D \left(\frac{\partial C_{Li}}{\partial L} \right)_{L=0} \tag{6.16}$$

With D: chemical diffusion coefficient of Li^+

C_{Li}: Li^+ concentration

L: distance from the surface of the inserted Li^+ to be considered

Thus, to determine the distance, over which diffusion will drive the redox reaction, one can write:

$$C_{Li} = \frac{x}{S.L} \tag{6.17}$$

Integration of (6.15) and (5.16) with inclusion of (6.17), leads to:

$$L = \sqrt{2.D.t} \tag{6.18}$$

With t: time of the process, related to the discharge rate imposed to the electrochemical system.

The chemical diffusion coefficient of the lithium ions, in a solid host lattice, is about 10^{-11} $cm^2.s^{-1}$ (mean value on different cathodic materials).

Thus, when the discharge rate is slow ($t \gg L^2/D$), the length of the lithium diffusion pathway is important (the whole particle is active), but when the discharge rate is fast ($t \ll L^2/D$), the distance traveled by the Li^+ ions is small (only a fraction of the particle is working), as illustrated by Fig. 6.9.

So, if D is very small, to access high rates (faster charge and/or discharge of the battery), it is crucial to work with nanoparticles.[38–40] This explains why a lot of work is now conducted in order to synthesize nanomaterials.

As previously recalled in this study, for TiO_2 or V_2O_5 composite electrodes with microsized particles that exhibit less than 1 $m^2.g^{-1}$ for their specific surface areas (i.e., actually the common value for most oxides tested in composite electrodes: $LiCoO_2$ exhibits 0.48 $m^2.g^{-1}$ and $LiNiO_2$ 0.89 $m^2.g^{-1}$), no quantitative adsorption phenomenon is observed, both on impedance spectra or on cyclic voltammetry experiments. But very recent investigations have shown than thin films of titanium dioxide[41] or vanadium pentoxide[42,43] react similarly as $LiCoO_2$ or $LiNiO_2$ electrodes with appearance of the same huge distortion in the impedance diagrams that suggests an identical physical process at the interface electrolyte/electrode.

Although it was demonstrated that for d^0 oxides, the adsorption is less effective, the specific surface developed by the thin films (66 $m^2{\cdot}g^{-1}$ for TiO_2 and 155–280 $m^2{\cdot}g^{-1}$ for V_2O_5) is so much greater than that exhibited by composite electrodes that it is possible to imagine a quantitative adsorption.[44]

This is a major issue, since nanomaterials will be intensively used for next generation of secondary lithium batteries. New electrode architectures and/or the use of additives/inhibitors in the electrolytes would be surely of great interest in order to alleviate degradation (ageing) or slowdown mechanisms inside the battery system.

6.5 Summary/Perspectives

Electrochemical ac or direct current (dc) pulse techniques applied on the simple electrochemical system Li/Li^+, PC/TiS_2 (where PC stand for propylene carbonate) initially corroborated the Randles model, that describes the lithium insertion as a dissolution reaction of the pair (Li^+, e^-) in the material host. By taking into account the mass transport kinetics of the lithium in the oxide, this famous model has permitted to suggest that the observed electrochemical behavior was correlated to the structure of the host material. However, this model is not complex enough to describe the phenomena that occur at numerous other electrode/electrolyte interfaces. In particular, the responses obtained by electrochemical impedance spectroscopy (EIS) on cathodes in $LiNiO_2$ or $LiCoO_2$ question the universality of this model and suggest a more complex mechanism that may contain passivation or adsorption steps. From this result, we can establish a large set of candidate mechanisms, where the aforementioned steps can also be followed by chemical or electrochemical insertion for instance.

The aim of this study was to lead simultaneously a double approach, experimental and theoretical, to build a phenomenological model able to describe the data collected on positive electrodes (mainly $LiCoO_2$ and $LiNiO_2$). What is at stake is considerable: isolating the relevant parameters that permit to describe and govern the interfacial mechanisms and more specifically the adsorption and insertion mechanisms. This will permit later on to stipulate the conditions of the optimization of the performances of the Lithium batteries (structure, morphology, interface, insertion rate, number of cycles, time, etc.).

Moreover, we have demonstrated that the size of the active material grains is of great importance for the understanding of the interfacial behavior, because of the larger specific surface area exhibited by the nanometer-sized materials.

References

1. K. Mizushima, P. C. Jones, P. J. Wiseman and J. B. Goodenough, *Mat. Res. Bull.*, 15 (1980) 783.
2. M. G. S. R. Thomas, W. I. F. David and J. B. Goodenough, *Mat. Res. Bull.*, 20 (1985) 1137.
3. W. Weppner and R. A. Huggins, *J. Solid State Chem.*, 22 (1977) 297.
4. C. Ho, I. D. Raistrick and R. A. Huggins, *J. Electrochem. Soc.*, 127 (1980) 343.
5. B. Garcia, J. Farcy, J.-P. Pereira-Ramos and N. Baffier, *J. Electrochem. Soc.*, 144 (1997) 1179.
6. S. Yamada, M. Fujiwara and M. Kanda, *J. Power Sources*, 54 (1995) 209.
7. V. Bianchi, S. Bach, C. Belhomme, J. Farcy, J.-P. Pereira-Ramos, D. Caurant, N. Baffier and P. Willmann, *Electrochimica Acta*, 46 (2001) 999.
8. D. Aurbach, M. D. Levi, E. Levi, H. Teller, B. Markovsky, G. Salitra, U. Heider and L. Heider, *J. Electrochem. Soc.*, 145 (1998) 3024.

9. P. Soudan, *PhD thesis*, Université Paris, 6, (1998).
10. Z. Lu, M.D. Levi, G. Salitra, Y. Gofer, E. Levi and D. Aurbach, *J. Electroanal. Chem.*, 491 (2000) 211.
11. M.D. Levi, Z. Lu and D. Aurbach, *Solid State Ionics*, 143 (2001) 309.
12. J. O. Besenhard, J. Heydecke, E. Wudy, H. P. Fritz and W. Foag, *Solid State Ionics*, 8, (1983) 6.
13. J. Farcy and R. Messina, unpublished results.
14. A. Kuhn, R. Amandi and F. Garcia-Alvarado, *J. Power Sources*, 92, (2001) 221.
15. M. G. S. R. Thomas, P. G. Bruce and J. B. Goodenough, *J. Electrochem. Soc.*, 132 (1985) 1521.
16. V. E. Henrich and P. A. Cox, in *The Surface Science of Metal Oxides*, Cambridge University Press (1996), London.
17. V. N. Ageev and S. M Solovev, *Surface Science*, 480 (2001) 1.
18. N. Sahai and D. A. Sverjensky, *Geochimica et Cosmochimica Acta*, 61 (1997) 2827.
19. S. Franger, S. Bach, J.-P. Pereira-Ramos and N. Baffier, *J. Electrochem. Soc.*, 147 (2000) 3226.
20. S. Franger, S. Bach, J.-P. Pereira-Ramos and N. Baffier, *J. Power Sources*, 97-98 (2001) 344.
21. S. Franger, S. Bach, J. Farcy, J.-P. Pereira-Ramos and N. Baffier, *Electrochimica Acta*, 48 (2003) 891.
22. R. Kostecki, F. Kong, Y. Matsuo and F. MacLarnon, *Electrochimica Acta*, 45 (1999) 225.
23. T. Fukushima, Y. Matsuda, H. Hashimoto and R. Arakawa, *Electrochem. Solid State Lett.*, 4 (2001) A127.
24. K. Kanamura, S. Toriyama, M. Ohashi and Z. Takehara, *J. Electroanal. Chem.*, 419 (1996) 77.
25. K. Kanamura, T. Umegaki, M. Ohashi, S. Toriyama, S. Shiraishi and Z. Takehara, *Electrochimica Acta*, 47 (2001) 433.
26. B. Garcia, J. Farcy, J-P. Pereira-Ramos, J. Périchon and N. Baffier, *J. Power Sources*, 54 (1995) 373.
27. Y. I. Jang, B. J. Neudecker and N. J. Dudney, *Electrochem. Solid State Lett.*, 4 (2001) A74.
28. V. Bianchi, PhD thesis, Université Paris, 6, (1998).
29. E. Laviron, *Electroanal. Chem.*, 12 (1982) 53.

30. F. Bedioui, *Techniques de l'Ingénieur*, traité Analyse & Caractérisation - P2 127 (1999) 16.
31. F. Bedioui, *Techniques de l'Ingénieur*, traité Analyse & Caractérisation - P2 128 (1999) 8.
32. M. Froment, M. Garreau, J. Thevenin and D. Warin, *J. Micro. Spectro. Electron.*, 4 (1979) 111.
33. J. Thevenin and R. H. Muller, *J. Electrochem. Soc.*, 134 (1987) 273.
34. P. R. Sorensen and T. Jacobsen, *Electrochemica Acta*, 27 (1982) 1671.
35. C. D. Robitaille and D. Fauteux, *J. Electrochem. Soc.*, 133 (1986) 315.
36. C. R. Pals and J. Newman, *J. Electrochem. Soc.*, 142 (1995) 3274.
37. R. E. Wetton, D. B. James and F. P. Warner, *Ions in Polymer*, Eds. A. Eisenberg (1980).
38. P. P. Prosini, M. Carewska, S. Scaccia, P. Wisnieski and M. Pasquali, *Electrochimica Acta*, 48 (2003) 4205.
39. A. Singhal, G. Skandan, G. Amatucci, F. Badway, N. Ye, A. Manthiram, H. Ye and J. J. Xu, *J. Power Sources*, 129 (2004) 38
40. S. Franger, F. Le Cras, C. Bourbon, C. Benoit, P. Soudan and J. Santos-Peña, *Recent Research Developments in Electrochemistry*, TRN Editions, Kerala (ISBN : 81-7895-183-5).
41. P. Krtil, D. Fattakhova, L. Kavan, S. Burnside and M. Grätzel, *Solid State Ionics*, 135 (2000) 101.
42. W. Dong, D. R. Rolison and B. Dunn, *Electrochem. Solid State Lett.*, 3 (2000) 457.
43. M. J. Parent, S. Passerini, B. B. Owens and W. H. Smyrl, *J. Electrochem. Soc.*, 146 (1999) 1346.
44. H. J. Freund and E. Umbach, *Adsorption on Ordered Surfaces of Ionic Solids and Thin Films*, Springer Series in Surface Sciences, vol. 33 - Berlin (1993).

Chapter 7

Electron Spin Resonance Studies of Lithium-Ion Battery Materials

Nail Suleimanov,[a] S. R. S. Prabaharan,[b] Dinar Abdullin,[a] and M. S. Michael[c]

[a] *Zavoisky Physical-Technical Institute of Russian Academy of Sciences, 420029 Kazan, Russian Federation*
[b] *School of Electronics Engineering, VIT University, Chennai Campus, Chennai, Karnataka, India*
[c] *Chemical Sciences Research Centre, SSN College of Engineering, Anna University Affiliated, Kalavakkam, Chennai, India*
suleiman@kfti.knc.ru

Electron spin resonance (ESR) is a sensitive spectroscopic method for the investigation of local crystal structure and electronic properties of paramagnetic species in materials. Such species carrying the magnetic moments are often employed as active electrode materials for lithium-ion batteries and their electrochemical properties substantially determined by their structural features. In this chapter, application of ESR method to study the electronic structure of layered and NASICON structured cathode materials is presented.

Nanotechnology in Advanced Electrochemical Power Sources
Edited by S. R. S. Prabaharan and M. S. Michael

ISBN 978-981-4241-43-4 (Hardcover), 978-981-4303-05-7 (eBook)
www.panstanford.com

7.1 Introduction

Lithium-ion batteries became the inalienable and quickly developing part of portable electronics due to their higher working voltage and capacity in comparison with other types of batteries. Generally, lithium-ion rechargeable batteries consist of lithiated Co, Ni, Mn oxides as positive electrodes(cathodes), various carbonaceous composites as negative electrodes(anodes), and lithium-bearing salt in organic solvents or polymer gels as electrolytes. The development of lithium-ion batteries is closely associated with the use of new electrode materials and electrolytes. Compounds used as electrode materials or electrolyte in lithium-ion batteries should satisfy a number of requirements concerning their physical and electrochemical properties. For example, a cathode material should have high electrode potential with respect to metal lithium, high ionic and electronic conductivities, and high specific capacity. In fact, the transition metal ions present in the cathode materials have a significant influence on the electrochemical characteristics of the cathodic materials. The operating voltage of a lithium-ion battery is determined by the transition metal ions present in the cathode materials. Moreover, the local environment of transition metal ions forms the pathways for the electron and lithium ions transport in the structure of cathode materials. These facts indicate that electrochemical properties of electrode material are substantially determined by their structural features, and hence, it is essential to study the transition metal ions and their local coordination to evaluate their electrochemical performance.

Today, a variety of physical methods are used to investigate the structural features of electrode materials: X-ray diffraction, neutron diffraction, XANES spectroscopy, NMR, ESR, Raman spectroscopy, and so on. The presence of transition metal ions (Co, Ni, Mn, V, Fe) in majority of the cathode materials stimulates their study by means of electron spin resonance (ESR) technique as well. The ESR is the sensitive method to study the details of local electronic and crystal structure and allows determining the valence state of magnetic ions in a crystal structure.

The majority of cathode materials employed in lithium-ion batteries have metal-oxide layers in which Li^+ can move in and

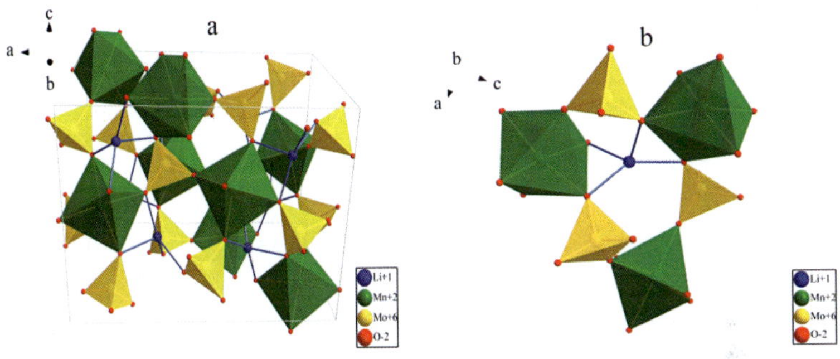

Figure 7.1 (a) NASICON structure of $Li_2Mn_2(MoO_4)_3$ and (b) interstitial positions of Li^+ ions in this structure.

out of their layers. Among them, the lithium cobaltates and lithium nickelates with the composition $LiMO_2$ (M = Co, Ni) are of significant research and technological interest. However, in real performances, these layered structures have shown at most 50% of the theoretical estimations[1] due to the certain types of structural imperfection.

Recently, a few members of NASICON family with the chemical formula $Li_2M_2(MoO_4)_3$ (M = Ni, Co, Mn) have been investigated as electrode materials for Li-ion batteries.[2–8] X-ray diffraction analysis of these compounds has shown that they adopt orthorhombic structure, which consists of MO_6 octahedrons and MoO_4 tetrahedrons sharing oxygen vertices (Fig. 7.1a). Lithium ions occupy interstitial positions in the framework formed by oxygen ions (Fig. 7.1b). These interstitial positions form 3-D pathways for the lithium transport in the NASICON structure and it has high theoretical lithium insertion/extraction capacity.[8] But the studies on the use of $Li_2M_2(MoO_4)_3$ compounds as cathode material showed rapid capacity decay during the cycling of Li-ion cell.[5–7] This capacity decay might be associated with the structural changes of these materials occurring during the cycling.

The electrochemical studies on different cathode materials indicate that along with crystal structure, electronic structure has an important influence on the processes of intercalation/extraction of lithium ions to/from lithiated compounds.

In this chapter, ESR study on layered cathode materials, $LiCoO_2$, and its substituted system $LiM_yCo_{1-y}O_2$ (M = Al, Ga) is discussed.

ESR investigations of NASICON structured $Li_2Mn_2(MoO_4)_3$ synthesized at micro and nanoscale compounds are also presented.

7.2 Electron Spin Resonance Phenomenon

The phenomenon of ESR spectroscopy can be explained by considering the behavior of a free electron. According to quantum theory, an electron has a spin leading to the magnetic moment $\mu_S = -g \cdot \mu_B \cdot S$, where S is the spin of electron ($S = 1/2$), μ_B is Bohr magneton, and g is Lande factor or g-factor, which is equal to 2.0023 for a free electron. If the electron is subjected to a steady magnetic field, $\boldsymbol{B}$ the energy of the spin system will depend upon the projection of the spin along the magnetic field. Quantum theory stipulates that only two projections $S_z = \pm 1/2$ in the applied field $\boldsymbol{B} \parallel z$ are permitted for electron spin. Consequently, the magnetic moment and the energy levels reduce to the two values $\mu_S = \pm 1/2 \cdot g \cdot \mu_B$ and $E_{\pm} = \pm 1/2 \cdot g \cdot \mu_B \cdot B$, respectively.

If an electromagnetic radiation is applied at a frequency ν that corresponds to the separation between the energy levels $\Delta E = E_{+} - E_{-}$, the amount of energy absorbed from the electromagnetic field by the spin system is given as (Fig. 7.2):

$$\Delta E = g \cdot \mu_B \cdot B = h \cdot \nu, \tag{7.1}$$

In continuous wave (CW) ESR experiment, the spectrum is obtained by sweeping in the magnetic field B at fixed microwave frequency ν. According to Eq. 7.1, the position of resonance signal in the scale of magnetic field is determined by the certain value of g-factor. As the g-factor is a characteristic feature of paramagnetic center, it is used in ESR to identify the paramagnetic species in the sample.

ESR spectrum becomes more complicated in the case of transition elements with unpaired 3d electrons, which act as paramagnetic species. As 3d electrons have a nonzero orbital magnetic momentum, the g-factor of transition elements differs from pure spin value of free electron. Moreover, the ESR spectrum of transition elements is often strongly affected by their surrounding ions also called as ligand ions. Firstly, the symmetry of the structural

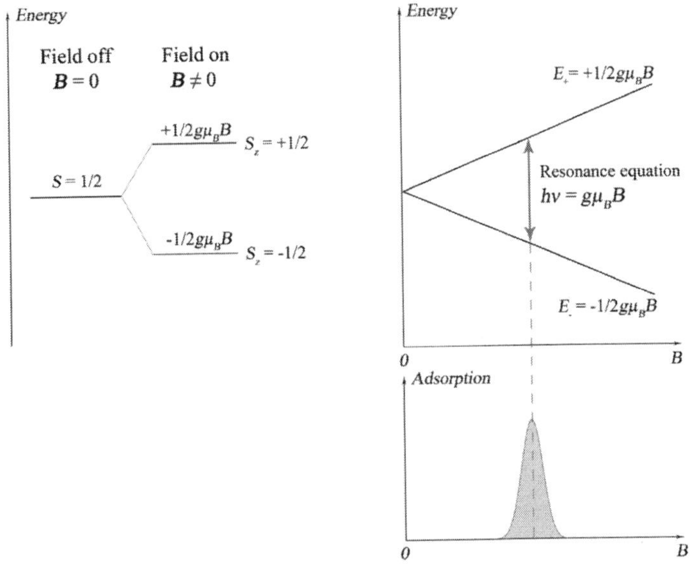

Figure 7.2 ESR phenomenon in case of free electron with spin $S = 1/2$.

unit formed by ligand ions around the transition element determines the anisotropy of g-factor, that is, its dependence on the direction of applied magnetic field. Secondly, the interaction of paramagnetic ion with electric crystal field created by surrounding ions causes the zero-field splitting of its ground energy levels, which result in the appearance of so-called fine structure in the ESR spectra. In addition, the ESR spectrum of transition elements with a nonzero nuclear spin contains a superfine structure resulting from the interactions between electron and nuclear spins. The zero-field splitting and hyperfine interaction constants can also be anisotropic.

Let us consider that a transition metal ion in a crystal is subjected to an external magnetic field. The energy states of the system are calculated using the so-called the spin Hamiltonian approach:

$$H = \mu_B \cdot \vec{B} \cdot \hat{g} \cdot \vec{S} + \vec{S} \cdot \hat{A} \cdot \vec{I} + D \cdot \left\{ S_z^2 - \frac{1}{3} \cdot S \cdot (S+1) \right\} + E \cdot \{ S_x^2 - S_y^2 \}, \tag{7.2}$$

where the Zeeman term $\mu_B \cdot \boldsymbol{B} \cdot \hat{g} \cdot \boldsymbol{S}$ represents the interaction of the electronic spin with the magnetic field, and hyperfine term $\boldsymbol{S} \cdot \hat{A} \cdot \boldsymbol{I}$ represents the electron-spin–nuclear-spin interaction, and the other two zero-field terms, namely D, representing the axial part, while E represents the deviation from axial symmetry. Thus, the identity of paramagnetic species can be determined by measurements of g-factor, zero-field splitting constants, hyperfine constants, and other resonance parameters.

7.3 Applications of ESR to Study Layered Cathode Materials

7.3.1 *Structural Imperfections of $LiCoO_2$*

ESR method has been applied as a tool to identify the impurities and deficiencies in the layered structured cathode materials with general formula, $LiMO_2$(M = Co, Ni, or Mn). It has been found that the final structures of materials can be strongly influenced by their synthetic and annealing temperatures, and oxygen and/or Li-deficient sites often exist in synthesized materials. Figure 7.3 shows room temperature ESR signals obtained for $LiCoO_2$ samples prepared at different synthetic temperatures.[9] The shapes of ESR lines are same irrespective of temperature of synthesis, but their intensity decreases as the synthetic temperatures increases.

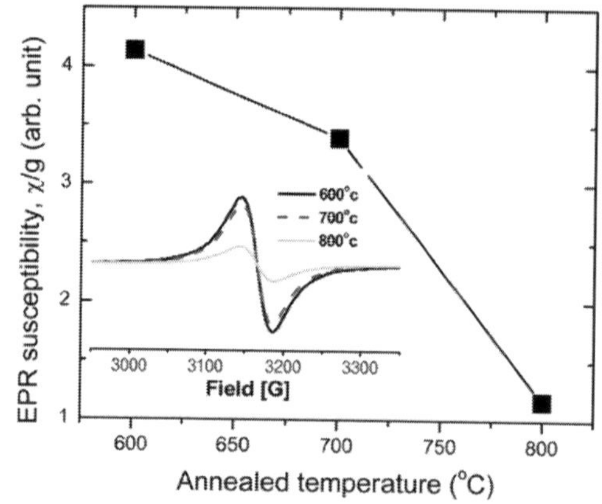

Figure 7.3 Synthetic temperatures dependence of ESR susceptibilities (Inset: the electron paramagnetic resonance (EPR) signals obtained for each $LiCoO_2$). Reprinted with permission from Ref. 10. Copyright 2004 American Chemical Society.

In ESR method, the integrated intensity, that is, the area beneath the resonance signal is proportional to the concentration and susceptibility of paramagnetic species in the sample. It is established that an Li-deficient lattice imperfection can lead to the appearance of Co^{4+} ions with $S = 1/2$, which give rise to observed single ESR spectrum with $g \sim 2$. Thus, the results of ESR investigations on $LiCoO_2$ indicate that the higher the synthetic temperatures, the less the lattice imperfection as well as the spin defects.

7.3.2 ESR Investigation of $LiM_yCo_{1-y}O_2$ (M = Al, Ga) System

Recently, it has been observed that the partial substitution of Co ions in $LiCoO_2$ by elements such as Al, Ga, and Mg increases the working voltage of lithium-ion cells.[11,12] The high working voltage of substituted system is attributed to the participation of oxygen in the redox processes, that is, the intercalation potential is mainly determined by participation of O ions in charge exchange during the cycling,[12–14] which has been supported by the result of ESR study of $LiM_yCo_{1-y}O_2$ (M=Al, Ga) compounds.[10,15,16]

As Co^{3+} ions adopt a low-spin diamagnetic configuration in $LiM_yCo_{1-y}O_2$ there is no contribution to ESR absorption by Co^{3+}. In order to carry out the ESR investigations of this system, a part of nonmagnetic Co^{3+} was substituted by magnetic Ni^{3+} ions, which play the role of spin probes.[15,16] These Ni^{3+} ions occupy octahedral cobalt positions in the host structure. The ESR spectra of Ni^{3+} doped $LiCoO_2$ and $LiAl_yCo_{1-y}O_2$ consist of single Lorentzian absorption line with $g = 2.142$ at 300 K. Below 230 K the anisotropic resonance signal with $g_\perp = 2.1935$ and $g_\| = 2.0535$ appears in Alsubstituted oxides which corresponds to the tetragonally distorted NiO_6 octahedra. Figure 7.4 shows the dependence of the g-factors of Ni^{3+} on the number of Al neighbors in nearest coordination sphere of the transition metal. The decrease in $g_\perp$ indicates that the Ni–O bond becomes more covalent and the enhancement of covalent character of Ni–O bond is associated with the appearance of additional electron density on the oxygen ions, which, as noted above, causes an increase in the electrode potential of $LiCoO_2$.

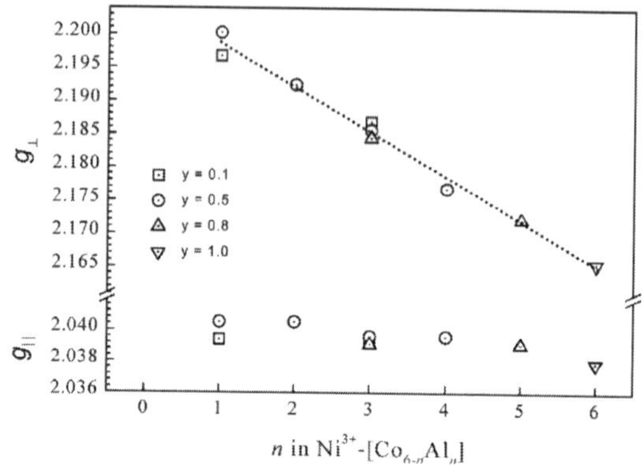

Figure 7.4 Values of the g-factor for Ni^{3+} in tetragonal crystal field *vs.* the number of Al ions included in the next metal $[Al_nCo_{6-n}]$ environment.

7.4 ESR Investigation of NASICON Structured $Li_2Mn_2(MoO_4)_3$

7.4.1 *Experiment*

$Li_2Mn_2(MoO_4)_3$ has been synthesized both in micro and nanoscale by employing two different synthesis protocols involving aqueous solutions of the respective nitrates with a fuel for mild combustion (we have named it as soft-combustion) and low temperature solvothermal process as described below.

Powders of micro-scaled $Li_2Mn_2(MoO_4)_3$ were synthesized by the low temperature combined glycine-nitrate aqueous soft-combustion procedure at low temperature as previously described.[4] Aqueous solutions of lithium nitrate and appropriate metal nitrates (Maganese nitrate in this case) were mixed with aqueous hexa ammonium heptamolybdate (Aldrich, USA) in an appropriate molar ratio (Li/Mn/Mo stoichiometry) and then mixed with aqueous glycine solution (aminoacetic acid) that acted as a mild combustion agent (fuel). The stoichiometric composition of glycine was twice the molar fraction of the starting materials in the above mixture solution. Purple colored solution thus obtained was heated to boiling at about 100°C, which became dark purple colored paste-like substance after evaporation of water. Further, the paste-like substance was allowed to decompose at 250°C in ambient air. Following heat disintegration, the reaction was duly ignited by glycine, yielding precursors of the final product, which upon further calcination (at 500°C and 600°C/6 h) in ambient air yielded a single-phase final product. The morphology of the samples obtained was investigated by scanning electron microscopy. It was found that samples consist of loosely agglomerated rod-like particles with submicrometer diameter size.

Samples of nanoscaled $Li_xMn_2(MoO_4)_3$ with $x = 0, 0.2, 0.8, 1.2, 1.8, 2.0$ were obtained by low-temperature solvothermal process. Stoichiometric amounts of lithium nitrate (Aldrich), manganese(II) nitrate (Aldrich), and hexa ammonium heptamolybdate (Aldrich, USA) were used as starting materials in our modified synthesis protocol, namely citric acid based solvothermal process to obtain single-phase products of $Li_xMn_2(MoO_4)_3$. The stoichiometric quantities of the above starting materials were dissolved in appropriate quantity of deionized water/isopropanol mixtures (80/20) separately and sonicated using an ultrasonic processor (UP200S 200 W, 24 kHz Hielscher Ultrasonics GmbH, Germany) for 10 min. Around 100 ml of saturated solution of citric acid was added to the above mixture solution containing stoichiometric amounts of Li/Mn/Mo. The resulting turbid solution was stirred continuously at 90°C until the clear solution was obtained. The heating was continued until the clear solution turned to a sticky paste-like substance. At this, the beaker containing the paste-like precursor was transferred to a heating mantle pre-heated at 350°C. This process yielded an as-prepared final product after the organic moieties were decomposed due to internal combustion of citric acid. The product thus formed has fine powders of the as-prepared final product, which was then calcined at 500°C and 600°C for 6 h. This kind of synthesis allows a better control of the size, morphology, and texture of solid particles. Moreover, the samples were prepared with various Li concentration of $Li_xMn_2(MoO_4)_3$ where $x = 0, 0.2, 0.8, 1.2, 1.8, 2.0$.

ESR measurements were carried out by means of Bruker BER-418S spectrometer at frequency 9.2 GHz (X-band) in the 7300 K temperature interval.

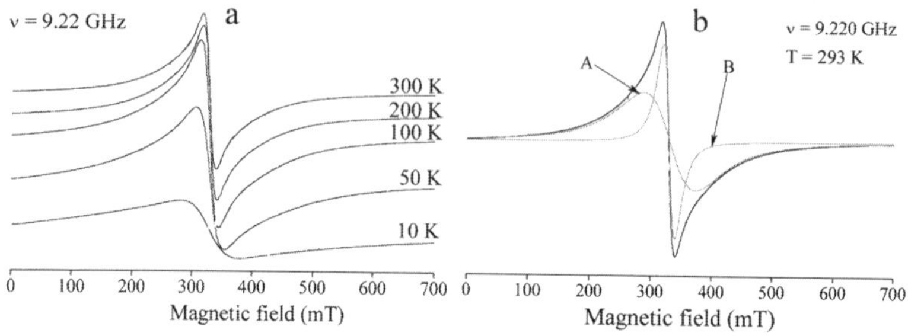

Figure 7.5 (a) ESR spectra of $Li_2Mn_2(MoO_4)_3$ at temperatures 10–300 K. (b) ESR spectra of $Li_2Mn_2(MoO_4)_3$ compound at 293 K and its approximation by two overlapping signals with Lorentzian line shape, which are denoted as line A and line B.

7.4.2 *ESR Analysis*

7.4.2.1 Micro-scaled $Li_2Mn_2(MoO_4)_3$

ESR spectra of $Li_2Mn_2(MoO_4)_3$ compound at different temperatures are presented in Fig. 7.5a. In order to determine the parameters of ESR spectrum, the approximation for observed absorption line was carried out with Gaussian and Lorentzian curves. Indeed, the best fit for experimental spectra was achieved by using two Lorentzian lines with different g-factors and linewidths (Fig. 7.5b). Let us denote the broad resonance line as line A and the narrower one as line B. On examining carefully the shape of the absorption line, one can see the small bend on its wings that suggests the existence of two overlapping resonance signals.

Figure 7.6 shows the temperature dependence of g-factors corresponding to the lines A and B. As seen in the Fig. 7.6, the g-factor of line B, $g_B = 2.000 \pm 0.006$, is constant within investigated temperature range, whereas the significant deviation of g-factor from its high-temperature value $g_A = 1.99 \pm 0.03$ is observed for line A at temperatures below 100 K.

The temperature dependences of the linewidths for both lines are presented on Fig. 7.7. The considerable broadening of resonance lines with the lowering of temperature below 100 K is observed.

Another important ESR parameter is the integral intensity, I, defined as the area under the ESR absorption curve. In the

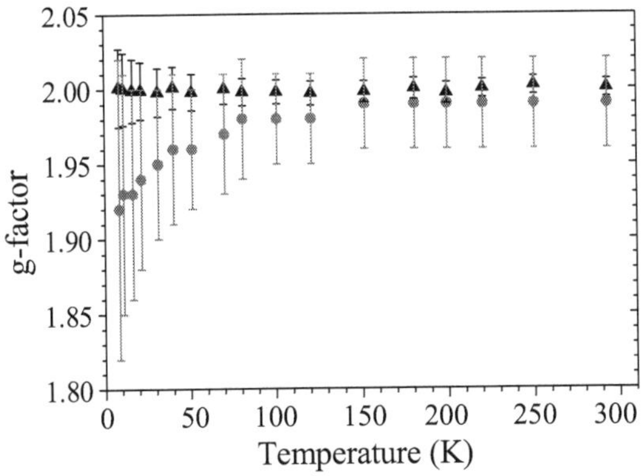

Figure 7.6 Temperature dependence of g-factors corresponding to lines A (circles) and B (triangles) in the ESR spectra of $Li_2Mn_2(MoO_4)_3$.

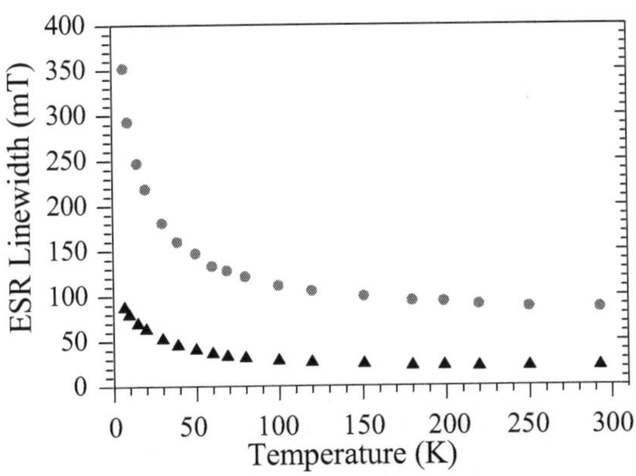

Figure 7.7 Temperature dependence of linewidths corresponding to lines A (circles) and B (triangles) in the ESR spectra of $Li_2Mn_2(MoO_4)_3$.

paramagnetic regime, I is proportional to the static magnetic susceptibility of spins participating in the resonance, and hence it follows the Curie–Weiss law, $I \sim C\ /\ (T - \theta)$, where C and θ are the Curie constant and Weiss constant, respectively. The temperature dependence of a reciprocal value of integral intensity I of observed resonance lines in $Li_2Mn_2(MoO_4)_3$ is shown in Fig. 7.8. The obtained values of Weiss constant are also presented in Fig. 7.8.

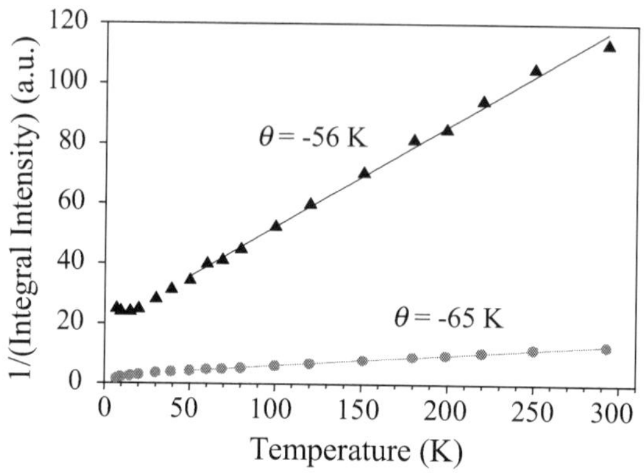

Figure 7.8 Temperature dependence of the reciprocal value of integral intensities corresponding to lines A (circles) and B (triangles) in the ESR spectra of $Li_2Mn_2(MoO_4)_3$. The Weiss constants obtained from these dependences are also presented.

The comparative analysis of the integral intensities has shown that the intensity of signal B is about 8% of the intensity of signal A. Thus, considering the broad resonance line as the main signal, the parameters of line A have been discussed first.

7.4.2.1.1 Resonance parameters of line A

The high-temperature value of g-factor 1.99 ± 0.03 is close to the g-factor of Mn^{2+} ion in octahedral crystal field.[17,18] This result is consistent with the structural data of $Li_2Mn_2(MoO_4)_3$ according to which the manganese ions reside in octahedral coordination of oxygen ions.

The Mn^{2+} ion has the electronic configuration $3d^5$ and ground state $^6S_{5/2}$. As the ground state corresponds to zero value of orbital magnetic momentum, the g-factor of Mn^{2+} ion is determined by spin value only. The admixture from higher states of the $3d^5$ configuration to the ground state under the action of both the crystal field and spin-orbit interaction causes the splitting of Mn^{2+} ground state with $S = 5/2$ into the three doublets with magnetic quantum numbers $M_S = \pm 5/2, \pm 3/2, \pm 1/2$. This results in the appearance of five fine-structure lines in the ESR spectra of Mn^{2+} ions. In turn, each fine structure line splits into six hyperfine-structure lines due to the hyperfine interaction between electron spin and nuclear spin of ^{55}Mn, $I = 5/2$. Moreover, it has been found that ESR spectra of Mn^{2+} ions in octahedral crystal field consists of the single resonance line without visible traces of both fine and hyperfine structures.[19,20] Such behavior is characteristic for the magnetically concentrated systems in which exchange interactions between magnetic ions can strongly affect ESR spectra. Exchange interaction causes a hopping of spins between the fine and hyperfine structure energy levels that leads to the merging of the fine and hyperfine absorption lines into a single resonance line located at the center of gravity of the spectrum.

As $Li_2Mn_2(MoO_4)_3$ compound is a magnetically concentrated system, fine and hyperfine structures of Mn^{2+} ion in its ESR spectra are absent due to exchange interaction between Mn^{2+} ions. According to the ESR theory, for magnetically concentrated systems, the dipole–dipole and exchange interactions basically determine the ESR linewidth when compared with that for isolated spins. Following the method of moments, the contribution of dipole–dipole interaction between the magnetic moments to the resonance linewidth can be estimated by the expression:[21]

$$H_{\mathrm{dd}}^2 = \frac{3}{4} \cdot g^2 \cdot \beta^2 \cdot S \cdot (S+1) \cdot \frac{1}{N} \cdot \sum_{k>j}^{N} \frac{(1 - 3 \cdot \cos^2 \theta_{jk})^2}{r_{jk}^6}, \qquad (7.3)$$

where g is the g-factor of magnetic ion, β is the Bohr magneton, S is the spin of magnetic ion, N is the number of magnetic ions, r_{jk} is the distance between j and k magnetic ions, and θ_{jk} is the angle between radius vector connecting j and k magnetic ions and the external magnetic field. Taking into account spherical angle in the case of powder samples, main contribution to the dipole linewidth is determined by the number of nearest neighboring magnetic ionsz; hence, Eq. 7.5 can be reduced to:

$$H_{\mathrm{dd}}^2 = \frac{3}{5} \cdot g^2 \cdot \beta^2 \cdot S \cdot (S+1) \cdot \sum_{k}^{z} \frac{1}{r_k^6}. \qquad (7.4)$$

From the structural data of $Li_2Fe_2(MoO_4)_3$[22], analogue of $Li_2Mn_2(MoO_4)_3$, it has been found that the number of nearest neighboring Mn^{2+} ions is 8, and the distance between magnetic ions varies between 5.1489 Å and 6.1446 Å. Substitution of these values to Eq. 7.6 gives the value of dipole linewidth $H_{\mathrm{dd}} = 827$ mT.

This value is considerably larger than the experimentally calculated value $\Delta H_{pp} = 86$ mT. It is because of exchange interaction between Mn^{2+} ions that narrows the dipole-broadened resonance line, which occurs in $Li_2Mn_2(MoO_4)_3$. In this case, the ESR linewidth can be roughly estimated by following expression:[23]

$$\Delta H_{1/2} = \frac{H_{dd}^2}{H_{ex}}, \tag{7.5}$$

where $\Delta H_{1/2}$ is the half-width of ESR line at half-power and H_{ex} is the exchange field, which is expressed through the value of exchange integral J according to[24] Eq. 7.8:

$$H_{ex} = \sqrt{\frac{8}{3} \cdot S \cdot (S+1) \cdot z \cdot J}. \tag{7.6}$$

It should be noted that Eq. 7.7 is valid only for the case when the resonant frequency ν_0 ($\sim$9.2 GHz) is greater than the exchange frequency J/h. Otherwise, H_{dd}^2 should be increased in Eq. 7.7 by the factor of 10/3.[23] Taking into account the relationship between half-width at half-power $\Delta H_{1/2}$ and the peak-to-peak linewidth ΔH_{pp} for the Lorentzian line, the exchange integral for the Mn^{2+} ions $|J| =$ 3.9 K was estimated using the expressions 7.7 and 7.8.

As In^{2+} ions are situated far enough from each other as deduced from structural data, the direct exchange interaction between them is negligibly small. But there is a possibility for exchange interaction between Mn and O based on its electronic structure. The electronic structure of $Li_2Mn_2(MoO_4)_3$ compound is governed by a strong hybridization between the 3d orbitals of transition metal, Mn and 2p orbitals of oxygen when In^{2+} ions are in octahedral coordination with oxygen ions (Fig. 7.9) d_{z2} and d_{x2-y2} atomic orbitals of manganese ions directly overlap with the p_x, p_y, and p_z orbitals of oxygen ions along the octahedral directions. This σ-overlap creates the e_g bands, of which the bonding band e_g^b has a predominant oxygen p character and the antibonding band e_g^* has a predominant metallic d character. The remaining d_{xy}, d_{yz}, and d_{xz} orbitals point away from oxygen and have only a slight π-overlap, giving rise to the nonbonding t_{2g} bands. Similarly, the overlap of O 2p with Mn 4p and O 2p with Mn 4s orbitals gives rise to t_{1u} and a_{1g} bands, respectively.

In this situation, the interactions between manganese ions can arise through super exchange interaction of neighboring oxygen

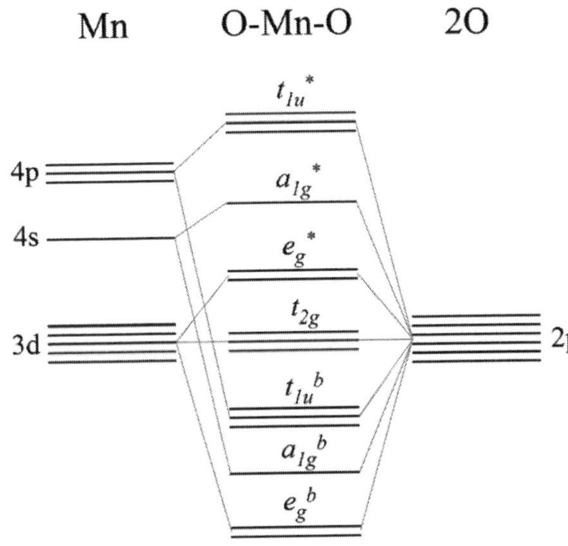

Figure 7.9 Schematic of the band structure expected for oxides with octahedrally coordinated In^{2+} ions.

anions. This exchange mechanism was first observed in manganese oxide by Anderson.[25,26] Finite overlapping between d and p orbitals of Mn and O ions allows at least one p-electron from oxygen go into the d orbital on In^{2+}. Then, remaining p-electron of oxygen can participate in exchange interactions with other neighbor In^{2+}, and it results in appearance of exchange spin coupling of In^{2+} ions. Structure analysis of $Li_2Mn_2(MoO_4)_3$ (Fig. 7.1a,b) shows that the superexchange is possible through two oxygen ions belonging to one MoO_4 tetrahedron and two neighbor MnO_6 octahedrons.

The negative value of Weiss constant $\theta = -69$ K suggests that the antiferromagnetic character arises due to superexchange interactions between In^{2+} ions. The exchange integral J can be estimated by means of the following expression:[23]

$$\frac{3}{2} \cdot k_B \cdot \theta = z \cdot J \cdot S \cdot (S+1), \tag{7.7}$$

where k_B is the Boltzmann constant, z is the number of nearest neighboring In^{2+} ions ($z = 8$), and S is the spin of In^{2+} ion ($S = 5/2$). The value obtained from the experiment, $J = -1.4$ K, matches satisfactorily with value of exchange interaction

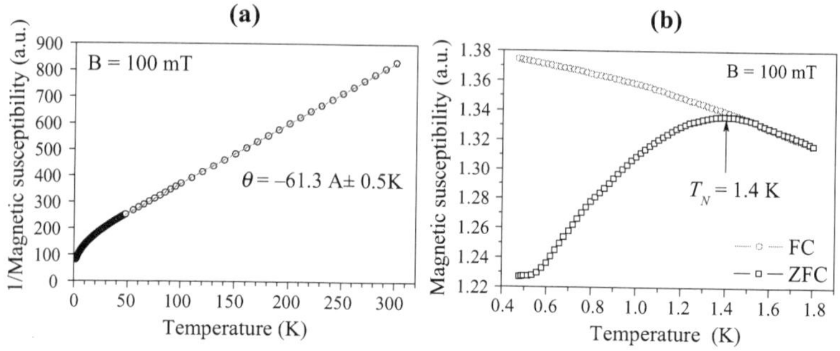

Figure 7.10 Temperature dependence of the magnetic susceptibility of $Li_2Mn_2(MoO_4)_3$. The Weiss constant obtained from approximation of these dependence with Curie–Weiss law is equal to -61.3 ± 0.5 K. (a) Temperature dependence of the magnetic susceptibility of $Li_2Mn_2(MoO_4)_3$ at ZFC and FC regimes. (b) Neel temperature is observed at 1.4 K.

calculated above. The increase of line width at low temperatures indicates the appearance of short magnetic ordering of In^{2+} ions in $Li_2Mn_2(MoO_4)_3$ compound.

Figure 7.10a shows the temperature dependence of magnetic susceptibility of $Li_2Mn_2(MoO_4)_3$, which was measured at temperature interval 0.4–300 K. The magnetic susceptibility obeys Curie–Weiss law, $\chi = C/(T - \theta)$, with Weiss constant $\theta = -61.3$ K in temperature range 4.2–300 K. Antiferromagnetic phase transition is observed at 1.4 K, which is shown in Fig. 7.10b. The temperature dependence of magnetic susceptibility in zero field cooled (ZFC) and field cooled (FC) regimes is shown as well in Fig. 7.10b. There is a difference in values of FC and ZFC susceptibility below the Neel temperature. This behavior indicates the existence of an antiferromagnetic arrangement in the sample, with the appearance of a remanent magnetization below the ordering temperature. The remanent magnetization could be explained by an incomplete compensation of the collinear antiferromagnetic arrangement of magnetic moments (weak ferrimagnetism). The presence of weak ferrimagnetism may be associated with the residence of some Mn^{2+} ions in antisite disorder positions that changes the distribution of exchange interactions in the system.

7.4.2.1.2 Resonance parameters of line B

Analysis of resonance parameters of line B indicates that this line has much in common with line A. First, g-factor of line B is close to the g-factor of line A at high temperatures. Second, both lines have the similar temperature dependences: at high temperatures, the linewidths do not depend on temperature, whereas at low temperatures, significant broadening in resonance signals takes place. It was shown that for the line A, such behavior is associated with the presence of exchange interaction between Mn^{2+} ions through the oxygen ions. In this connection, it is pertinent to suggest that magnetic ions giving the signal B are involved in structure of existing exchange interactions as well. Moreover, the lines A and B have the close values of Curie–Weiss constants. All these analogies clearly indicate that the line B can be attributed to the Mn^{2+} ions.

It was found that in some cathode materials, the so-called antisite defects are formed where some of the transition metal ions reside on lithium sites.[27–30] For instance, up to 10% of lithium interstitial positions in the $LiFePO_4$ can be substituted by Fe ions.[30] The appearance of antisite defects is mainly associated with synthesis method. Thus, the signal B is assigned to the Mn^{2+} ions residing on tetrahedrally coordinated Li sites in $Li_2Mn_2(MoO_4)_3$ structure. By comparing intensities of resonance lines A and B, it has been found that the concentration of such defects would be about 8%.

7.4.2.2 Nanoscaled $Li_xMn_2(MoO_4)_3 (x = 0–2.0)$

A symmetric absorption line is observed for $Li_xMn_2(MoO_4)_3$ for all x values in the ESR spectrum. The resonance parameters g-factor and line width are deduced from the absorption line and plotted against temperature and lithium concentration and given in Figs. 7.11 and 7.12.

The g-factor of the resonance line remains constant with composition within the limits of experimental error equal to 2.00 in temperature range 100–300 K as shown in Fig. 7.11 a, b. But below 100 K, g-factor changes with concentrations of Li as well as temperature and the change does not follow any trend as clearly seen in Fig. 7.11b. For $Li_xMn_2(MoO_4)_3$ with $x = 1.8$ and 2.0, g-

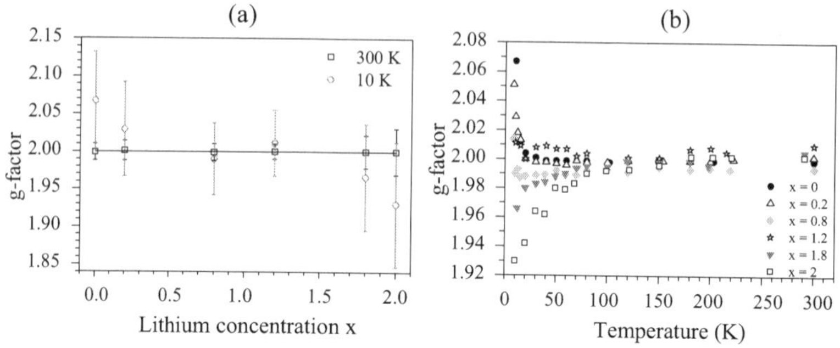

Figure 7.11 (a) The g-factor of the resonance line in the ESR spectrum of $Li_xMn_2(MoO_4)_3$ compounds as a function of lithium concentration x. (b) Temperature dependence of the g-factor of the resonance line in the ESR spectrum of $Li_xMn_2(MoO_4)_3$ compounds.

factor decreases with a decrease in temperature due to shift of resonance signal to higher magnetic fields for these samples. The g-factor decreases with an increase in temperature for samples with $x = 0$ and $x = 0.2$ that corresponds to the shift of resonance signal to the smaller magnetic fields. The samples with intermediate concentrations do not show any significant change in the g-factor with temperature.

The dependence of the width of resonance line with temperature and concentration is presented in Fig. 7.12a, b. As seen in these figures, the temperature behavior of linewidth is same as that of g-factor of $Li_xMn_2(MoO_4)_3$, that is, at high temperatures, the linewidth remains constant with composition. As shown in Fig. 7.12a, the width of resonance line changes with x value. The greatest linewidth, 106 mT, is observed in stoichiometric sample with $x = 2.0$ at room temperature. This value decreases to the lowest value, 34 mT, for the samples with $x = 1.2$.

7.4.3 *Comparison of Resonance Parameters of Microscaled and Nanoscaled $Li_2Mn_2(MoO_4)_3$*

Comparing the parameter, g-factor of ESR spectra of microscaled and nanoscaled $Li_2Mn_2(MoO_4)_3$ samples, it has been found that nanoscaled $Li_xMn_2(MoO_4)_3$, where $(x = 0 - 2)$, has the ESR

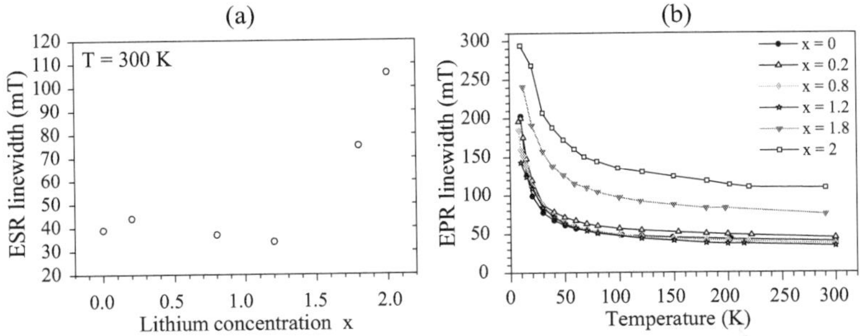

Figure 7.12 (a) The linewidth of the resonance line in the ESR spectrum of $Li_xMn_2(MoO_4)_3$ compounds as a function of lithium concentration x. (b) Temperature dependence of the linewidth of the resonance line in the ESR spectrum of $Li_xMn_2(MoO_4)_3$ compounds.

parameters identical to that of line A in the ESR spectra of microscaled $Li_2Mn_2(MoO_4)_3$. It indicates that the ESR signals have same origin in both cases, due to Mn^{2+} ions in octahedral crystal field created by oxygen ions in $Li_2Mn_2(MoO_4)_3$. But the line widths of resonance signals differ significantly, as valence state of Mn ions changes with lithium concentration of $Li_xMn_2(MoO_4)_3$. As it was indicated, the stoichiometric $Li_2Mn_2(MoO_4)_3$ contains manganese ions in the 2+ valence state, and in the $Mn_2(MoO_4)_3$ compound, all manganese ions are expected to be in the 3+ valence. But for samples with the lithium concentration $x < 2$, part of the Mn ions is in the +3 valance state, which is connected with the different values of spins of the Mn^{2+} ion ($S = 5/2$) and of the Mn^{3+} ion ($S = 2$). However, as the g-values of Mn^{2+} and Mn^{3+} ions are close to each other and both ions are involved in a common system of exchange interaction between Mn^{2+} and Mn^{3+}, it is difficult to distinct these ions in the ESR spectra of $Li_xMn_2(MoO_4)_3$.

It is important to note that the samples of $Li_xMn_2(MoO_4)_3$, nanoscaled samples, synthesized by solvothermal method are free from the anti-site disorder resonance signal, which is observed in micro-scaled $Li_2Mn_2(MoO_4)_3$ samples. The appearance of anti-site disorder defects is associated with location of Mn^{2+} in lithium interstitial positions. It impedes the diffusion of Li^+ in crystal structure of cathodic materials that have the negative influence on their electrochemical performance. Thus, the ESR investigations show that the use of solvothermal method for the preparation of cathodic material $Li_2Mn_2(MoO_4)_3$ is preferable when compared with low-temperature soft-combustion method.

7.5 Summary and Outlook

The results of ESR investigations on layered transition metal oxides, $LiMO_2$(M = Co, Ni, or Mn), have shown that lattice imperfection as well as the spin defects are the main causes for less achievable (50%) theoretical capacity and the ESR studies on $LiCoO_2$ indicate that the higher the synthetic temperatures, the less the lattice imperfection as well as the spin defects. The ESR analysis could explain the increase in electrode potential of substituted $LiMO_2$ system $\{LiM_yCo_{1-y}O_2$ (M = Al, Ga)$\}$ by means of enhancement of covalent character of Ni-O bond, which is associated with the appearance of additional electron density on the oxygen ions. ESR measurements of micro-scaled $Li_2Mn_2(MoO_4)_3$ confirm that manganese is in the 2+ valence state and the absence of ESR signal from molybdenum allows us to assume that the valence state of molybdenum is 6+, as in this state, molybdenum has no unpaired electrons so that it is nonmagnetic. The two ESR signals with Lorentzian shape obtained for micro-scaled $Li_2Mn_2(MoO_4)_3$ are assigned to Mn^{2+} in octahedral environmental and to the Mn^{2+} in anti-site disorder positions. Considerable overlapping between 3d orbitals of transition metal and 2p orbitals of oxygen has been experimentally established in $Li_2Mn_2(MoO_4)_3$ and $Li_xMn_2(MoO_4)_3$ compounds that leads to the superexchange interaction between magnetic ions. The overlapping between d orbitals of transition metal and 2p oxygen orbitals plays an important role in Li-electron transport properties of $Li_2Mn_2(MoO_4)_3$ and $Li_xMn_2(MoO_4)_3$ compounds. ESR spectra of $Li_xMn_2(MoO_4)_3$ compounds with $x = 0$, $x = 0.2$, $x = 0.8$, $x = 1.2$, and $x = 1.8$ demonstrate the single resonance line with g-factor corresponding to the Mn^{2+}/Mn^{3+} ions located in octahedral crystal field of oxygen ions. No anti-site disorder resonance signal was observed in $Li_xMn_2(MoO_4)_3$ samples synthesized by low-temperature solvo-thermal method, which explains the better performance of nanoscaled $Li_xMn_2(MoO_4)_3$ synthesized by this method.

Acknowledgments

N. Suleimanov and D. Abdullin would like to thank the Russian Foundation for Basic Research (RFFI) for the support through the grants 12-08-97064 and 12-08-00755.

References

1. M.M. Thackeray, *J. Electrochem. Soc.* **142**, 2558 (1995).
2. D. Morgan, G. Ceder, M.Y. Saidi, J. Barker, J. Swoyer, H. Huang, G. Adamson, *Chem. Mater.* **14**, 4684 (2002).
3. A. Manthiram, J.B. Goodenough, *J. Solid State Chem.* **71**, 349 (1987).
4. K.M. Begam, M.S. Michael, S.R.S. Prabaharan, in Next generation lithium ion batteries for electrical vehicles, ed. by C.R. Park (InTech, Vukovar, 2010), p. 93.
5. K.M. Begam, M.S. Michael, Y.H. Taufiq-Yap, S.R.S. Prabaharan, *Electrochem. Solid St.* **7**, A242 (2004).
6. S.R.S. Prabaharan, S. Ramesh, M.S. Michael, K.M. Begam, *Mater. Chem. Phys.* **87**, 318 (2004).
7. S.R.S. Prabaharan, M.S. Michael, K.M. Begam, *Electrochem. Solid St.* **7**, A416 (2004).
8. M.S. Michael, K.M. Begam, M. Cloke, S.R.S. Prabaharan, *J. Solid State Electr.* **12**, 1025 (2008).
9. C.-H. Lee, *J. Adv. Engineering Technol* **1**, 59 (2008).
10. R. Stoyanova, E. Zhecheva, R. Alcantara, J.L. Tirado, *J. Phys. Chem. B* **108**, 4053 (2004).
11. R. Stoyanova, A. L. Barra, E. Zhecheva, R. Alcantara, G. Ortiz, J. L. Tirado, *Inorg. Chem.* **48**, 4798 (2009).
12. C. Julien, G. A. Nazri, A. Rougier, *Solid State Ionics* **135**, 121 (2000).
13. G. Ceder, Y.-M. Chang, D. R. Sadoway, M. K. Aydinol, Y. I. Jang, B. Huang, *Nature* **392**, 694 (1998).
14. M. V. Koudriachova, N. M. Harrison, *J. Mater. Chem.* **16**, 1973 (2006).
15. G. Ceder, M. K. Aydinol, A. F. Kohan, *Comput. Mater.* Sci. **8**, 161 (1997).
16. E. Zhecheva, R. Stoyanova, R. Alcantara, J.L. Tirado, *J. Phys. Chem. B* **107**, 4290 (2003).

17. J.I. Wertz, J.R. Bolton, Electron spin resonance: Elementary theory and practical applications (McGraw-Hill, New York, 1972), 548 p.
18. A. Abragam, B. Bleaney, Electron paramagnetic resonance of transition ions (Clarendon Press, Oxford, 1970), 899 p.
19. J. Deisenhofer, M.V. Eremin, D.V. Zakharov, V.A. Ivanshin, R.M. Eremina, H.-A. Krug von Nidda, A.A. Mukhin, A.M. Balbashov, A. Loidl, *Phys. Rev. B* **65**, 104440 (2002).
20. K.L. Nagy, B. Nafradi, N.D. Kushch, E.B. Yagubskii, E. Herdtweck, T. Feher, L.F. Kiss, L. Forro, A. Janossy, *Phys. Rev. B* **80**, 104407 (2009).
21. J.H. Van Vleck, *Phys. Rev.* **74**, 1168 (1948).
22. C.C. Torardi, E. Prince, *Mater. Res. Bull.* **21**, 719 (1986).
23. P.W. Anderson, P.R. Weiss, *Rev. Mod. Phys.* **25**, 269 (1953).
24. T. Morya, *Prog. Theor. Phys.* **16**, 23 (1956).
25. P.W. Anderson, *Phys. Rev.* **7**, 350 (1950).
26. P.W. Anderson, *Phys. Rev.* **115**, 2 (1959).
27. J. Chen, M.S. Whittingham, *Electrochem. Commun.* **8**, 855 (2006).
28. S.-Y. Chung, S.-Y. Choi, T. Yamamoto, Y. Ikuhara, *Phys. Rev. Lett.* **100**, 125502 (2008).
29. P. Jozwiak, J. Garbarczyk, F. Gendron, A. Mauger, C.M. Julien, *J. Non-Cryst. Solids* **354**, 1915 (2008).
30. J. Chen, J. Graetz, *ACS Applied Materials & Interfaces* **3**, 1380 (2011).

Chapter 8

Graphene and Graphene-Based Nanocomposites for Electrochemical Energy Storage

Marappan Sathish[a] and Itaru Honma[b]

[a]*Institute of Multidisciplinary Research for Advanced Materials, Tohoku University, 2-1-1, Katahira, Sendai, Miyagi 980-8577, Japan*

[b]*Functional Materials Division, CSIR-Central Electrochemical Research Institute, Karaikudi 630006, India*

marappan.sathish@gmail.com

Graphene, a two-dimensional sp^2-hybridized carbon, has received considerable attention due to their high surface area, often-cited desirable physical and electrochemical properties. Various efforts have been made on the preparation of the graphene via a number of physical and chemical methods. Some of these methods provide high-quality graphene and have opened up new possible routes to address the challenges in preparation and molecular engineering of high-quality processable graphene. The high restacking rate of graphene sheets is the major drawback to attain the forecasted high surface area of 3620 m^2/g and associated electrochemical properties. As high surface area carbon materials are promising for enhanced energy storage, attempts have been made to exploit graphene and graphene-based nanocomposite electrode materials

Nanotechnology in Advanced Electrochemical Power Sources
Edited by S. R. S. Prabaharan and M. S. Michael

ISBN 978-981-4241-43-4 (Hardcover), 978-981-4303-05-7 (eBook)
www.panstanford.com

for supercapacitor, Li-ion battery, and Li–air battery applications. This chapter highlights the fundamentals and recent progress in graphene and graphene-based nanocomposite materials as electrodes for supercapacitor, Li-ion battery, and Li–air battery. And, a brief description of the synthesis and modification of graphene and graphene-based nanocomposite materials is presented.

8.1 Introduction

Preparation of diverse nanomaterials with unique crystalline structures, size, morphologies, and the effect of various parameters on the energy storage have been discussed in the previous chapters. As the readers would have noted, other than the crystalline nature, the carbon coating or enhancing the electronic conductivity has been identified as a key to attain desired storage capacity with superior cyclability. Carbon materials have long been incorporated into the electrodes of energy-storage devices as electro-conductive additives, supports for active materials, electron transfer catalysts, intercalation hosts, substrates for current leads, and as agents for the control of heat transfer, porosity, surface area, and capacitance. Carbon-based nanomaterials, particularly graphite, are well known for electrochemical applications as an electrode and support for various active electrode materials. Carbon materials alone are widely used for energy storage applications such as supercapacitors, Li-ion batteries, and so on because of their low cost and versatile existing forms. And, carbon is considered unusual in the number of its allotropic structures and the diversity of structural forms, as well as in its broad range of physical properties. Of all the carbon nanostructures relevant for charge storage, graphite intercalation compounds have been studied extensively and the lithium-intercalated graphite can be considered as the first example of lithium storage in a carbon-based material.[1] Similarly, carbon, in its various forms, is the most extensively examined and widely utilized electrode material in supercapacitors[2] owing to the high surface area and currently pore size (nanometer to micrometer range). Considerable research is presently being directed toward the development of carbon materials with a tailored pore-size, high surface area, good electrical conductivity, and desired surface functionalities for enhanced electrochemical energy storage.

At this point, it would be appropriate to introduce graphene or graphene-based nanocomposites as potential candidate for electrochemical energy storage applications. Graphene nanosheets either have a single layer or a few layers of carbon atoms in a hexagonal network and are free of defects, with all carbon atoms linked together by strong and flexible bonds. Thus, it has unusual electronic, optical, mechanical, and thermal properties. The outstanding properties of graphene reported so far include Young's modulus (~1.0 Tpa), fracture strength (125 Gpa), high thermal conductivity (~5000 Wm/K), high intrinsic mobility, quantum Hall effect at room temperature, an ambipolar electric field effect along with ballistic conduction of charge carriers, tunable band gap, and specific surface area of 3620 m^2/g and so on.[3–8] Indeed, graphene provides infinite possibilities for the modification and fictionalization of its carbon backbone. Thus, it is believed that graphene nanosheets act as a potential candidate for various applications including electrochemical energy storage. Numerous attempts have been made in recent days to explore the full potential of graphene nanosheets in electrochemical energy storage systems such as supercapacitors, Li–ion battery, and Li–air batteries.[7–10] However, only limited success was attained so far due to the various issues such as preparation of defect-free monolayer graphene nanosheets in bulk scale and fast restacking of graphene nanosheets.

Recently, many efforts have been made on the preparation of the graphene via a number of physical and chemical methods.[10–12] Some of these methods provide high-quality graphene and have opened up new possible routes to address the challenges in the preparation and molecular engineering of high-quality process-able graphene at low cost and large scale. Attempts have also been made to prepare grapene nanosheets with various organic functional molecules, which enhance the use of graphene nanosheets for various applications, owing to the integration of properties. For the similar reason, the incorporation of inorganic nanostructures, such as metal, metal ions, metal oxides, and metal complexes, has also gained attraction due to their widespread applications. With this background, the idea of incorporating metal or metal oxides into the graphene nanosheets has been just conceived and few attempts have been initiated recently. In this chapter, more attention has been paid on the possible preparation methods, characterization, and potential application of metal, metal oxide, and other functional molecules incorporated in graphene nanocomposites in electrochemical energy storage applications.

8.2 Graphene Nanosheets and Functionalized Graphene Nanosheets Preparation

Although the synthesis of mono-layered and multi-layered graphite was shown by thermal decomposition of carbon on single crystal Pt substrates by B. Lang *et al.* in 1975,[13] due to lack of consistency between properties and failure to identify the beneficial applications of the product, the process was not studied extensively. Recently, Novoselov *et al.* in 2004[14] examined the single-layer graphene nanosheets preparation by micromechanical cleavage in which highly oriented pyrolitic graphite is pealed using scotch-tape and deposited on to a silicon substrate. Although this method is not suitable for largescale synthesis of single-layer graphene or of few-layer graphene nanosheets, it reopens the graphene nanosheets research and their possible applications in various fields are under examination. Thus, several alternate preparation strategies have been reported recently.[10–12] Depending on the requirement and quality of the graphene, one can choose appropriate method for their use. It is well known that the graphite is stacked with many layers of graphene sheets, bonded together by weak van der Waals force. Thus, in principle, it is possible to produce graphene sheets from high purity graphite, if these bonds can be broken or weakened by chemical or mechanical force. Basically, all these graphene nanosheets preparation methods use mechanical or chemical energy to break these weak bonds and separate out individual graphene sheets. The first attempt in this direction was proposed by Viculis *et al.*,[15] wherein the potassium metal was used to intercalate a pure graphite sheet and then exfoliate it with ethanol to form dispersion of C sheets. The most common preparation methods used for graphene nanosheets preparation

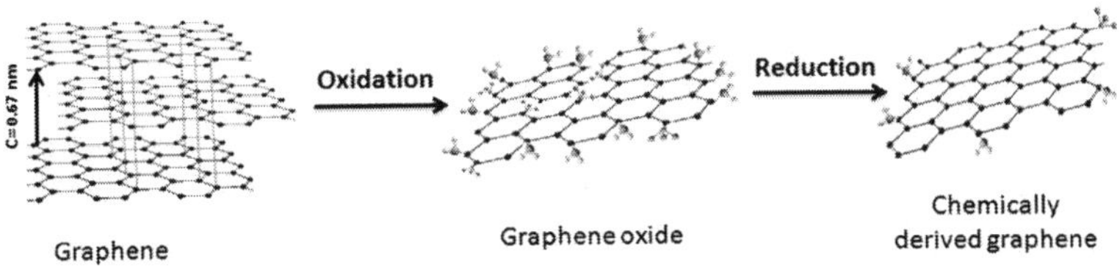

Figure 8.1 An illustration of the chemical route for synthesis of graphene.

are solution phase chemical oxidation, Hummers method, modified Hummers, and Offeman's method,[16,17] in which the graphite is used as starting material and strongly oxidized by mixture of $KMnO_4$ and H_2SO_4. The resulting graphene oxide (GO) can be easily exfoliated into GO nanosheets by simple ultrasonication. Further, it can be easily reduced to reduced graphite oxide (RGO) using conventional reducing agents. The general solutionbased chemical oxidation method is shown in Fig. 8.1. The RGO layers are to be considered as chemically modified graphene as it was not fully reduced and they generally contain some oxygencontaining functional groups, such as OH or COOH groups. Consequently, the resulting RGO will comprise many structural defects with limited physicochemical properties forecasted and this chemical oxidation method can be used for bulk production (few grams) of graphene nanosheets. In the case of functionalization, the GO may be used as such to make chemical bonding between the functional group and graphene surface, and the unreacted functional groups can be reduced later.

Recently, varieties of solutionbased graphene exfoliation methods have been developed.[9,12,18] Unzipping multi-walled carbon nanotubes (MW-CNTs) longitudinally will produce graphene nanosheets or nano-ribbons with various widths and lengths up to several micrometers.[19] The unzipping process was carried out using chemical oxidation and thermal treatment. However, the solution phase exfoliation of graphene suffers from lot of structural defects, due to oxidation and reduction processes, leading to much poor electrical properties of graphene. The future inventions need to concentrate on controlling the number of layers and minimizing impurity levels to lead this process into an industrial scale production level.

Chemical vapor deposition (CVD) and thermal decomposition based methods open a new avenue in the synthesis of high-quality graphene. There are number of reports on the synthesis of graphene through thermal CVD and plasma-enhanced CVD (PECVD) methods.[10,11] One of the highly upcoming techniques of graphene growth is thermal decomposition of Si on the (0001) surface plane of single crystal of 6H-SiC. Graphene sheets were found to be formed when H_2-etched surface of 6H-SiC was heated to temperatures of 1250°C–1450°C, for a short time (1–20 min). These methods offer good quality defect-free large area (centimeter scale) graphene sheets with controlled layer thickness on the substrate. These developments are necessary to realize the application of graphene in electronic devices such as photovoltaic, field effect transistors, sensors, and so on.

8.3 Graphene Nanosheets for Supercapacitors

Electrochemical capacitors, also called supercapacitors, store energy using either ion adsorption (electrochemical double layer capacitors, EDLCs) or fast surface redox reactions (pseudocapacitance). It is well known that the EDLCs store the opposite charge electrostatically using reversible adsorption of ions of the electrolyte on active materials surface that are electrochemically stable and have high accessible surface area. As there is no chemical reaction involved in the storage mechanism, the process is highly reversible for millions of cycles and results long life for the capacitors. And, the specific capacitance depends on the available active specific surface area of the active materials. Whereas, in the pseudo capacitance, surface or nearsurface redox reactions occur during the charge storage mechanism, resulting in high specific capacitance with relatively short cycle life compared with EDLCs. Thus, the development of high capacitive energy storage systems with optimum cycle life, low-cost, and environmentally friendly materials is essential to meet the energy demands of modern society and emerging environmental concerns. Carbons and carbon-based composite materials are the most widely used owing to their high surface area, moderate cost and eco-friendly nature. Varieties of carbon morphologies with different surface area and chemical nature such as carbon nanotubes, carbon nanofibers, carbon fibers, onions, nanohorns and so on have been investigated.[20,21] Similarly, activated carbons, mesoporous carbon, template carbon and chemically derived carbon have been examined. The carbons used in EDLC are generally pre-treated to remove moisture and most of the surface functional groups are present on the carbon surface to improve stability during cycling. The presence of functional group will result in increased serious resistance and capacitance fading during aging. The double layer capacitance of activated carbon reaches 100–120 F/g in organic electrolytes; this value can exceed 150–300 F/g in aqueous electrolytes, but at a lower cell voltage because the electrolyte voltage window is limited by water decomposition.[20]

The key to reaching high capacitance by charging the double layer is in using high surface area carbon materials with electronically conducting electrodes. Graphitic carbon satisfies all the requirements for this application, including high conductivity, electrochemical stability and open porosity that offer huge surface area.[21] Thus, graphenebased supercapacitors are under intensive investigation as potential alternatives to the activated carbon that are used in the current supercapacitor electrodes. The effective surface area of graphene materials should highly depend on the number of layers, that is, single or few layered graphene sheets with less agglomeration might be expected to exhibit higher effective surface area. The recent research reports on clean graphene materials with specific capacitance ranging from 120 to 250 F/g. The chemical nature of the graphene nanosheets and its purity greatly depends on the method of preparation and subsequent processing of the resulting graphene sheets. As mentioned earlier, the presence of functional groups on the graphene surface greatly influences its capacitance. In general, the presence of oxygen-containing functional groups diminishes the capacitance and cycle life.[1] Y. Wang *et al.*[22] report a maximum specific capacitance of 205 F/g with excellent long cycle life along with nearly 90% specific capacitance retained after 1200 cycle tests. Also, it is confirmed that the interfacial capacitance of multilayer graphene sheets is found to depend on the number of layers. In addition to the graphene quality, the fabrication of electrodes and its structure also influences the performance of the resulting supercapacitors. Recently, Ajyan *et al.* report an "in-plane" fabrication approach for ultrathin supercapacitors based on electrodes composed of pristine graphene and multilayer-reduced GO.[23] And, this approach allows for the formation of an efficient electrical double layer by utilization of the maximum electrochemical surface area and results in a maximum specific capacitance of 247 F/g. The research on supercapacitors using graphene nanosheets is under tremendous progress; combination of high-quality graphene sheets with suitable electrode fabrication technique will lead to the commercial production of EDLC capacitors with high capacitance in near future.

The large specific pseudocapacitance of faradaic electrodes (typically 300–1000 F/g) exceeds that of carbon-based materials using double-layer charge storage, resulting in great interest in these systems. Specific capacitance of more than 600 F/g has been reported for RuO_2 based system owing to its good conductivity and fast and reversible electron transfer together with an electro-adsorption of protons on the surface. However, the Ru-based aqueous electrochemical capacitors are expensive, and the 1 V voltage window limits their applications to small electronic devices.[20] Thus, pseudocapacitive transition metal oxides such as MnO_2, NiO, and redox polymers such as polyanilines, polypyrroles, and polythiophenes could be used to make electrodes, because they are predicted to have a high capacitance for storing electrical charge, inexpensive, and not harmful to the environment. The poor conductivity and lack of stability during cycling are the major drawbacks associated with these materials for their usage in supercapacitors. Thus, numerous efforts have been made to use these metal oxides successfully in supercapacitors by making composite with conductive support. As the graphene nanosheets have vast surface area with excellent conductivity, they will be appropriate candidates to accommodate large amount of metal oxides. And, the EDLC behavior of graphene nanosheets can also be enhanced, which contributes to the total capacitance of the resulting composite. Recently, graphene-metal oxide nanocomposite systems have been developed by various researchers, and high specific capacitances with good cycling performance have been reported.[9,11,20,21,24] Also, layered double hydroxides (LDHs) materials containing transition metals have been reported to be promising electrode materials for supercapacitors because of their relatively low cost, high redox activity, and environmentally friendly nature. Z. Gao *et al.* report the preparation of graphene-Ni/Al LDH nanocomposite and a maximum specific capacitance of 781.5 F/g with an excellent cycle life.[25] The observed capacitance is almost 1.5 times higher than the pure LDH electrodes. The larger capacitance for GNS/LDH may be caused by the combination of electric double-layer capacitance and faradic pseudocapacitance. At the same time, the open structure system of GNS/LDH improves the contact between the electrode materials and the electrolyte and thus makes full use of the electrochemical active material contribution to the overall capacitance. Development of suitable GNS-metal oxide or LDH nanocomposites with appropriate ratio will lead to high capacitance owing to the combination of EDLC and pseudocapacitance from GNS and metal oxide or LDH, respectively.

G. Yu *et al.*[26] demonstrated the solution-processed graphene/MnO_2 nanostructured textiles for high-performance electrochemical capacitors applications. In their study, solution-exfoliated graphene nanosheets (~5 nm thickness) were conformably coated on three-dimensional, porous textiles support structures, and pseudo-capacitive MnO_2 nanomaterials were deposited by controlled electro-deposition (Fig. 8.2). This technique offers high loading of active electrode materials and facilitates the easy access of electrolytes to those materials. The hybrid graphene/MnO_2-based textile yields high-capacitance performance with specific capacitance up to 315 F/g. Also, they have fabricated asymmetric electrochemical capacitors with graphene/MnO_2-textile as the positive electrode and single-walled carbon nanotubes (SWNTs)-textile as the negative electrode in an aqueous Na_2SO_4 electrolyte solution. These devices exhibit promising characteristics with a maximum power density of 110 kW/kg, an energy density of 12.5 Wh/kg, and excellent cycling performance of nearly 95% capacitance retention over 5000 cycles. These kind of low-cost, high-performance energy textiles based nanostructures offer great promise to realize the future large-scale energy storage devices.

Three-dimensional (3D) conductive wrapping (Fig. 8.3) of graphene/MnO_2 nanostructured electrodes with CNTs or

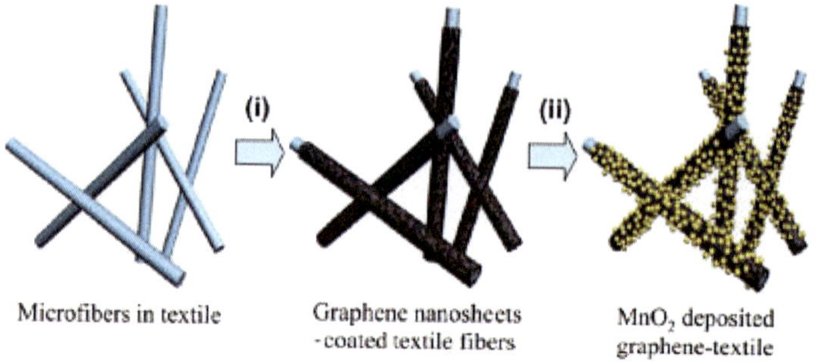

Figure 8.2 Schematic illustration of key steps for preparing hybrid graphene/MnO_2-nanostructured textiles, high-performance EC electrodes. (i) Conformal coating of solution-exfoliated graphene nanosheets (gray color) onto textile fibers. (ii) Controlled electrodeposition of MnO_2 nanoparticles (yellow dots) on graphene-wrapped textile fibers. Reprinted with permission from Ref. 26. Copyright 2011 American Chemical Society.

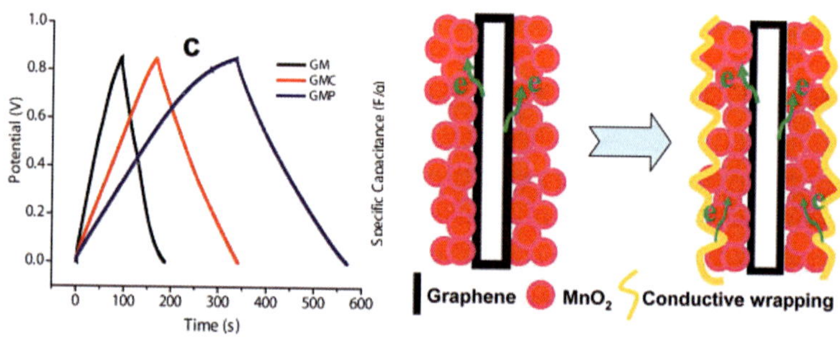

Figure 8.3 (a) Schematic illustration showing conductive wrapping of graphene/MnO2 (GM) to introduce an additional electron transport path. (b) Galvanostatic charge–discharge curves for GM, GMC, and GMP composite electrodes at current density of 0.5 mA/cm^2. Reprinted with permission from Ref. 27. Copyright 2011 American Chemical Society.

conducting polymer was developed by Yu *et al.*[27] for better electrochemical utilization of highly insulating MnO_2 materials and greatly enhance their supercapacitor performance. The supercapacitance of carbon nanotubes and conducting polymer (PEDOT:PSS) wrapped graphene/MnO_2 substantially increased by nearly 20% and nearly 45%, respectively, with values as high as nearly 380 F/g. In addition, excellent cycling performance with more than 95% capacitance retention over 3000 cycles was achieved. Although GMP system offers higher specific capacitance due to large pseudocapacitance contributed from conductive polymer, the wrapping with CNTs can be advantageous in those applications that require high operation voltages owing to their better electrochemical stability across a large voltage range and environmental safety. Such a 3D conductive wrapping approach represents an effective and convenient technique to improve the specific capacitance and rate capability of oxide-based supercapacitors and can be applicable to a wide range of insulating energy storage electrode materials.

To exploit the potential of graphene-based materials for supercapacitor applications, graphene-conducting polymer nanocomposite was prepared by several preparation routes, and electrochemical capacitance was reported in the range of 233~500 F/g.[21] However, the capacitance was mainly dominated by the pseudocapacitance from the polymer films coated on the graphene paper surface, and the EDLC from the graphene sheet was less utilized due to the agglomerated layer like structure in the graphene paper. Among the conductive polymers, carbon (or) graphene–PANI composites have been extensively studied and well documented in literature for supercapacitor applications.[28–30] And, flexible nanoelectrodes have also been developed using carbon nanotube/PANI or graphene/PANI nanocomposites for supercapacitors applications.[31,32] Typically, in all these studies, the aniline polymerization on graphene surface was carried out using oxidants such as ammonium persulphate ($(NH_4)_2S_2O_8$) or ferric chloride ($FeCl_3$), and the experimental strategy plays a vital role in the morphology of the graphene/PANI composite and their electrochemical response. Recently, we showed the preparation of graphene-polyaniline nanocomposite electrodes via oxidative polymerization of aniline by MnO_2. And, a superior supercapacitive performance (641 F/g, ~15–40% enhancement than the reported capacitance for graphene-polyaniline) has been observed.[33] As mentioned earlier, the method of polymerization plays a vital role on the property of materials. In our preparation, GO/MnO_2 composite was prepared by mixing appropriate amount of MnO_2 nanosheets and GO nanosheets. Then, appropriate amount of aniline was added to the above composite and the chemical oxidation polymerization of aniline was initiated by reduction of

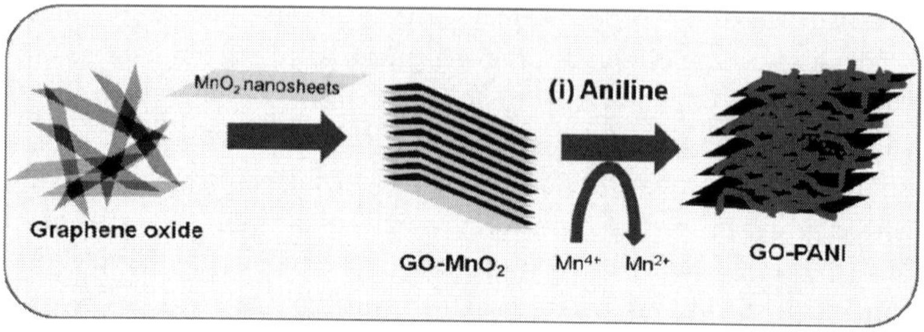

Figure 8.4 Schematic representation of polyaniline formation on graphene surface via oxidative polymerization of aniline by MnO_2. Reproduced from Ref. 33 by permission of the Royal Society of Chemistry.

Mn^{4+} ion and the resulting Mn^{2+} ions will go into the solution. This process enables the formation of slow and uniform polyaniline nanofibers on the graphene surface with significant porosity, which enables the impulsive peculation of electrolyte to access large surface area.[33] Thus, the graphene surface has also been used for the EDLC in addition to the psudocapacitance form polyaniline. Figure 8.4 shows the schematic representation of polyaniline formation on the graphene sheets.

In the row of various chemical modification processes, doping of hetero-atoms, such as nitrogen, sulfur, and boron into graphene backbone, is another possible route. Also, the hetero-atom doping in carbon materials (CNTs, graphene) and metal oxides has always created excitement in material chemistry, as the materials property can be tuned significantly. Recently, studies have been focused in this direction and several possible routes have been identified for the effective N- or B-doping in graphene sheets.[34,35] And, significant enhancement in capacitance has been reported for the hetero-atom doped graphene sheets and the following possible mechanisms have been proposed for the capacitance enhancement: (i) The improvement of electrode wetability due to the increase in the number of hydrophilic polar sites; (ii) The decrease of equivalent series resistance (ESR) of a capacitor cell by the increase of the carbon electric conductivity; (iii) The occurrence of space charge layer capacitance in carbon by the increase of its electron density; (iv) The occurrence of pseudocapacitance through faradic charge transfer, because the nature of carbon becomes electron donor. Although it is difficult to point out one particular effect for the capacitance enhancement, it is believed that the pseudocapacitance through faradic charge transfer is the most important factor to enhance the capacitance of N- or B-doped graphene sheets. And, the research in this direction is in infant stage; additional experiments and theoretical understating is necessary. The mechanism of understanding and further developments in the preparation method are expected to have an important role in the improvement of supercapacitor performance in the coming years.

Asymmetric or hybrid systems offer an attractive alternative to conventional pseudocapacitance or EDLCs by combining a battery-like electrode (energy source) with a capacitor-like electrode (power source) in the same cell.[20] The high specific capacitances, cell voltage, and charge/discharge rates offered by such hybrid structures make them promising candidates as electrodes in supercapacitors. MnO_2–graphene composite electrodes have been developed for high-voltage hybrid electrochemical capacitor based on graphene as negative electrode and a MnO_2–graphene composite as positive electrode in a neutral aqueous Na_2SO_4 solution as electrolyte.[36] These ECs can be cycled reversibly in the high voltage region of 0–2.0 V (Fig. 8.5a). The resulting energy density, 30.4 Wh/kg, is much higher than those of symmetric ECs based on graphene/graphene (2.8 Wh/kg) (Fig. 8.5b) and MnO_2-graphene/MnO_2-graphene (5.2 Wh/kg) (Fig. 8.5c) and higher than

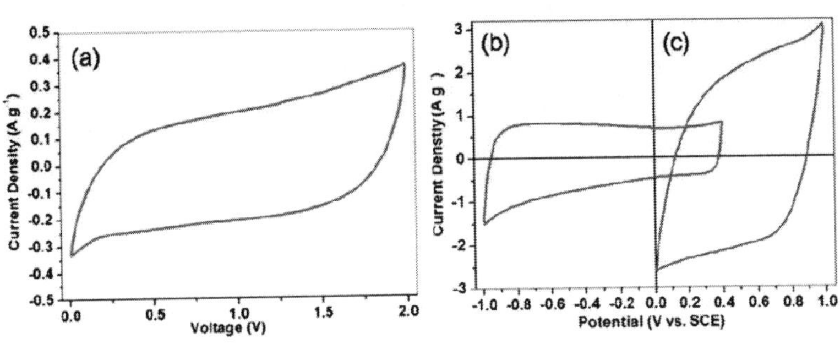

Figure 8.5 Schematic representation of polyaniline formation on graphene surface via oxidative polymerization of aniline by MnO_2. Reprinted with permission from Ref. 36. Copyright 2010 American Chemical Society.

those of other MnO_2-based asymmetric ECs. These findings open up the possibility of graphene-based composites for applications in safe aqueous electrolyte-based high-voltage hybrids systems with high energy and power densities.

8.4 Graphene Nanosheets for Li-Ion Battery

Intercalation of lithium into graphite proceeds via a mechanism known as staging in which the lithium fully intercalates into very distant graphene layers before occupying neighboring graphene layers[37] (Fig. 8.6a). And, the staging phenomenon is characterized by different potential steps in the experimental galvanostatic discharge curve as shown in Fig. 8.6b. The maximum specific lithium insertion capacity for graphite corresponding to the formation of first-stage graphite intercalation compounds (LiC_6) is 372 mAh/g. During the intercalation process, lithium transfers its 2s electrons to the

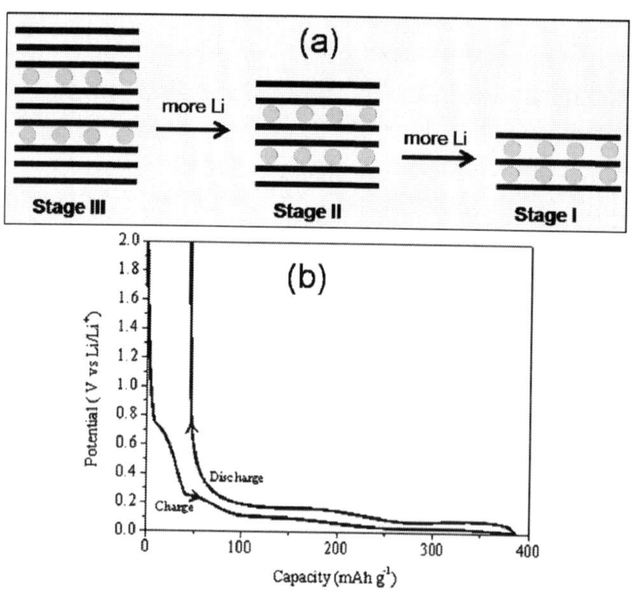

Figure 8.6 (a) Schematic of stage formation during intercalation of lithium into graphite and (b) Li-ion charge–discharge profile of graphite. Adapted from Refs. 1 and 37.

carbon host and is situated between the carbon sheets. There are many excellent reviews available for detailed understating in the fundamentals and mechanism aspects of the Li-ion storage in graphite or carbon materials.[1,38–40]

The forecasted large surface-to-volume ratio and highly conductive nature of graphene nanosheets suggests the possibilities of replacing the graphite in lithium-ion batteries to attain enhanced reversible lithium storage. This is because lithium ions could be bound not only on both sides of graphene nanosheets but also on the edges and covalent sites of the graphene nanosheets. The high chemical diffusivity of Li, nearly 10^{-7}–10^{-6} cm^2/s, on a graphene plane also contributes to its high-power applications.[41,42] Therefore, it is expected that graphene could overtake its 3D counterpart (graphite) for enhanced lithium storage in lithium-ion batteries. It has been proposed by various researchers that lithium ions can be adsorbed on both sides of the graphene nanosheets leading to two layers of lithium for each graphene sheet, with a theoretical capacity of 744 mAh/g through the formation of Li_2C_6.[43–45] In addition, the nanocavities existing between graphene nanosheets due to scrolling and crumpling could also contribute to the lithium storage. According to the micropore mechanism, the extraction of lithium from the nanocavities has to go through the "way" of graphene crystallites. The interaction between lithium atoms and nanopores results in an appreciable voltage hysteresis during the charging process.[46] The charge/discharge profiles of graphene anode in the first cycle and the 100th cycle are shown in Fig. 8.7. Inset of Fig. 8.7 shows the Dahn's model of Li-ion adsorption on the two sides of graphene nanosheets.[47]

It could be seen clearly that a specific capacity of 945 mAh/g was delivered at the initial discharge and a reversible capacity of 650 mAh/g was attained at the first charge.[42] The irreversible capacity (loss) could be associated with the formation of the SEI layer in the first cycle. The shape of the discharge and charge curves is typical of nanosize carbonaceous materials. The largest part of discharge curve (lithium insertion) falls in the region below 0.5 V and the inserted lithium ions were removed during the charge process (lithium extraction) in a wide voltage range of 0.05–3.0 V. After 100 cycles, the graphene anode still maintained a specific

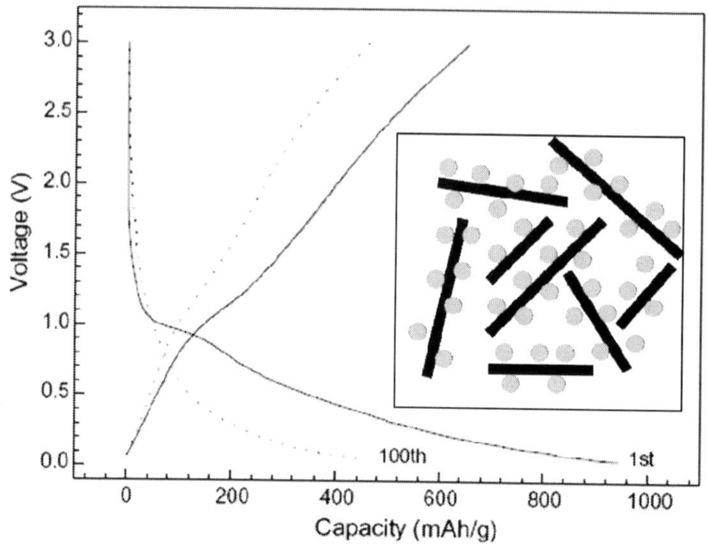

Figure 8.7 Charge–discharge profile of graphene nanosheets. Inset Dahn's model of Li-ion adsorption on the two sides of graphene nanosheets. Reprinted from Ref. 43, Copyright 2009, and Ref. 45, Copyright 1996, with permission from Elsevier.

capacity of 460 mAh/g, which is much enhanced performance than that of graphite anodes.[42] Chemically derived graphene exhibits a high reversible capacity up to 1264 mAh/g at a low rate (such as 50–100 mA/g) but is rate-limited at a high charge/discharge rate (500 mA/g or higher) with capacity fluctuations.[46–50] This capacity fluctuation is closely related to the complex surface chemistry and electrochemical behavior of graphene at the interface of the electrolyte and electrode at a high starting current density. Therefore, graphene nanosheet electrode exhibited lithium storage behavior that was typical of both soft graphitized carbon and hard carbon. Using graphene nanosheets instead of 3D graphite enhances both the Li-ion storage capacity and the charge/discharge rate of the Li-ion battery.

Recently, various attempts have been made to explore alternative materials (other than carbon nanostructures) for succeeding high-capacity Li-ion battery applications.[11] Metal or metal oxides such as Si, Sn, SnO_2, Fe_2O_3, TiO_2, Co_3O_4, and so on nanostructures are promising alternative anode materials for reversible Li-ion storage due to their high theoretical gravimetric capacity, low cost, and eco-friendly property. The major drawback associated with these materials is high volume changes during charge–discharge cycling that result in rapid deterioration and low retention of capacity. However, attempts have been made to resolve this issue significantly by reducing the particle size that offers high surface-to-volume ratio, high surface area, enhanced electron transport, and reduced strain associated with intercalation process. Thus, various synthetic strategies have been introduced for the preparation of nanoparticle, nanosheets, and nanorods to achieve high Li-ion storage with improved cyclic performance. However, the cyclic performance of the above materials is limited due to the poor electronic conductivity and volume extension/contractions. The recent progress in the preparation of high surface area GNS with high conductivity offers GNS as better support for the dispersion of nanometer scale metal or metal oxide nanostructures. Further, the GNS can also contribute to the Li-ion storage and the nanoparticle decoration on their surface hinders the actual restacking of GNS that offers the advantage of additional hidden surface to be available for Li-ion storage. Thus, it is believed that a few nanometer size nanoparticle decorations on GNS would greatly enhance the Li-ion storage capacity of GNS and metal oxide due to the synergic effect. Thus, the development in preparation of metal or metal oxide decorated GNS for Li-ion storage is highly warranted for both fundamental understanding and applications point of view.

It is well known that tin oxide (SnO_2) could be a good substitute for the carbon anode in LIB owing to its high theoretical Li^+ storage capacity of 782 mAh/g. It shows very large volume change of about 300% during charge/discharge process, which causes crumbling and cracking of electrode, leading to electrical disconnection from current collectors. Therefore, most of SnO_2 electrode materials suffer from the rapid fading of capacity. Although several attempts have been made to prevent capacity fading, only limited success was attained so far in enhancing the cyclability. Recently, Paek *et al.*[51] report the preparation of 3D flexible graphene–SnO_2 nanocomposite without any deterioration of fundamental electrochemical properties of both components. As shown in Fig. 8.8, SnO_2/GNS

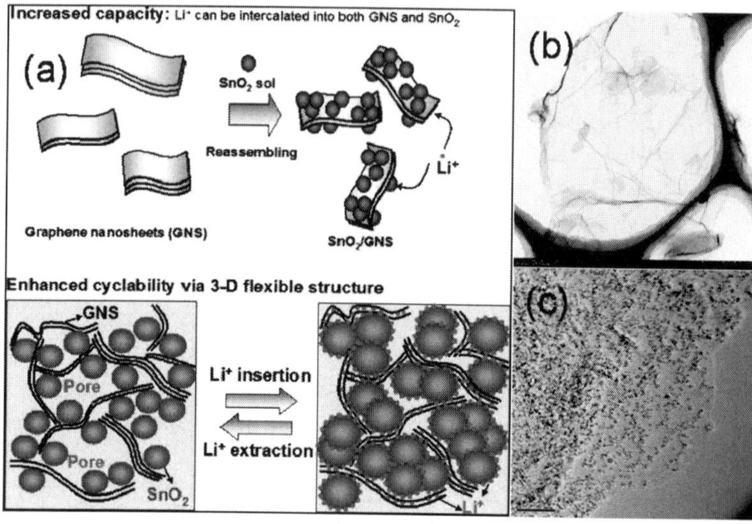

Figure 8.8 (a) Schematic representation of synthesis and structure of SnO_2 decorated GNS. (b) TEM image of few layer graphene nanosheets. (c) TEM images of SnO_2 decorated GNS. Reprinted with permission from Ref. 51. Copyright 2009 American Chemical Society.

nanostructured electrodes, in which the dimensional confinement of tin oxide nanoparticles by the surrounding GNS limits the volume expansion upon lithium insertion, and the developed nanopores between SnO_2 and GNS could be used as buffered spaces during charge/discharge, resulting in the superior cyclic performances of 70% of reversible capacity after 30 cycles. Similarly, Co_3O_4 nanoparticles anchored on conducting graphene were synthesized using various facile preparation strategies for advanced high-performance lithium-ion batteries.[52] The obtained composite materials consist of homogeneously depressed, few nanometer size nanoparticle on graphene sheets. The Co_3O_4 nanoparticles act as spacers to keep the neighboring sheets separated. This Co_3O_4-graphene nanocomposite displays superior Li-battery performance with large reversible capacity, excellent cyclic performance, and good rate capability. Also, Fe_3O_4 decorated graphene nanosheets have been reported for high Li-ion storage capacity. Also, hierarchical 3D nano-assemblies have been reported as possible alternative to the graphite electrodes.

As mentioned earlier, in all these cases, graphene is used not only as a support for metal oxide nanostructures, but also graphene sheets contribute to the Li-ion storage significantly. Thus, there is no limitation for the loading of metal oxide nanostructures. Graphene-metal oxide nanocomposite with divers weight ratio has been reported in the literature for enhanced Li-ion storage.

Recently, graphene nanocomposites have been investigated for cathode materials in Li-ion battery. In cathode materials, graphene is used as a conducting support for the cathode materials. Ding *et al.* prepared a $LiFePO_4$ (LFP)/graphene composite using coprecipitation method.[53] However, there was no significant improvement in rate performance compared with conventional carbon-coated LFP, because the LFP nanoparticles were only loosely loaded on graphene sheets, thereby resulting only in a limited enhancement in electron conductivity. Later, X. Zhou *et al.* report graphene-modified LFP composite cathode materials with excellent high rate capability and cycling stability by creating a 3D network of graphene wrapping on LFP nanoparticles.[54] Graphene layers wrapping homogeneously around the surface of LFP nanoparticles serve as a fast path for electron migration during charge/discharge processes. And, it can deliver a capacity of 70 mAh/g at 60 C discharge rate and showed a capacity decay rate of less than 15% when cycled under 10 C charging and 20 C discharging for 1000 times.

8.5 Graphene Nanosheets for Li–Air Battery

The lithium air (Li–air) battery is one of the most promising technologies among various electrochemical energy storage systems, with a theoretical energy density nearly 10 times that of conventional lithium-ion batteries. This is because lithium metal as an anode has a capacity 10 times higher than that of conventional graphite anodes, and oxygen as the cathode of a Li–air battery can be absorbed freely from the environment leading to a significant reduction in the weight and the cost of the battery. However, the lithium-air battery is far from practical use at this stage and there are many obscurities to be revealed for its potential applications in energy storage. The performance of Li–air batteries is affected by many factors such as oxygen partial pressure, relative humidity, choice of catalysts, electrolyte composition, the macrostructure of the air electrode, and cell designs. To understand the fundamental mechanism and challenges in Li–air batteries, refer Chapter 4 elsewhere in this book (Michael *et al.*) and the references therein. This chapter focuses on the use of graphene and graphene-based materials as possible air cathode material in the Li–air batteries.

The electrochemical reaction for Li–air battery in a nonaqueous electrolyte is widely assumed to be

$$2Li^+ + O_2 + 2e- \leftrightarrow Li_2O_2(\text{solid})$$

The precipitation of reaction products (Li_2O_2) formed in the cathode during the discharge process eventually completely filled in the nano-pores in the cathode materials and blocks the oxygen pathway. Consequently, the oxygen flow will be reduced or stopped, which limits the capacity of the Li–air batteries. The capacity of the cathode is assumed to depend on its maximum ability to accommodate the Li_2O_2 without blocking the O_2 flow. And, it is estimated by its pore volume, with the assumption that the product (Li_2O_2) fills that volume. Hence, there is a critical need to design an optimum air electrode containing micrometer-sized open porosity for rapid oxygen diffusion and substantial nanoporosity (2–50 nm) to catalyze Li-O_2 reactions while preventing excessive growth of the discharge products that block chemical pathways.[55] Thus, materials with high porosity such as mesoporous, micro-porous, and nanoporous carbon are generally used for this purpose. Also, during the charge process, the complete decomposition of Li_2O_2 is essential to achieve excellent recharge capacity. Thus, a catalyst is needed to be used in a positive electrode of the Li–air battery to improve the electrochemical reversibility and charge/discharge efficiency of the Li–air battery. Recently, graphene and graphene-based materials have been used as cathode materials with or without catalyst. It is believed that the GNS prepared by a chemical method that readily produces exfoliated graphene sheets from graphite has many edge sites and defect sites located on the GNS surface. Such edge and defect sites are considered to serve as active sites for chemical reactions. Furthermore, the two-dimensional structure of GNS provides a pathway for access by oxygen gas from both sides

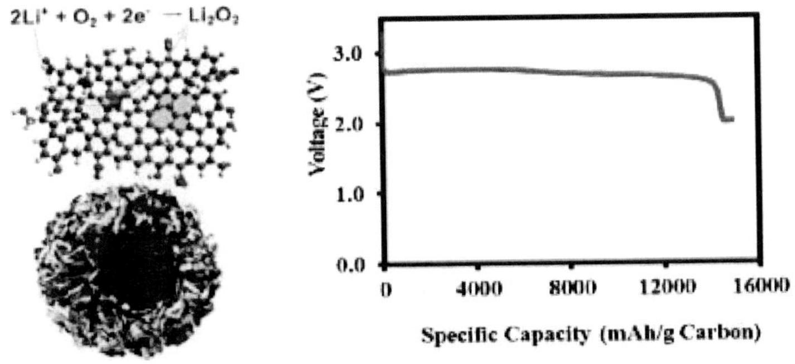

Figure 8.9 An ideal bimodal porous structure of functionalized graphene nanosheets and its discharge curve as the air electrode in a Li–O_2 cell. Reprinted with permission from Ref. 55. Copyright 2011 American Chemical Society.

of the nanosheet. With these characteristics, GNS were expected to be useful novel catalysts for Li–air batteries. Experimentally, there are few studies that have been already reported for Li–air batteries using graphene or graphene-based electrodes.[55–57]

J. Xiao *et al.*[55] demonstrated air electrode consisting of an unusual hierarchical arrangement of functionalized graphene sheets (contains lattice defects and hydroxyl, epoxy, and carboxyl group with no catalyst), which delivers an exceptionally high capacity of 15000 mAh/g in Li–O_2 batteries, which is the highest value ever reported in this field. This excellent performance is attributed to the unique bimodal porous structure of the electrode, which consists of microporous channels facilitating rapid O_2 diffusion, while the highly connected nanoscale pores provide a high density of reactive sites for Li–O_2 reactions. The hierarchically ordered porous structure (Fig. 8.9) in bulk graphene enables its practical applications by promoting accessibility to most graphene sheets in this structure. Further, it was shown that the defects and functional groups on graphene favor the formation of isolated nanosized Li_2O_2 particles and help prevent air blocking in the air electrode. The modeling by density functional theory and electron microscopy characterization suggests that lattice defect sites on the functionalized graphene play a critical role in the formation of small, nanometer-sized discharge products (Li_2O_2).

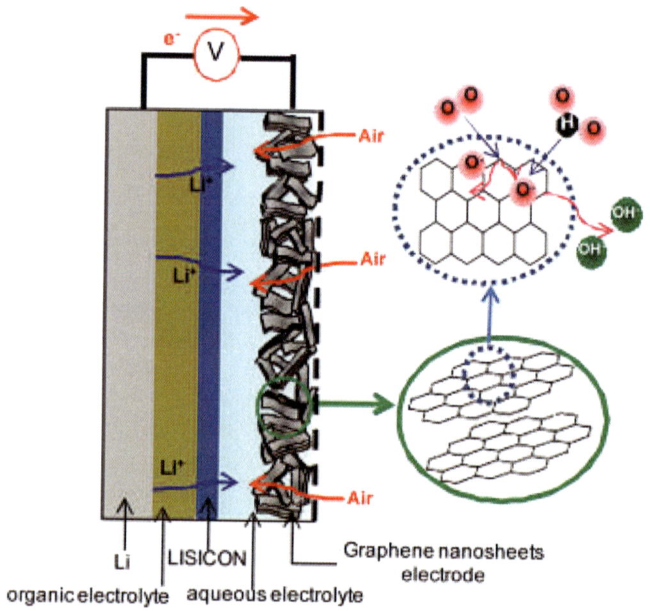

Figure 8.10 Structure of the rechargeable Li–air battery based on GNS as an air electrode. Reprinted with permission from Ref. 56. Copyright 2011 American Chemical Society.

Similarly, metal-free, heat-treated GNS was examined by Yoo *et al.*[56] as air electrode (Fig. 8.10) in a Li–air battery with a hybrid electrolyte and the results were compared with GNS and 20 wt % Pt/carbon black sample. At 0.5 mA/cm, the heat-treated GNS showed a high discharge voltage that was near that of the 20 wt % Pt/carbon black. This was ascribed to the presence of sp^3 bonding associated with edge and defect sites in GNS. These edge and defect sites can effectively serve as active sites for specific chemical reactions and act as catalysts for reduction of oxygen. In comparison with the stacked structure of graphite, the two-dimensional structure of GNS permits ready access of oxygen gas from both sides of the nanosheet onto defect sites and edges. Moreover, the heat-treated GNS not only provided a similar catalytic activity in reducing oxygen in the air but also showed a much more stable cycling performance than GNS when used in a rechargeable Li–air battery. This improvement resulted from removal of adsorbed

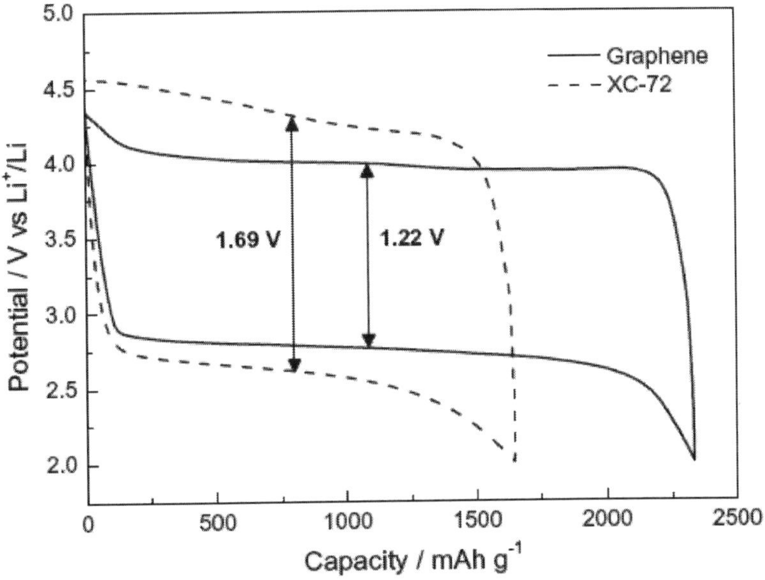

Figure 8.11 Charge/discharge voltage profiles (third cycle) of the as-prepared GNS and Vulcan XC-72 carbon. Capacities are per gram of carbon in the electrode. Cycling was carried out at a current density of 50 mA/g in 1 atm O_2 atmosphere at room temperature (20°C). The cut voltage ranges were 2.0–4.4 V for the GNS electrode and 2.0–4.6 V for the Vulcan carbon electrode. Reprinted from Ref. 57, Copyright 2012, with permission from Elsevier.

functional groups and from crystallization of the GNS surface into a graphitic structure on heat treatment. Moreover, the difference in the discharge voltage between the first and the 50th cycle was as little as 0.07 V. Thus, the improved charge–discharge cycling performance suggests that heat-treated GNS have a good durability when used in Li–air rechargeable batteries.

Sun *et al.*[57] investigated the graphene nanosheets as cathode catalysts for Li–air batteries with alkyl carbonate electrolyte. The electrocatalytic activity of graphene nanosheets was examined in Li–air cells and compared with Vulcan XC-72 carbon (Fig. 8.11). The discharge capacity of the as-prepared GNSs electrode is 2332 mAh/g, while Vulcan XC-72 carbon electrode shows a lower discharge capacity of 1645 mAh/g. During the discharge process, the GNS electrode shows a discharge potential plateau of 2.75 V *vs.* Li/Li^+, which is higher than that of Vulcan XC-72 carbon electrode (2.60 V *vs.* Li/Li^+). During the charge process, the charge potential plateau of GNS is 3.97 V *vs.* Li/Li^+ (half charge capacity), which is much lower than that of the Vulcan XC-72 carbon (4.29 V *vs.* Li/Li^+). The GNS electrodes exhibited a much better cycling stability and lower overpotential than that of the Vulcan XC-72 carbon. The reduced overpotential shows that the as-prepared GNSs are more electrochemically active toward that reaction in alkyl carbonate electrolyte than that of the Vulcan XC-72. The significant difference of the electrochemical performance between the GNS and the Vulcan XC-72 carbon in Li–air cells could be ascribed to the differences of their catalytic activities that arise from the presence of carbon vacancies and defects on the surface. It is well known that these vacancies and defects are very high in GNS owing to the preparation process by strong chemical oxidation and reduction of graphite and exfoliated graphite oxide sheets, respectively. The carbon atoms at the edge of graphene nanosheets are considered to be "partially radical," which can offer special chemical reactivity, compared with nonedge carbon atoms. Thus, the sp^3-bonded carbon atoms in GNS contribute to edge states and defects and might serve as active sites for the electrochemical reaction of oxygen. The detailed mechanism for the oxygen-involved electrochemical reaction in alkyl carbonate electrolytes is still not completely clear and is under intensive investigation.[57]

8.6 Summary

This chapter has focused on the preparation and electrochemical energy storage applications of graphene and graphene-based nanocomposites electrode materials. A number of preparation methods have been developed for the GNS such as chemical oxidation, CVD, unzipping CNTs, and so on. However, preparation of large quantity, few layered GNS with defect-free structure is still challenging. The chemical exfoliation, cleavage, oxidation methods can yield large quantity GNS with large structural defects. And, physical methods, the CVD-based methods, can yield very good quality GNS with low quantities. Thus, it is essential to select appropriate preparation method on the basis of the use of graphene. It is expected that the development of GNS preparation on a large scale, with defect-free and desired structural properties, may become more attractive.

GNS and GNS-based nanocomposites are under extensive investigation for electrochemical energy storage devices such as super capacitors, Li-ion batteries, and Li–air batteries. GNS alone has been shown as better energy storage material than other carbon materials. Graphene-based nanocomposites such as GNS–MnO_2, GNS–polyaniline, and so on have been shown for enhanced energy storage in supercapacitors. Similarly, GNS-metal oxide nanocomposites are shown for enhanced capacity, cyclic performance, and better C-rate in Li-ion storage. The recent investigations on graphene and graphene-based materials with or without catalyst show good performance in Li–air batteries as cathode materials. It is believed that the edge and defect sites are considered to serve as active sites for chemical reactions and the two-dimensional structure of GNS provides a pathway for good access to oxygen gas. Thus, GNS and GNS-based nanocomposites are promising candidate for electrochemical energy storage application and a number of investigations are highly warranted to explore the potential of these materials.

References

1. Kaskhedikar, N. A. and Maier, J., *J. Adv. Mater.*, **21**, pp. 2664–2680 (2009).
2. Pandolfo, A. G. and Hollenkamp, A. F. *J. Power Sources*, **157**, pp. 11–27 (2006).
3. Allen, M. J., Tung, V. C. and Kaner, R. B. *Chem. Rev.*, **110**, pp. 132–145 (2010).
4. Lee, C., Wei, X., Kysar, J. W. and Hone, *Science*, **321**, pp. 385–388 (2008).
5. Balandin, A. A., Ghosh, S., Bao, W., Calizo, I., Teweldebrhan, D., Miao, F. and Lau, C. *Nano Lett.*, **8**, pp. 902–907 (2008).
6. Bolotin, K. I., Sikes, K. J., Jiang, Z., Klima, M., Fudenberg, G., Hone, J., Kim, P. and Stormer, H. L, *Solid State Commun.*, **146**, pp. 351–355 (2008).
7. Stoller, M. D., Park, S., Zhu, Y., An, J. and Ruoff, R. S., *Nano Lett.*, **8**, pp. 3498–3502 (2008).

8. Zhang, Y., Tan, Y. W., Stormer, H. L. and Kim, P., *Nature,* **438**, pp. 201–204 (2005).
9. Zhang, L. L., Zhou R. and Zhao, X. S., *J. Mater. Chem.,* **20**, pp. 5983–5992 (2010).
10. Choi, W., Lahiri, I., Seelaboyina, R. and Kang, Y. S., *Solid State Mater. Sci.,* **35**, pp. 52–71 (2010).
11. Huang, X., Qi, X., Boey, F. and Zhang, H. *Chem. Soc. Rev.,* **41**, pp. 666–686 (2012).
12. Park, S. and Ruoff, R. S., *Nature Nanotech.*, **4**, pp. 217–224 (2009).
13. Lamg, B. A., *Surface Sci.,* **53**, pp. 317–329 (1975).
14. Novoselov, K. S., Geim, A. K., Morozov, S. V., Jiang, D., Zhang, Y., Dubonos, S. V., Grigorieva, I. V. and Firsov, A. A., *Science,* **306**, pp. 666–669 (2004).
15. Viculis, L. M., Mack, J. J. and Kaner, R. B. A., *Science,* **299**, pp. 1361 (2003).
16. Hummers, W. S. and Offeman, R. E., *J. Am. Chem Soc.,* **80**, 1339 (1958).
17. Cote, L. J., Kim, F. and Huang, J., *J. Am. Chem. Soc.,* **131**, pp. 1043–1049 (2009).
18. Tung, V. C., Allen, M. J., Yang, Y. and Kaner, R. B., *Nature Nanotech.,* **4**, pp. 25–29 (2009).
19. Kosynkin, D. V., Higginbotham, A. L., Sinitskii, A., Lomeda, J. R., Dimiev, A., Price, B. K. and Tour, J. M., *Nature,* **458**, pp. 872–877 (2009).
20. Simon, P. and Gogotsi, Y., *Nature Mater.,* **7**, pp. 845–854 (2008).
21. Zhang, L. L. and Zha, X. S., *Chem. Soc. Rev.,* **38**, pp. 2520–2531 (2009).
22. Wang, D. W., Li, F. Wu, Z. S., Ren, W. and Cheng, H. M., *Electrochem. Commun.,* **11**, pp. 1729–1732 (2009).
23. Yoo, J. J., Balakrishnan, K., Huang, J., Meunier, V., Sumpter, B. G., Srivastava, A., Conway, M., Reddy, A. L. M., Yu, J., Vajtai, R. and Ajayan, P. M., *Nano Lett.,* **11**, pp. 1423–1427 (2011).
24. Chen, D., Tanga, L. and Li, J., *J. Mater. Chem.,* **20**, pp. 5983–5992 (2010).
25. Gao, Z., Wang, J., Li, Z., Yang, W., Wang, B., Hou, M., He, Y., Liu, Q., Mann, T., Yang, P., Zhang, M. and Liu, L., *Chem. Mater.,* **23**, pp. 3509–3516 (2011).
26. Yu, G., Hu, L., Vosgueritchian, M., Wang, H., Xie, X., McDonough, J. R., Cui, X., Cui, Y. and Bao, Z., *Nano Lett.,* **11**, pp. 2905–2911 (2011).
27. Yu, G., Hu, L., Liu, N., Wang, H., Vosgueritchian, M., Yang, Y., Cui, Y. and Bao, Z., *Nano Lett.,* **11**, pp. 4438–4442 (2011).
28. Wang, D. W., Li, F., Zhao, J. P., Ren, W. C., Chen, Z. G., Tan, J., Wu, Z. S., Gentle, I., Lu, G. Q. and Cheng, H. M., *ACS Nano,* **3**, pp. 1745–1752 (2009).
29. Zhang, K., Zhang, L. L., Zhao, X. S. and Wu, J., *Chem. Mater.,* **22**, pp. 1392–1401 (2010).

30. Yan, J., Wei, T., Shao, B., Fan, Z., Qian, W., Zhang, M. and Wei, F., *Carbon,* **48**, pp. 487–493 (2010).
31. Meng, C., Liu, C., Chen, L., Hu, C. and Fan, S., *Nano Lett.,* **10**, pp. 4025–4031 (2010).
32. Wu, Q., Xu, Y., Yao, Z., Liu, A. and Shi, G., *ACS Nano,* **4**, pp. 1963–1970 (2010).
33. Sathish, M., Mitani, S., Tomai, T. and Honma, I., *J. Mater Chem.*, **21**, pp. 16216–16222 (2011).
34. Hulicova, D., Kodama, M. and Hatori, H., *Chem. Mater.*, **18**, pp. 2318–2326 (2006).
35. Kwon, T., Nishihara, H., Itoi, H., Yang, Q. H. and Kyotani, T., *Langmuir,* **25**, pp. 11961–11968 (2009).
36. Wu, Z. S., Ren, W., Wang, D. W., Li, F., Liu, B. and Cheng, H. M., *ACS Nano,* **4**, pp. 5835–5842 (2010).
37. Winter, M., Besenhard, J. O., Spahr, M. E. and Novák, P. I., *Adv. Mater.,* **10**, pp. 725–763 (1998).
38. Besenhard, J. O. and Fritz, H. P., *Chem. Int. Ed. Engl.,* **22**, pp. 950–975 (1983).
39. Yamamoto, O., Imanishi, I., Takeda, Y. and Kashiwagi, H., *J. Power Sources,* **54**, pp. 72–75 (1995).
40. Kasuh, T., Mabuchi, A., Tokumitsu, K. and Fujimoto, H., *J. Power Sources,* **68**, pp. 99–101 (1997).
41. Persson, K., Sethuraman, V. A., Hardwick, L. J., Hinuma, Y., Meng, Y. S., Ven, A., Srinivasan, V., Kostecki, R. and Ceder, G., *J. Phys. Chem. Lett.,* **1**, pp. 1176–1180 (2010).
42. Uthaisar, C. and Barone, V., *Nano Lett.,* **10**, pp. 2838–2842 (2010).
43. Wang, G., Shen, X., Yao, J. and Park, J., *Carbon,* **47**, pp. 2049–2053 (2009).
44. Dahn, J. R., Zheng, T., Liu, Y. H. and Xue, J. S., *Science,* **270**, pp. 590–593 (1995).
45. Liu, Y. H., Xue, J. S., Zheng, T. and Dahn, J. R., *Carbon,* **34**, pp. 193–200 (1996).
46. Wu, Y. P., Wan, C.R., Jiang, C. Y., Fang, S. B. and Jiang Y., *Carbon,* **37**, pp. 1901–1908 (1999).
47. Zheng, T., Xing, W. and Dahn, J. R., *Carbon,* **34**, pp. 1501–1507 (1996).
48. Lian, P., Zhu, X., Liang, S., Li, Z., Yang, W. and Wang, H., *Electrochim. Acta,* **5**, pp. 3909–3914 (1996).
49. Wallace, G. G., Wang, C. Y., Li, D. and Too, C. O., *Chem. Mater.,* **21**, pp. 2604–2606 (2009).

50. Pan, D. Y., Wang, S., Zhao, B., Wu, M. H., Zhang, H. J., Wang, Y. and Jiao, Z., *Chem. Mater.*, **21**, pp. 3136–3142 (2009).
51. Paek, S. M., Yoo, E. J. and Honma, I., *Nano Lett.*, **9**, pp. 72–75 (2009).
52. Kim, H., Seo, D. H., Kim, S. W., Kim, J. and Kang, K., *Carbon*, **49**, pp. 326–332 (2011).
53. Ding, Y., Jiang, Y., Xu, F., Yin, J., Ren, H., Zhuo, Q., Long, Z. and Zhang, P., *Electrochem. Commun.*, **12**, pp. 10–13 (2010).
54. Zhou, X., Wang, F., Zhu, Y. and Liu, Z., *J. Mater. Chem.*, **21**, pp. 3353–3358 (2011).
55. Xiao, J., Mei, D., Li, X., Xu, W., Wang, D., Graff, G. L., Bennett, W. D., Nie, Z., Saraf, L. V., Aksay, I. A., Liu, J. and Zhang, J. G., *Nano Lett.*, **11**, pp. 5071–5078 (2011).
56. Yoo, E. and Zhou, H., *ACS Nano*, **5**, pp. 3020–3026 (2011).
57. Sun, B., Wang, B., Su, D., Xiao, L., Ahn, H. and Wang, G., *Carbon*, **50**, pp. 727–733 (2012).

Chapter 9

Carbon Nanotubes for Energy Storage Application

Elzbieta Frackowiak, Grzegorz Lota, and Krzysztof Fic

Institute of Chemistry and Technical Electrochemistry,
Poznan University of Technology, 60-965 Poznan, Piotrowo 3, Poland
Elzbieta.Frackowiak@put.poznan.pl

Carbon-based materials are of interest in developing supercapacitors. Hitherto, variety of carbonaceous materials have been investigated, developed, and commercialized as supercapacitor electrodes. As power and energy densities have become imperative for the development of supercapacitors, recently attention has been focused to introduce carbon nanotubes (CNTs) as a three-dimensional support for materials exhibiting pseudocapacitive properties, for example, electrically conducting polymers (PANI, PPy, PEDOT) or transition metal oxides, for example, MnO_2. It has been confirmed that CNTs play a vital role, forming conducting matrix around the electrode-active materials forming nano-composite electrodes. This chapter is devoted to studies centered on CNTs as support for various pseudocapacitive materials and it explains the merits of having CNTs as conductive additive support in the context of enhancing the achievement of active capacitances.

Nanotechnology in Advanced Electrochemical Power Sources
Edited by S. R. S. Prabaharan and M. S. Michael

ISBN 978-981-4241-43-4 (Hardcover), 978-981-4303-05-7 (eBook)
www.panstanford.com

9.1 Introduction

Carbon nanotubes (CNTs), because of their unique physicochemical properties,[1,2] become an interesting component of different energy storage devices (supercapacitors, fuel cells, accumulators). Especially, conducting and microtextural properties of CNTs are noteworthy for these various electrochemical applications. The presence of more or less continuous graphitic layers, cylindrically rolled, together with a tight entanglement of nanotubes supply a conducting and well accessible carbon network with a moderate surface area (approximately 100–400 m^2/g). The supercapacitor performance is connected with the accumulation of charges in the electrode/electrolyte interface;[3–5] hence, such a nanotubular material in the form of an easy available web is of great interest. However, apart from mesopores that are perfect for ion transportation, accessible micropores are required for the accumulation of ions in the electrical double layer. Generally, for this target, a highly developed specific surface area is demanded.[6,7] Therefore, pure carbon nanotubes as capacitor electrode material can only supply very moderate capacitance values depending on their microtexture and catalyst impurities.[8] It has been found that the more defected outer walls of CNTs, the higher values of capacitance. The presence of a thin amorphous carbon layer on their walls also supplies better conditions for charge accumulation. In addition, catalyst impurities of nanotubes can be a source of redox reactions giving pseudocapacitance effects. For the supercapacitor application, CNTs can play a perfect role as a conducting support of active materials such as conducting polymers,[9–20] metallic oxides[18–27] as well as they can be an excellent component of carbon/carbon composites.[28–31] Apart from capacitors, a small amount of carbon nanotubes might be used as a percolating additive in electrode active mass of typical accumulators. On the contrary, due to their mesoporous character, carbon nanotubes could constitute a good support of catalytic particles for the application in fuel cells. In this chapter, various electrochemical applications of nanotubular material (capacitors, fuel cells, accumulators as well as hydrogen storage) are presented and a profitable role of nanotubes is underlined.

9.2 Carbon Nanotubes/Conducting Polymers Composites for Capacitor

The values of capacitance for pure nanotubes are very moderate, that is, from 5 F/g to 40 F/g being strongly dependent on CNTs preparation method.[8] Generally, the more developed microporosity, the better capacitance properties. It is well known that nanotubes posses mesoporous character with a very limited micropores amount; hence, enhancement of CNTs capacitance can be reached by an increase of their surface area, for example, by activation process. However, a drastic capacitance improvement might be realized by introducing pseudocapacitance effects through electrically conducting polymers (ECPs) wherein apart from the pure electrostatic attraction of ions that takes place in electric double layer, additional faradaic reactions can proceed.[9–20] ECP/CNTs composite can be prepared by chemical and electrochemical polymerization and 20–30% content of nanotubes is a sufficient proportion. Figure 9.1 shows chemically elaborated polyaniline (80%) on the nanotubular network (20%) wherein a perfect template effect between two composite components is observed.

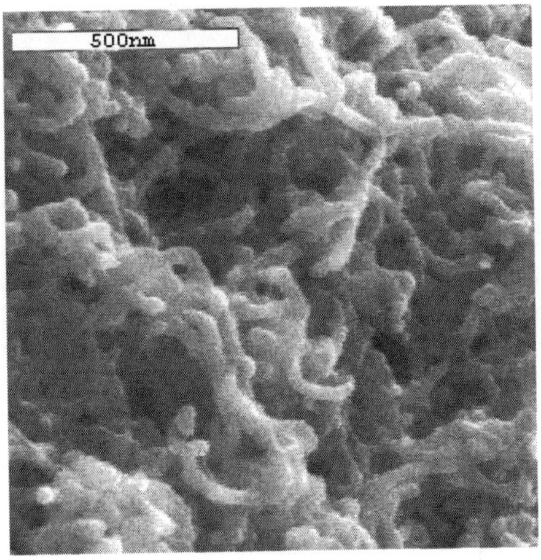

Figure 9.1 Scanning electron micrograph showing the morphology of a polyaniline/CNT composite (80% of polyaniline). Reprinted from Ref. 12, Copyright 2005, with permission from Elsevier.

Generally, a synergy effect is obtained between the components of composite, that is, CNTs and ECP. A nanocomposite gives better results than its components alone. The most attractive ECP as supercapacitor materials are polyaniline (PANI), polypyrrole (PPy), and 3,4-ethylene dioxythiophene (PEDOT) because of the quick redox reactions (example of PPy):

$$[PPy^{+}A^{-}] + e^{-} \leftrightarrow [PPy] + A^{-}$$

During oxidation process (p-doping), ions from electrolyte are inserted to the polymer backbone, and upon reduction they are released back into the electrolytic solution. The doping/dedoping process takes place in the bulk of the electrode, giving high values of capacitance. However, insertion/deinsertion of counter ions is at the origin of volumetric changes of ECPs during cyclic performance and progressive degradation of electrode. Hence, the application of a moderate amount of carbon nanotubes improves mechanical properties of electrode. Swelling and shrinkage of ECPs does not occur during doping/dedoping of ECP material with nanotubes, because they adapt to volumetric changes and protect electrode from mechanical destruction. The good electrochemical performance of ECP/CNTs composite is observed during cycling, especially if the suitable operating voltage range of capacitor is selected. Generally, the operating voltage of a symmetric capacitor based on ECP electrodes cannot exceed 0.6–0.8 V due to oxygen evolution in the positive range of potentials and switching to an insulating state in the region of negative values.[12] The best application of ECP composites is their use in asymmetric configuration, for instance, selecting different type of ECPs or other electrode materials (activated carbons, metal oxides). Such concept allows to reach higher energy as well as higher power of supercapacitors.[7,18,20]

9.3 Carbon Nanotubes/Metal Oxides Composites for Capacitor

Metal oxides (MnO_2, RuO_2, $Ni(OH)_2$, and so on) are the next and more environment-friendly group of pseudocapacitance materials used for capacitor application. In the case of hydrous oxides that

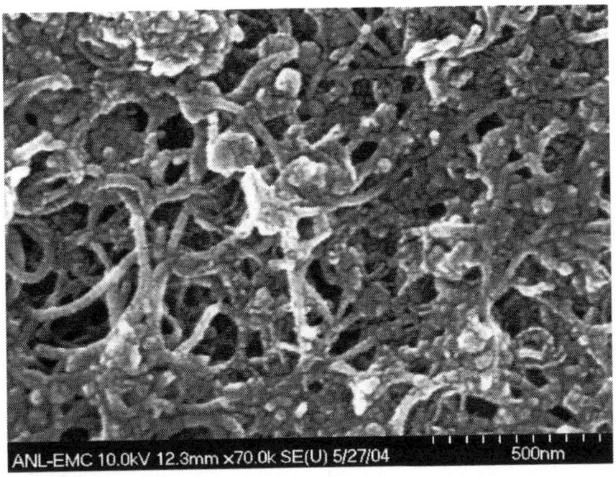

Figure 9.2 Chemically prepared composite CNTs/MnO_2 (Courtesy of M. Thackeray, ANL, USA).

have a significant resistivity, a well conducting percolator must be used. Hence, carbon nanotubes play a perfect percolator role in such composites.[21–27] Their electrical conductivity and mechanical properties (high resiliency) as well as mesoporous character are profitable for the formation of attractive composites with transition metal oxides.

A good adhesion of manganese oxide to entangled nanotubes framework forms perfect textural properties for an easy access of ions to the bulk of the active material (Fig. 9.2). A high conductivity and a good resiliency of the composite electrodes are observed. The pseudocapacitance properties of hydrous oxides are attributed to the redox exchange of protons and/or cations with the electrolyte according to the following reaction:[15,20,21]

$$MnO_x(OH)_y + \delta e^- + \delta H^+ \quad \leftrightarrow MnO_{x-\delta}(OH)_{y+\delta}$$

Impedance spectroscopy measurements (Fig. 9.3) show how the improved conducting properties indicate MnO_2/CNTs as a nanocomposite. Semicircle on the axis Z' (real part), which represents electronic properties of the system, is situated below 2 Ω.

In the symmetric capacitor in which both electrodes are built from the same MnO_2/CNT composite, the working potential of positive and negative electrode cannot be controlled. Taking into account the possibility of irreversible transformation of Mn(IV) into

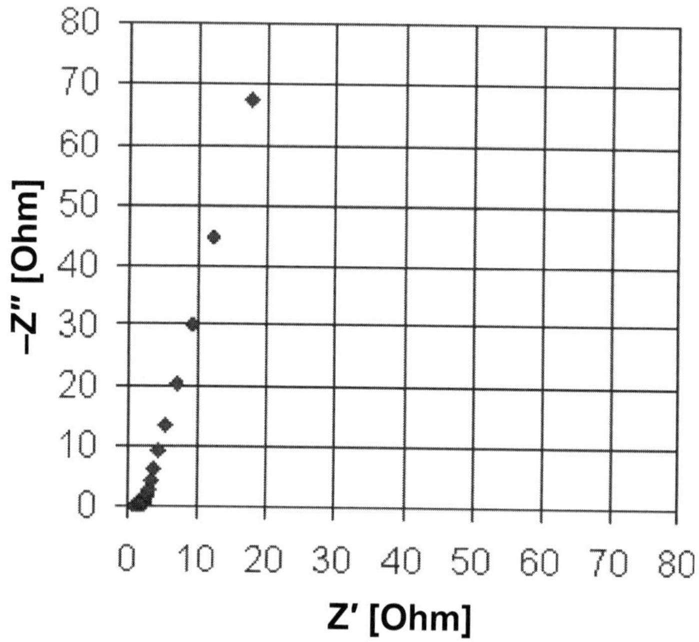

Figure 9.3 Impedance spectroscopy measurements (100 kHz–10 mHz) in symmetric two electrode system. Composition of electrode: 30 % MnO_2/70 % CNTs (Courtesy of K. Lota, PhD thesis).

Mn(II) at the negative electrode,[15] the asymmetric configuration has been realized wherein MnO_2/CNT composite is a positive electrode whereas the activated carbon, which is able to store reversibly hydrogen, played a role of negative electrode. Galvanostatic charge–discharge of such supercapacitor (+) MnO_2(30%)/CNTs(70%) composite/activated carbon (−) at two different current densities is shown in Fig. 9.4. Two high current loads 500 mA/g and 1000 mA/g were used.

For the asymmetric system, the values of capacitance are 160 F/g at 500 mA/g current load, whereas in voltammetry experiments for a low scan rate of potential, the capacitance reaches 220 F/g. The asymmetric configuration (2 V in aqueous solution) in addition can extremely increase the energy of such system being close to the values obtained in an organic medium.

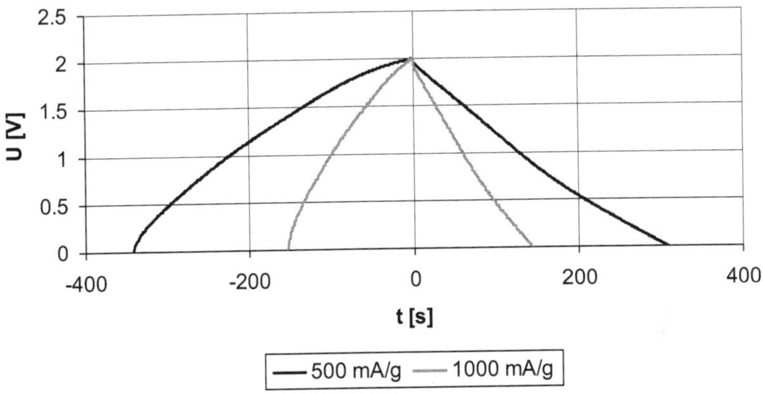

Figure 9.4 Galvanostatic charge–discharge characteristics of asymmetric supercapacitor (+) MnO_2(30%)/CNTs(70%) composite/activated carbon (−) at two different current densities (Courtesy of K. Lota, PhD thesis).

9.4 Pseudoeffects from Heteroatoms in the Carbon Network

Carbon nanotubes can also be used as a frame for carbon obtained from different polymeric precursors.[28–31] In this case, the nanotubes skeleton is very profitable to prevent shrinkage during the carbonization process, and consequently, an interesting CNT/carbon composite is obtained for supercapacitor application. Especially attractive is carbon/CNT composite rich in heteroatoms. Carbons rich in heteroatoms, even with a moderate surface area, can supply interesting capacitance values because of pseudocapacitance connected with quick faradaic reactions. The most often present heteroatoms in the carbon network are oxygen, nitrogen, hydrogen, and sulfur. Especially nitrogen and oxygen are of great interest for capacitance enhancement. The profitable effect of these heteroatoms in the carbon network has been already documented.[28,31] The type of surface groups is crucial for a good cycling performance of supercapacitors. In the case of nitrogen, it should be noted that various chemical states are present in the porous carbons. Pyridinic, pyrrolic, quaternary, pyridonic, and pyridine N-oxide nitrogen is commonly distinguished among N atoms substituted for carbon in the ring system. The variety of surface functionality results both from the position occupied in the ring system and from the extent of association with oxygen, which hardly can be avoided during synthesis. The surroundings of the nitrogen atom in a graphene layer obviously affect its charge, electron donor/acceptor properties, and the contribution to the delocalized π electron system. Generally, the presence of nitrogen in the carbon material can enhance capacitance not only due to faradaic redox reactions but also because of the modification of its electronic character as well as an improved electrode wettability.

The nitrogen enrichment can be realized by a few ways; however, the carbonization of suitable precursors rich in heteroatoms is the most popular procedure. It has been already proved that through the carbonization of organic precursors rich in nitrogen, carbon materials adapted for supercapacitor could be obtained.[28,31] The porosity of the final carbons depends often on the type and amount of gases evolved during carbonization; sometimes, a profitable autoactivation of carbon can take place; hence, in this case, adding nanotubes to the substrate before carbonization is useful and gives exceptional capacitance properties of such carbon/carbon composites.

The beneficial effect of nitrogen in the composite with an incorporated nanotubular backbone has been clearly demonstrated.[31] Melamine was the carbon precursor rich in nitrogen (45 wt%); hence, polymerized blends of melamine with formaldehyde allow various nitrogenated composites to be prepared. The carbon composites have been obtained by polymerization of melamine with formaldehyde (without any catalyst) in the presence of a controlled amount of multiwalled carbon nanotubes. Such a polymerized blend was carbonised at 750°C for 1 h under nitrogen flow. The final carbonization product was named M+F (i.e., melamine and formaldehyde without carbon nanotubes), whereas Nt+M+F means composite with carbon nanotubes; 2M and 3M stand for a twofold and threefold melamine proportion in the blend, respectively. The results of elemental analysis showed that the nitrogen content varied in the final product from 7.4 to 21.7 wt% (Table 9.1). The oxygen content was calculated by difference and its amount was comparable in all the samples varying from 5.9 to 7.8 wt%. The

Table 9.1 Physicochemical and electrochemical characteristics of nanocomposites rich in nitrogen. C stands for the capacitance estimated at 5 Ag^{-1} current load

Sample	S_{BET} [m^2g^{-1}]	V_{total} [cm^3g^{-1}]	V_{micro} [cm^3g^{-1}]	C [$F g^{-1}$]	C [μFcm^{-2}]	Nitrogen [wt. %]	Nitrogen [at. %]
M+F	329	0.162	0.152	4	1.2	21.7	24.8
Nt+3M+F	403	0.291	0.174	100	24.8	14.0	13.6
Nt+2M+F	393	0.321	0.167	126	32.1	11.7	12.7
Nt+M+F	381	0.424	0.156	83	21.8	7.4	7.9

physicochemical parameters of composites have been correlated with their electrochemical behaviour (Table 9.1).

TEM images of composites with various amount of nitrogen content are shown in Fig. 9.5. It is seen that the entangled morphology of the nanotubes with the presence of open mesopores is well preserved in the composite. On the contrary, the simple carbonization product of the melamine/formaldehyde (M+F) blend (Fig. 9.5a)—without nanotubes—gives an amorphous texture. In the case of the composite with nearly 7 wt% nitrogen (Nt+M+F) (Fig. 9.5b), the carbon nanotubes are hardly covered by the carbonization product. An ideal good adhesion of carbon homogeneously distributed on nanotubes is demonstrated in Fig. 9.5c (Nt+2M+F), whereas increasing the melamine content in the case of Nt+3M+F (Fig. 9.5d), the composite morphology gradually changes with a compact texture together with some agglomerates.

Three main N1s peaks have been observed by XPS analysis; however, pyridinic with a binding energy of 398.7 eV and quaternary (400.9 eV) are dominant. The amount of pyridinic groups is the same, that is, 4.9 at.% in the Nt+2M+F and Nt+3M+F composites, whereas it reaches 10.2 at.% for M+F and only 2.7 at% for the Nt+M+F composite. The quaternary nitrogen gradually decreases from 13.4 at% for the M+F composite to 4.8 at% for Nt+M+F, whereas it is almost the same, that is, 7.9 at% and 7.3 at% for Nt+3M+F and Nt+2M+F, respectively.

It is important to stress that the values from elemental analysis and from XPS are very comparable, which is a proof of the same perfect distribution of nitrogen in the bulk of the carbon network

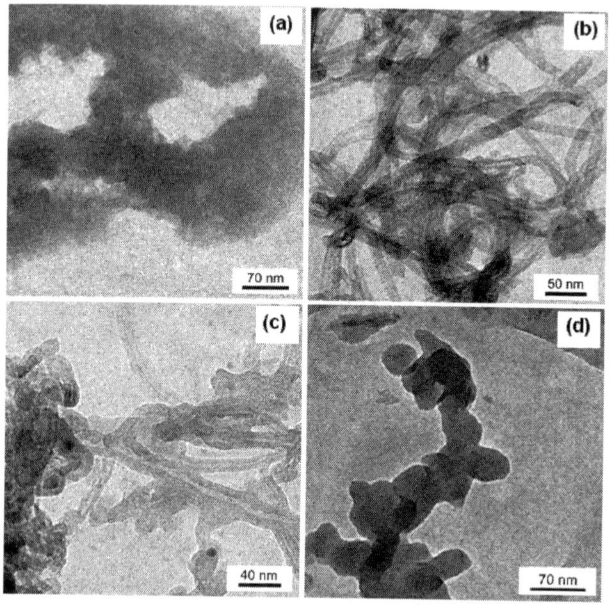

Figure 9.5 TEM images of composites from nanotubes and carbonised polymer blends (melamine and formaldehyde). (a) M+F with 21.7 N %, (b) Nt+M+F with 7.4 N %, (c) Nt+2M+F with 11.7 N %, (d) Nt+3M+F with 14.0 N %. Reprinted from Ref. 31, Copyright 2007, with permission from Elsevier.

as well as on the surface. On the contrary, the amount of oxygen was almost the same in all the samples giving values from 5.9 to 7.8 wt%.

The nitrogen adsorption/desorption isotherms allow the specific surface area, pore size distribution as well as the micro/meso ratio to be estimated. The total surface area is quite similar for the investigated samples and it ranges from 329 to 403 m^2/g being the most developed for the Nt+3M+F composite as shown in Fig. 9.6. The nitrogen adsorption isotherms showed that the carbon materials are typically mesoporous (apart from the material M+F, that is, without nanotubes), and the amount of micropores is very moderate. The micropore volume values for all the samples are comparable varying from 0.152 to 0.174 cm^3/g. The porosity characteristics of all the composites are illustrated in Table 9.1.

The electrochemical performance of the nitrogen-enriched composites used as electrodes in supercapacitors is determined by the

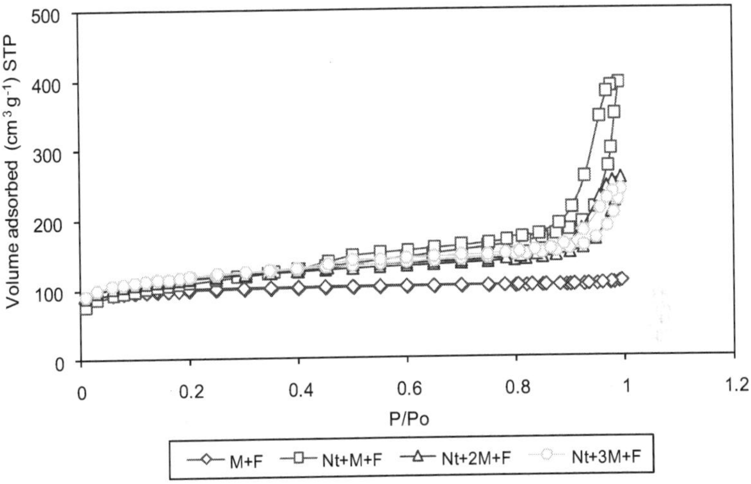

Figure 9.6 Nitrogen sorption isotherms at 77 K for nitrogenated composites. Reprinted from Ref. 31, Copyright 2007, with permission from Elsevier.

porosity development; however, the surface chemistry being related primarily to the type and concentration of nitrogen-containing groups also plays an important role. It is also well known that the presence of nitrogen improves the wettability of the carbon matrix; hence, it can be predicted that an aqueous electrolytic solution will be preferable. The electrochemical properties of the nitrogen-enriched composites have been carefully investigated. The voltammetry characteristics at a potential scan rate of 10 mV/s for all the composites in 1 mol/L H_2SO_4 show a mirror-like behavior typical for an ideal capacitor. Only the M+F composite without nanotubes containing 21.7 wt% nitrogen presents definitively worse characteristic, the most probably because of a higher resistance. The capacitance values estimated by the integration of charge varied from 86 F/g for the M+F composite up to 157 F/g for Nt+2M+F. The dependence of capacitance *vs.* current load for all the composites is very interesting from the practical point of view (Fig. 9.7.). The various current loads have been selected from 50 mA/g to 50 A/g. It is well visible how the ability of charge accumulation diminishes with the load, but the Nt+2M+F sample is still able to supply a capacitance of 60 F/g at extremely high 50 A/g current density.

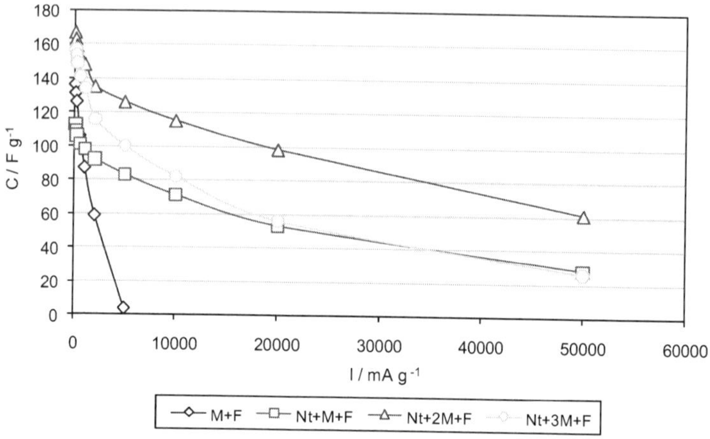

Figure 9.7 Capacitance *vs.* current load for all the composites investigated in acidic electrolytic solution.

Trials were undertaken to explain the electrochemical behavior of both electrodes separately. For that, a two-electrode capacitor was charged/discharged at 1 A/g current load by controlling simultaneously the potential for both electrodes. Such characteristics for the Nt+2M+F composite operating in acidic medium are shown in Fig. 9.8. It is interesting to note that the capacitance of the total capacitor is equal to 148 F/g, whereas the positive electrode shows 133 F/g and the negative one 166 F/g. (All values are expressed per electrode). It shows that both electrodes have a different contribution to the total capacitance.

The same procedure was done for the voltammetry experiments at a scan rate of 5 mV/s. During this investigation, a higher discrepancy between the capacitance behavior of both electrodes was found, and the positive electrode showed higher capacitance values. We could assume that different functional groups are involved in both cases. But other parameters should be also taken into account, for example, the electrical conductivity.

Impedance spectroscopy seems to be the most adapted method for explaining this behavior because it allows the study of the frequency effect on the electrode series resistance and on the specific capacitance. Studying the dependence of capacitance *vs.* frequency for the total two-electrode capacitor built from the Nt+2M+F

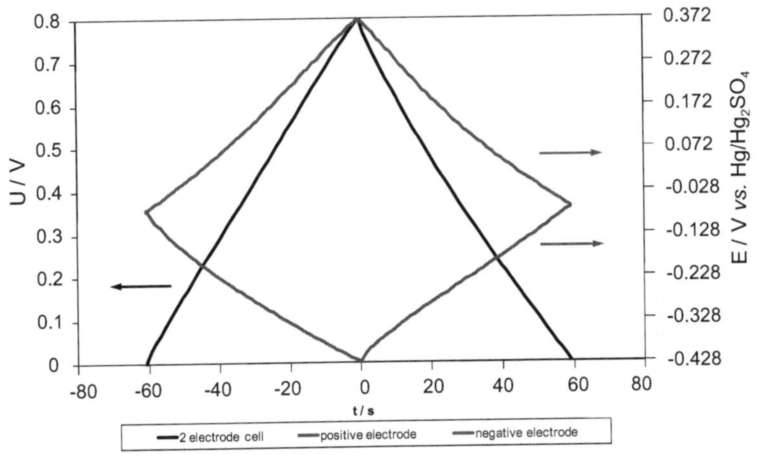

Figure 9.8 Galvanostatic charge/discharge of a capacitor (1 A/g) built from the Nt+2M+F composite with a potential control of the positive and negative electrodes, separately.

material and for both electrodes separately, an anomalous behavior is observed (Fig. 9.9). At 1 mHz frequency, the capacitance of the negative electrode is strikingly high (227 F/g) and the value for the positive electrode (151 F/g) is similar to the value for the total capacitor (162 F/g). On the contrary, an anomalous increase of capacitance is observed for the positive electrode in the range of higher frequency (from 0.1 Hz to 100 Hz). To explain this, very useful information can be obtained from the analysis of the impedance real part *vs.* frequency. The first drop in the range of frequency from 1 to 10 mHz is driven by the series and the parallel resistance. The latter is due to the electronic leakage current through the separator, to the charge redistribution inside the electrode and to the self-discharge. At very low frequency, the parallel resistance contribution is dominant. In the range of frequency from 10 mHz to 10 Hz, the information on the series resistance due to the electronic resistance in the conductors and to the ionic resistance in the electrolyte is supplied. Generally, in this frequency region, the parallel resistance is negligible.

At lower frequency, the ions are able to reach the electrode deeper in the pores of activated carbon. Then, from 10 Hz to

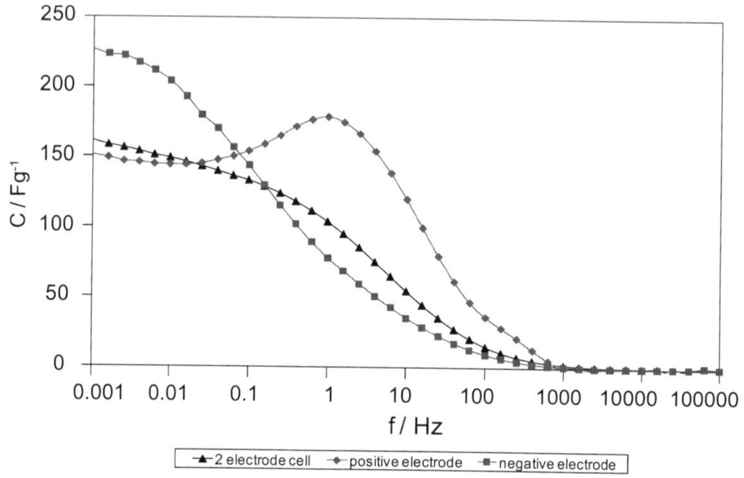

Figure 9.9 Capacitance *vs.* frequency dependence for a two-electrode cell built from the Nt+2M+F composite and for both electrodes (+) and (−) separately.

1 kHz, the various contributions are the electronic resistance of all the electrical connections, the contact resistance between the activated carbon and the current collector as well as the minimal resistance of the electrolyte. The region above 1 kHz reflects the supercapacitor inductance and the parasitic inductance of the electrical cables, which is negligible in our case. Taking into account all the above-mentioned explanations, some anomalies of nitrogen-rich composites could be reasonably elucidated. High capacitance at 1 mHz proves that redox reactions of nitrogenated groups on the negative electrode are extremely efficient; on the contrary, the electronic structure/density of electronic states due to nitrogen is dominant for the positive electrode. We can assume that both effects are responsible for the excellent charge propagation of the Nt+2M+F composite. It is also necessary to take into account that both electrodes operate together in one system; hence, if one electrode works very bad in the selected frequency range, the second one will operate contrarily (perfectly in this case) just to preserve equilibrium.

Apart from capacitance properties of the composite measured in acidic electrolyte, some attention has been also devoted to 6

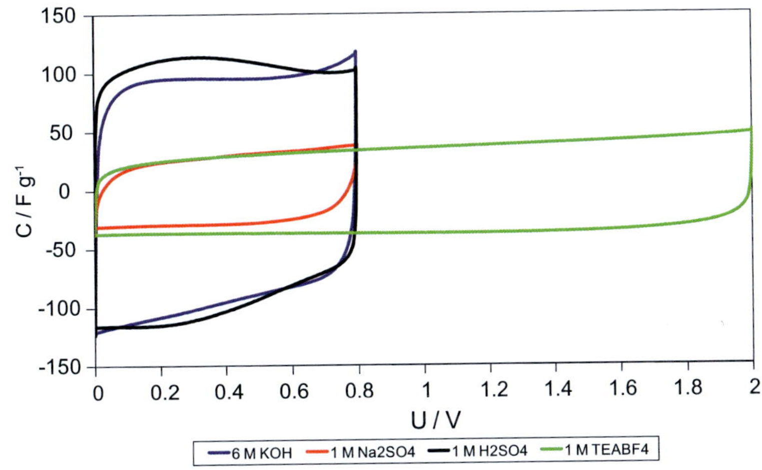

Figure 9.10 Voltammetry characteristics at 10 mV/s scan rate for the Nt+M+F nitrogen enriched composite. Various aqueous electrolytic solutions and organic medium.

mol/L KOH medium, 1 mol/L Na_2SO_4 neutral solution, and the organic electrolyte 1 mol/L^1 tetraethylammonium tetrafluoroborate ($TEABF_4$) in acetonitrile (ACN). Figure 9.10 shows the voltammetry characteristics at 10 mV/s in all these electrolytic solutions, using the Nt+M+F composite with 7.4 wt% of nitrogen. The best performance (101 F/g) was observed in acidic solution, slightly exceeding the alkaline one (92 F/g). As expected, in organic solution, capacitance diminished to 35 F/g. In this case, the lower values are justified by the lack of protons in the organic electrolyte, but also the size of ions in organic solutions is bigger. In the case of neutral electrolyte, the lowest values of 26 F/g were observed.

Surely, the nitrogen presence has a profitable effect on the capacitance values, as well as for good capacitor performance at drastic current loads. However, an excess of nitrogen (presumably over 15%) will definitively aggravate the conducting properties, and in turn, the capacitance characteristics and supercapacitor cyclability. The form in which N participates in the carbon network is especially important. It seems that the effect of $-NH_2$ groups outside of the matrix, with N chemically bound to the carbon network ("chemical nitrogen"), will be of less importance and most probably

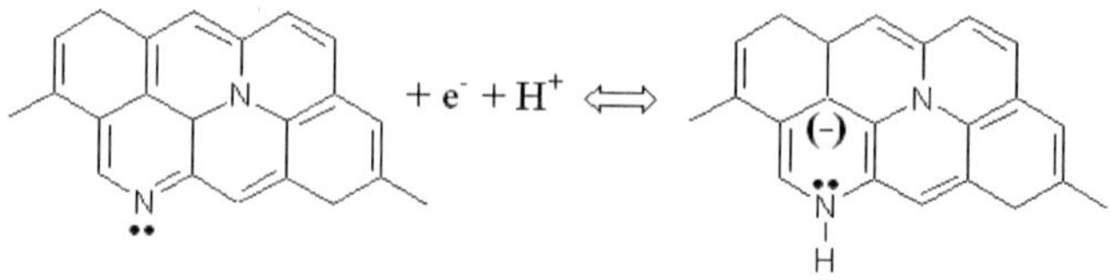

Figure 9.11 Scheme of reversible redox reactions of pyridinic groups.

such groups could block the entrance to the pores. Nitrogen substituted to carbon ("lattice nitrogen") in the periphery, in the form of pyridinic groups, could play some useful role according to the reversible redox reaction shown in the Fig. 9.11.

The donor properties are responsible for filling the conduction band by electrons; in turn, more ions can be sorbed in the electrical double layer, especially for composites in which carbon is optimally substituted by nitrogen. However, it was confirmed experimentally that a higher amount of nitrogen (over 15%) could lead to a significant aggravation of conductivity.

Some trials were undertaken to correlate N content in nitrogenated carbons with the electron density states from molecular quantum calculation. The energy gap, that is, a difference between highest occupied molecular orbital (HOMO) and lowest unoccupied molecular orbital (LUMO) is tightly connected with a conductivity of material. The calculation of the gap between conduction and valence band was based on the Hartree–Fock method and the Koopmans theory. A few different compositions were assumed with varying N% content. The results showed that for the compositions with nitrogen exceeding 14% N, the gap starts to increase. Model of carbon matrix with pyridinic nitrogen of 11.88% N is shown in Fig. 9.12. From molecular calculation, the energy gap is equal to 4.08 eV. HOMO and LUMO are presented in Fig. 9.12 as transparent, colored surfaces. Each molecular orbital is a mathematical function that describes the wave-like behavior of an electron in a molecule. This function can be used to calculate chemical and physical properties such as the probability of finding an electron in any specific region. The orbitals being charged positively are colorized in red, and the orbitals being charged negatively are colorized in green.

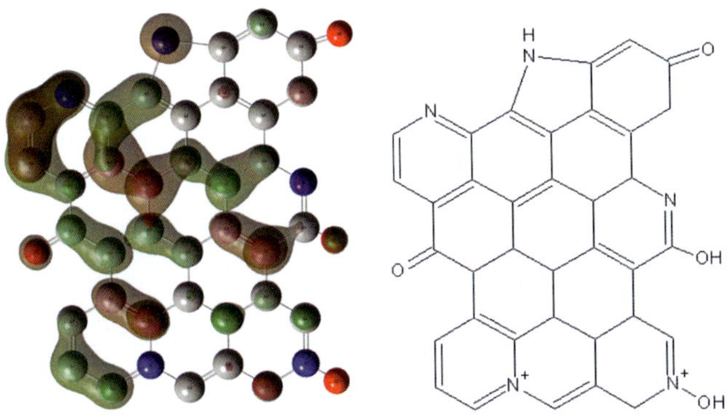

Figure 9.12 Model of carbon matrix with pyridinic nitrogen of 11.88% N.

The difference of the energies between HOMO and LUMO, termed the band gap, may be interpreted as material ability to electrical conductivity. However, not only the energy of band gap can inform about this property, but the shape and conformation of these orbitals are also very important. Nitrogen as whole but also type of nitrogen functionality strongly influenced and changed the primary conformation of HOMO/LUMO structure in investigated material. Generally, the more developed HOMO/LUMO structure in nitrogenated carbons, in a big simplicity, makes the electron/current flow much easier. For comparison, HOMO/LUMO gap for the same material without nitrogen is higher and equal to 5.0 eV.

The effect of different N functionality is shown in Fig. 9.13 in which the nitrogen content is very close (11.92%), but more quaternary nitrogen is present in the carbon matrix. Here, the HOMO-LUMO gap is smaller than for pyridinic groups and is 3.81 eV (better conductivity is expected).

On the contrary, for maximal amount of nitrogen, that is, 25.8% N, the band gap drastically increases to 4.63 eV (not shown here). Theoretical prediction matches with experimental data, and the loss of conductivity has been proved by electrochemical methods (Table 9.1).

Apart from the nitrogen effect, the high charge propagation can be partly explained by an exceptional electronic transport in carbon nanotubes, which are still preserved after the carbonization process.

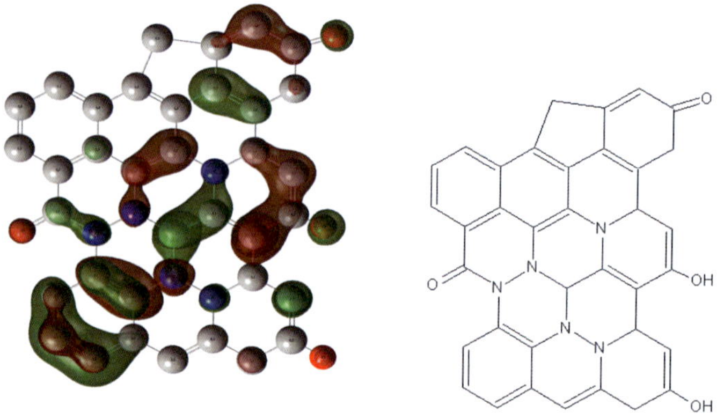

Figure 9.13 Model of carbon matrix with quaternary nitrogen of 11.88% N.

It should be also stressed that these composites have been prepared without activation process; hence, their moderate surface area gives profitable values of specific surface capacitance as well as attractive volumetric capacity.

9.5 Pseudocapacitance Effect Related with Hydrogen Electrosorbed in Carbon

The reversible electrosorption of hydrogen on the carbon network surface by cathodic decomposition of water is another example of pseudocapacitance phenomenon,[32–36] which affects the performance of negative electrode in supercapacitor. This method is very attractive because of its simple and safe conditions of use, and the fact that sorption of hydrogen takes place at room temperature. Moreover, from the point of view of energy efficiency, it is interesting to associate in a unique step the hydrogen production by water electrolysis and its simultaneous storage in the carbon substrate. The *in situ* produced hydrogen easily penetrates into the nanopores of carbon, wherein it is adsorbed due to the driving force of the negative polarization.[32,33] In alkaline solution, water is reduced according to the following reaction:

$$H_2O + e^- \rightarrow H + OH^-$$

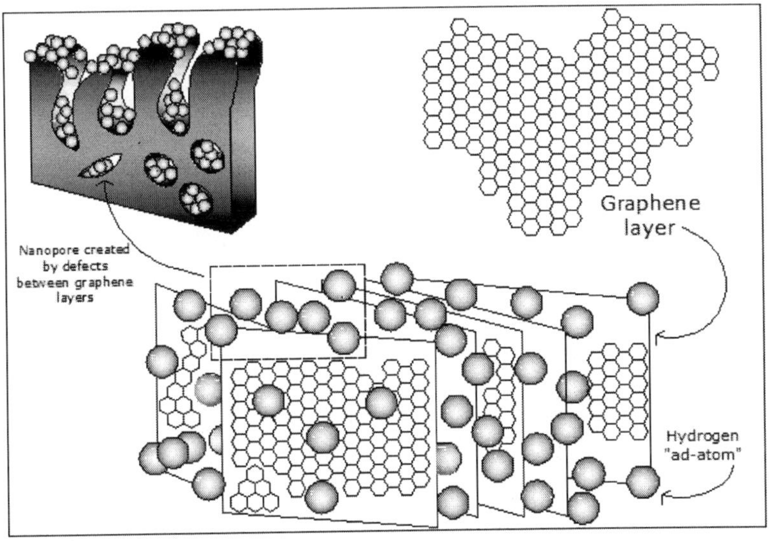

Figure 9.14 Model of hydrogen insertion into the carbon network.

and the formed nascent hydrogen is adsorbed on the surface of nanopores

$$<\mathrm{C}> + x\mathrm{H} \rightarrow <\mathrm{CH}_x>$$

where $<$C$>$ represents the carbon host, and $<\mathrm{CH}_x>$ hydrogen adsorbed in the later.

The overall charge/discharge phenomenon is summarized as follows:

$$<\mathrm{C}> + x\mathrm{H_2O} + x\mathrm{e}^- \leftrightarrow <\mathrm{CH}_x> + x\mathrm{OH}^-$$

In the process of hydrogen insertion into the carbon network, it seems that especially defective domains[34] can adsorb hydrogen more preferably (Fig. 9.14). However, the filling of graphene layers through an intercalation phenomenon is still controversial.

For understanding the mechanism of hydrogen electrosorption, various techniques must be used. Cyclic voltammetry is a well adapted electrochemical technique to elucidate the mechanism and kinetics of reversible hydrogen storage. An example of voltammetry characteristics using a microporous activated carbon cloth from viscose (AC; $S_{\mathrm{BET}} = 1390\ \mathrm{m^2/g}$) in 6 mol/L KOH is shown in Fig. 9.15.

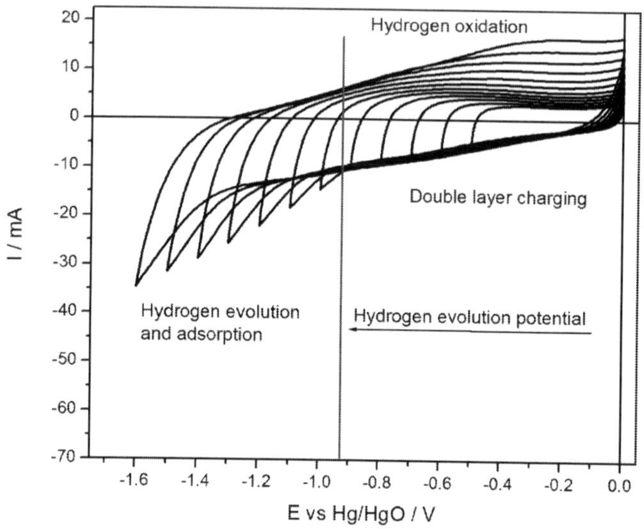

Figure 9.15 Voltammetry characteristics (5 mV/s) of carbon cloth (AC) in 6 mol/L KOH.

The minimum potential is gradually shifted for each cycle toward hydrogen evolution. When the electrode potential is higher than the thermodynamic value that corresponds to water decomposition (the theoretical value of the equilibrium potential in 6 mol/L KOH is −0.874 V *vs.* NHE), the voltammetry curves have the rectangular shape typical of charging the electrical double layer. From the fourth loop, as the potential becomes lower than this value, both the double layer is charged and hydrogen is adsorbed in the pores of carbon. During the anodic scan, the reactions run in an opposite direction and a peak corresponding to the electro-oxidation of adsorbed hydrogen is observed. This pseudo-capacity contributes to the total capacity in addition to the electrical double layer capacity. When the negative potential limit decreases, the anodic current increases, and the corresponding hump shifts toward more positive values of potential. This important polarization between the cathodic and anodic processes indicates that hydrogen electrochemically stored is trapped more energetically than for a classical physisorption and/or that there are important diffusion limitations.

The important polarization (overpotential) between charge/discharge indicates a strong interaction of adsorbed hydrogen with the carbon substrate. As it can be still desorbed in ambient conditions, one must conclude to a weak chemical bonding of hydrogen. The particularity of electrochemical storage compared with the typical adsorption under gas pressure is due to the production of nascent hydrogen through water electroreduction. This very reactive form of hydrogen interacts with the active sites of carbon, being very energetically trapped. From a practical point of view, such a state of hydrogen is particularly attractive, as it is stabilized in the carbon substrate (e.g., the self-discharge is not important and it can be desorbed in ambient conditions). Taking into account that the capacity is quite high (up to 500 mAh/g) and that the process occurs in the negative range of potentials, it proves that nanoporous carbons are very interesting as negative electrode of asymmetric supercapacitors.

For comparison, the voltammetry characteristics (5 mV/s) performed on the multiwalled carbon nanotubes in the same way as for the activated carbon cloth are shown in Fig. 9.16.

The more quantitative method is a galvanostatic charge/discharge (1 A/g and 50 mA/g). The amount of electrosorbed hydrogen can be easily estimated from the charge passed during oxidation (272 mAh/g = 1 wt% of hydrogen). Hydrogen storage ability for AC cloth and nanotubes in alkaline solution is shown in Fig. 9.17.

The amount of stored hydrogen in AC is $1.49\%_{wt}$ H, whereas for nanotubes it is only $0.27\%_{wt}$H because of mesoporosity character of multiwalled carbon nanotubes and lack of microporosity. On observing their electrochemical behavior by cyclic voltammetry, presented in Fig. 9.16, carbon nanotubes reveal extremely different shape of curves in cathodic area, without hysteresis loop between cathodic and anodic polarization. It suggests that hydrogen is evolved but is not being adsorbed on carbon nanotubes. The disturbances in voltammetric curves observed in cathodic area proved that recombination reactions are favored.

The capillary structure of carbon nanotubes could be in favor of electrochemical hydrogen storage, but probably closed ends of these materials generate the diffusion limitation. Electrodes made with

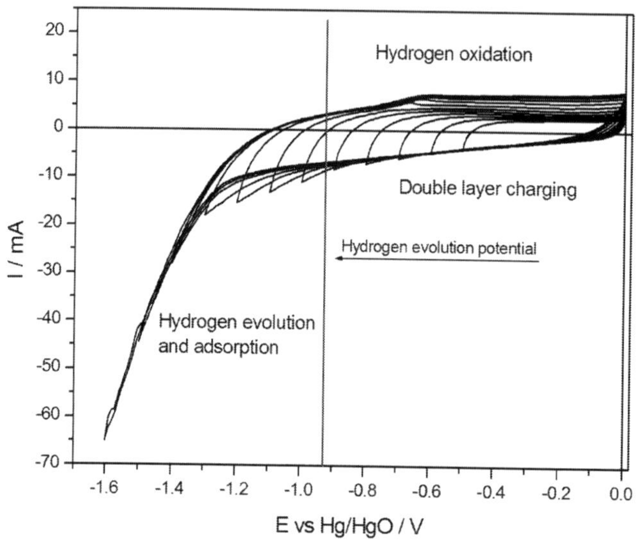

Figure 9.16 Voltammetry characteristics (5 mV/s) of nanotubular material (MWNTs) in 6 mol/L KOH.

high-purified and open SWCNTs by Rajalakshmi *et al.*[36] demonstrate the reversible capacity up to 800 mAh/g corresponding to 2.9 %wt of hydrogen. They also reported that this kind of electrodes reveals electrochemical behavior very similar to metal-hydride electrodes in Ni-MH batteries. Further literature study indicates that also the diameter of carbon nanotubes, especially for MWCNTs, may be the crucial factor.

An investigation made by Zhang *et al.*[37] reveals the difference in electrochemical hydrogen storage of pure multi-walled carbon nanotubes (MWNTs) with different diameters. They concluded that MWNTs with different diameters represented a great discrepancy in the electrochemical hydrogen storage capacity, even tested under the same experimental conditions. This clearly indicates that the tube diameter of CNTs is an important factor that influences its electrochemical hydrogen storage performance. For electrochemical hydrogen storage, short nanotubes with open tips are required to overcome the diffusion limitation. Different approaches have been developed to get shorter and open-ended CNTs. Ball milling appears

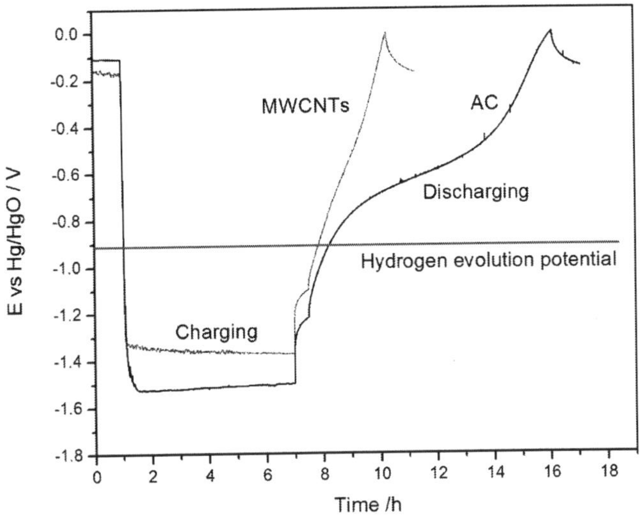

Figure 9.17 Comparison of galvanostatic hydrogen insertion/deinsertion into AC and nanotubes in 6 mol/L KOH.

to be the simplest and cleanest process for cutting and opening CNTs on a large scale without any chemical or thermal treatment. Yi Wang *et al.*[35] applied ball milling in cyclohexanone solution and obtained CNTs with open ends and much shorter than before milling. The results seem to be very interesting because CNTs after milling reveal reversible hydrogen storage capacity of approximately 700 mAh/g. In this work, authors discussed the role and influence of the most important factors, that is, specific surface area and micropore volume for hydrogen storage capacity and concluded that the first one seems to be more crucial. From this point of view, the application of modified carbon nanotubes as hydrogen storage materials is still an open question.

9.6 Carbon Nanotubes as a Support for Catalyst in Fuel Cells

The direct methanol fuel cell is one of the most promising energy sources for the nearest future. As methanol oxidation in acidic medium requires the application of noble metals, generally platinum and/or its binary alloys, an intensive research is devoted to decrease the catalyst loading. The kinetics and total efficiency of the process strongly depend on the catalyst dispersion, which is quite often determined by the support. Some investigations have already been performed with an application of carbon nanotubes as a promising catalyst support for cathodic and anodic reactions in fuel cell.[38–49] The nanotubular texture and/or structure were correlated with an electrochemical behavior during methanol oxidation.[38]

MWNTs served as an attractive support of catalytic particles for methanol electrooxidation in acidic medium. Pt-Ru (50:50) nanoparticles deposited on the different type of nanotubular materials were investigated to select the best support and to estimate the efficiency of CH_3OH oxidation. Pt-Ru supported on NTs graphitized at 2500°C supply the optimal voltage/current characteristics with a maximum current load of 2.7 A/g. Mesoporous network of nanotubes combined with their excellent conducting properties allows a quick transport and easy accessibility of the reagent molecules to the catalytic sites of electrode interface giving a good performance of methanol anode.

Binary Pt-Ru (50:50) catalyst particles from Alfa Aesar Johnson Matthey have been deposited on different kinds of nanotubes. To ensure the same distribution and amount of catalyst and to pronounce only effect of a support, the deposition of catalyst particles from ethanol suspension was applied for all the nanotubular materials. Entangled NTs prepared by catalytic decomposition of acetylene on a $Co_xMg_{(1-x)}O$ solid solution catalyst or stiff nanotubes obtained by propylene decomposition at 800°C using a porous alumina template were studied as a catalyst support. In order to estimate the influence of nanotubes texture on electrocatalytic behavior, some samples were further activated by KOH at 850°C during one hour (KOH:C= 4:1) or annealed at 2500°C for 1 h.

Methanol electro-oxidation in acidic medium (1 mol/L CH_3OH + 1 mol/L H_2SO_4) was investigated by voltammetry method (5 mV/s) and galvanostatic and impedance spectroscopy. For all the nanotubular carbon samples, the nitrogen adsorption/desorption isotherms reveal a mesoporous character (IV-type isotherm) due to the entanglement and/or the presence of the central canal. The

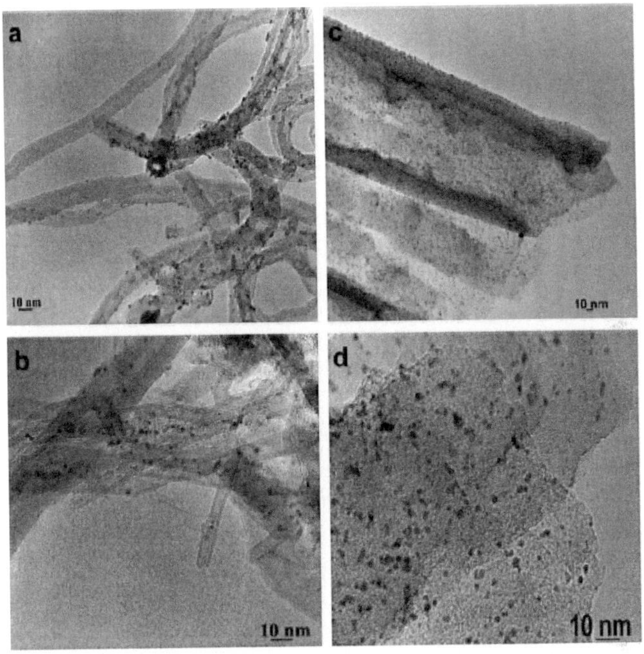

Figure 9.18 TEM images for nanotubes with catalysts Pt/Ru (a) nanotubes, (b) HTT-treated nanotubes, (c) templated carbon nanotubes, and (d) magnification of catalyst particles. Reprinted from Ref. 38, Copyright 2006, with permission from Elsevier.

BET-specific surface area of the $Co_xMg_{(1-x)}O$ catalytically grown nanotubes is 290 $m^{2/}g$ and it diminishes to 159 m^2/g after annealing, whereas KOH activation causes its enhancement to 402 m^2/g. For the nanotubes obtained by the template technique, it reaches only 15 m^2/g. Transmission electron microscopy images of Pt-Ru particles deposited on three types of nanotubes, that is, pristine $Co_xMg_{(1-x)}O$, graphitized, and template, are presented in Fig. 9.18a–c, respectively. The catalyst is well dispersed with a unique particle size of 2–3 nm estimated from Fig. 9.18d presenting magnification of template sample.

The first characteristic feature of catalytic activity for the methanol oxidation is a shift of the rest potential of working electrode into the negative direction. The most remarked shift was observed in the case of graphitized nanotubes with Pt-Ru nanoparticles. In this case, the value of rest potential reaches −0.464 V *vs.* Hg/Hg_2SO_4. Cyclic voltammetry characteristics (scan rate 5 mV/s) for a methanol oxidation in 1 mol/L H_2SO_4 on the Pt-Ru particles deposited on different NTs are shown in Fig. 9.19. A peak current responsible for the partial methanol oxidation was placed between approximately 0.1 V and 0.3 V *vs.* Hg/Hg_2SO_4. The highest values for this peak (approximately 9.5 A/g) were obtained in the case of graphitized NTs, whereas the lowest values of this electrochemical response were measured for KOH-activated nanotubes. The best electrochemical response in the case of graphitized NTs can be interpreted by the optimal conducting properties of these material, that is, during oxidation of methanol, electrons are more easily taken out from the reagent molecules and carried away to the main current collector and then to the electrical circuit. On the contrary, KOH-activated NTs, which are more defected and rich in oxygen, obviously present definitively worse conductivity as well as a negligible current response (max. value of 0.8 Ag^1). Similar voltammetry behavior (Fig. 9.19) was observed for catalytic and templated nanotubes, what could be interpreted

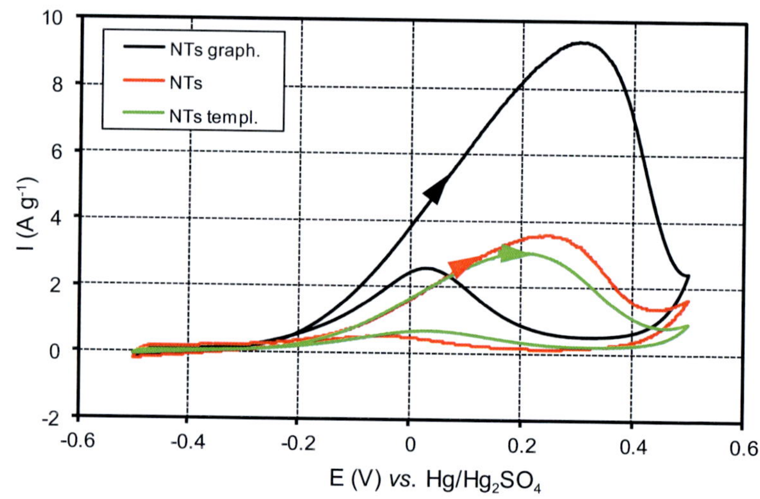

Figure 9.19 Voltammetry characteristics (5 mV/s) of methanol oxidation on Pt/Ru catalyst deposited on three types of NTs. Electrolytic solution: 1 mol/L CH_3OH + 1 mol/L H_2SO_4.

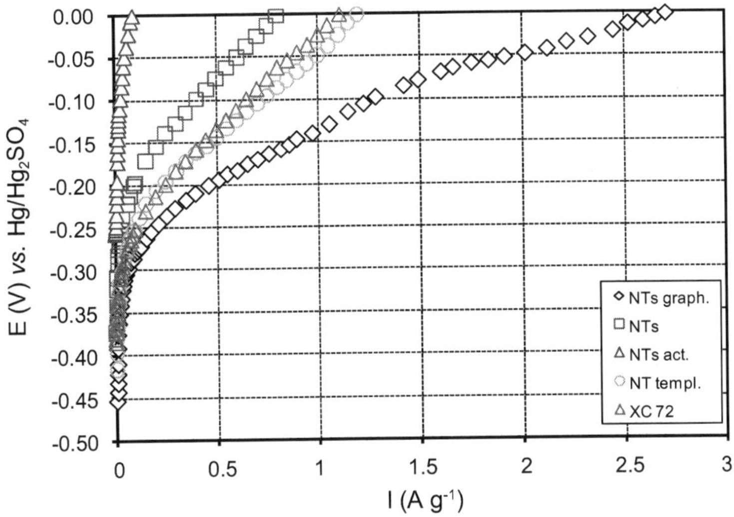

Figure 9.20 Current load/potential characteristics for methanol oxidation on Pt-Ru deposited on various nanotubular carbon materials.

that entangled or straight morphology does not affect significantly methanol oxidation. However, such behavior was remarked only during voltammetry experiments, which are not the most indicative for estimating the methanol anode performance.

The current–potential characteristics (Fig. 9.20), which are the most indicative for practical fuel cell performance, were also investigated to select the optimal and the most advantageous catalyst support from nanotubular material. It is clearly seen that KOH-activated nanotubes with a defected texture, enriched in oxygen and aggravated conductivity, show the worse characteristic, whereas graphitized NTs have the best performance. However, templated nanotubes show definitively better performance than raw catalytic nanotubes, most probably due to the wide central canal that accelerates diffusion of reagents as well as oxidation products.

One can summarize that nanotubular materials fulfil perfectly the conditions of conducting and well accessible mesoporous network for catalytic particles. Pt-Ru particles of 2–3 nm size were equally deposited on different nanotubular supports. The selection of optimal support is crucial for a performance of methanol anode and it was clearly shown that the best results of methanol oxidation are obtained for graphitized carbon nanotubes with deposited catalyst. This high electrocatalytic activity could be attributed to the peculiar texture of nanotubes accessible for reagents, perfect conductivity that improves electron transfer as well as to the specific interaction between catalyst particles and nanotubes.

Open mesopores of carbon nanotubes allow to disperse catalyst more homogeneously that can decrease a cost of fuel cell. Such support can be considered as a very useful component of the membrane DMFC.

9.7 Summary

Carbon nanotubes can serve as a three-dimensional support of materials with pseudocapacitive properties, for example, electrically conducting polymers (PANI, PPy, PEDOT) or transition metal oxides, for example, MnO_2. CNTs play a perfect conducting role in these composites.

In addition due to the high resiliency of carbon nanotubes, the supercapacitor electrodes can be charged/discharged many cycles without mechanical degradation. The exceptional conducting and mechanical properties of MWNTs can be used for preparation of unique carbon/carbon composites, with a comparable texture but variable nitrogen content. Pseudocapacitance effects related to the nitrogen presence in the carbon composite based on nanotubes have been efficiently used in supercapacitor electrodes. The useful effect of nitrogen is connected with faradaic reactions, partly not only due to the presence of pyridinic groups but also because of the presence of quaternary nitrogen, which could be responsible for a significant modification of the density of electronic states. An exceptionally good dynamics of charge was obtained for the composite with approximately 12 wt% of N but only in acidic and alkaline medium. Neutral and organic solutions are not adapted as electrolyte for nitrogenated carbons because of the lack of protons. A suitable amount of nitrogen without any conductivity aggravation is not only able to enhance the capacitance but also to improve the electrodes wettability in aqueous medium. Some trials were undertaken to correlate N content in carbons with the electron density states from molecular quantum calculation. The energy gap between HOMO and LUMO level affected by nitrogen content was correlated with conductivity and capacitance properties of composite.

Reversible electrosorption of hydrogen is an interesting pseudocapacitive contribution during charging/discharging a nanoporous carbon electrode in aqueous electrolyte. Carbon materials of suitable nanotexture, that is, high ultramicroporosity able to adsorb hydrogen by electrochemical way, are perfect electrodes of negative polarity. Carbon nanotubes are not sufficient candidate for H electrosorption, but the modification, for example, cutting, can improve hydrogen storage ability. Developing asymmetric supercapacitors with two different materials for negative and positive electrode should supply high energy and power devices due to the great extension of operating voltage.

Unique mechanical properties of CNTs allow using them as supercapacitor components in the entangled form but also aligned[50] or even stretchable.[51]

Apart from the supercapacitor application, it has been proved that carbon nanotubes can find a potential function as a support of catalyst for methanol oxidation. They are also well adapted as a catalyst support for reduction processes. It can have a great use in fuel cell technology. It is well known that for this target, generally mesoporous carbon material is the most convenient. It has been shown that carbon nanotubes, but especially additionally thermally treated nanotubes at 2600°C, are well adapted as catalyst support. For this application, a good conductivity enables a fast charge transfer, whereas mesopores play an important role for diffusion of reagents to the interface as well as diffusion of products to the electrolyte bulk.

In addition, CNTs can serve as an excellent percolating agent[52] in the electrode materials for many other electrochemical systems, for example, lead-acid or Li-ion batteries.

Acknowledgments

The authors would like to acknowledge the financial support from the Ministry of Science and Education (Poland), grant DS 31-203/2010.

References

1. S. Iijima, *Nature* 354 (1991) 56.
2. M.S. Dresselhaus, G. Dresselhaus, P.C. Eklund, *Science of fullerenes and carbon nanotubes*, Academic Press, San Diego, 1996.
3. B.E. Conway, *Electrochemical supercapacitors – scientific fundamentals and technological applications*, Kluwer Academic/Plenum, New York 1999.
4. R. Kotz, M. Carlen, *Electrochim. Acta* 45 (2000) 2483.
5. E. Frackowiak, F. Béguin, *Carbon* 39 (2001) 937.
6. D. Qu, H. Shi, *J. Power Sources* 74 (1998) 99.
7. E. Frackowiak, *Phys. Chem. Chem. Phys.* 9 (2007) 1774.
8. E. Frackowiak, F. Béguin, *Carbon* 40 (2002) 1775.
9. K. Jurewicz, S. Delpeux, V. Bertagna, F. Béguin, E. Frackowiak, *Chem. Phys. Lett.* 347 (2001) 36.
10. K.H. An, K.K. Jeon, J.K. Heo, S.C. Lim, D.J. Bae, Y.H. Lee, *J. Electrochem. Soc.* 149 (2002) A1058.
11. K. Lota, V. Khomenko, E. Frackowiak, *J. Phys. Chem. Solids*, 65 (2004) 295.
12. V. Khomenko, E. Frackowiak, F. Béguin, *Electrochim. Acta*, 50 (2005) 2499.
13. C. Peng, J. Jin, G.Z. Chen, *Electrochim. Acta* 53 (2007) 525.
14. B. Dong, B.L. He, C.L. Xu, H.L. Li, *Mater Sci. Engineer. B* 143 (2007) 143.
15. E. Frackowiak, F. Béguin, *Recent Advances in Supercapacitors*, ed. V. Gupta Transworld Research Network, Kerala, India 2006; 79–114.
16. C. Meng, C. Liu, S. Fan, *Electrochem. Commun.* 11 (2009) 186.
17. H. Zhang, G. Cao, W. Wang, K. Yuan, B. Xu, W. Zhang, J. Cheng, Y. Yang, *Electrochim. Acta* 54 (2009) 1153.
18. E. Frackowiak, *Dekker encyclopedia of nanoscience and nanotechnology*, Taylor and Francis, New York, (2008) 707–721.
19. H. Zhang, G. Cao, Y. Yang, *Energy Environ Sci.* 2 (2009) 932.
20. V. Khomenko, E. Raymundo-Pinero, E. Frackowiak, F. Béguin, *Appl. Phys. A* 82 (2006) 567.
21. E. Raymundo-Pinero, V. Khomenko, E. Frackowiak, F. Béguin, *J. Electrochem. Soc.*152 (2005) A229.
22. S.B. Ma, K.W. Nam, W.S. Yoon, X.Q. Yang, K.Y. Ahn, K.H. Oh, K.B. Kim, *Electrochem. Commun.* 9 (2007) 2807.
23. J.H. Park, J.M. Ko, O.O. Park, *J. Electrochem. Soc.* 159 (2003) A864.

24. G.X. Wang, B.L. Zhang, Z.L. Yu, M.Z. Qu, *Solid State Ionics* 176 (2005) 1169.
25. Y. Shan, L. Gao, *Mater. Chem. Phys.* 103 (2007) 206.
26. J. Yan, Z. Fan, T. Wei, J. Cheng, B. Shao, K. Wang, L. Song, M. Zhang, *J. Power Sources* 194 (2009). 1202.
27. P.-C. Chen, G. Shen, S. Sukcharoenchoke, C. Zhou, *Appl. Phys. Lett.* 94 (2009).
28. F. Béguin, K. Szostak, G. Lota, E. Frackowiak, *Adv. Mater.* 17 (2005) 2380.
29. T. Liu, T.V. Sreekumar, S. Kumar, R.H. Hauge, R.E. Smalley, *Carbon* 41 (2003) 2440.
30. Ch. Emmenegger, Ph. Mauron, P. Sudan, P. Wenger, V. Hermann, R. Gallay, A. Zuttel, *J. Power Sourc.* 124 (2003) 321.
31. G. Lota, K. Lota, E. Frackowiak, *Electrochem. Commun.* 9 (2007) 1828.
32. E. Frackowiak and F. Béguin, *Carbon* 40 (2002) 1775.
33. K. Jurewicz, E. Frackowiak, F. Béguin *Appl. Phys. A* 78 (2004) 981.
34. D. Qu, *J. Power Sources* 179 (2009) 309.
35. Y. Wang, W. Deng, X. Liu, X. Wang, *Int. J. Hydrogen Energy* 34 (2009) 1437.
36. N. Rajalakshmi, K.S. Dhathathreyan, A. Govindaraj, B.C. Satishkumar, *Electrochim. Acta* 45 (2000) 4511.
37. H.Y. Zhang, X.J. Fu, J.F. Yin, C. Zhou, Y.M. Chen, M.H. Li, *Phys Lett A* 339 (2005) 370.
38. E. Frackowiak, G. Lota, T. Cacciaguerra, F. Béguin, *Electrochem. Commun.* 8 (2006) 129.
39. G. Che, B.B. Lakshmi, C.R. Martin, E.R. Fisher, *Langmuir* 15 (1999) 346.
40. G. Girishkumar, K. Vinodgopal, P.V. Kamat, *J. Phys. Chem. B* 108 (2004) 19960.
41. Z. Liu, J.Y. Lee, W. Chen, M. Han, L.M. Gan, *Langmuir* 20 (2004) 181.
42. W. Li, C. Liang, J. Qiu, W. Zhou, H. Han, Z. Wei, G. Sun, Q. Xin, *Carbon* 40 (2002) 791.
43. B. Rajesh, K.R. Thampi, J. M. Bonard, N. Xanthopoulos, H. J. Mathieu, B. Viswanathan, *J. Phys. Chem. B.*,107 (2003) 2701.
44. Z. He, J. Chen, D. Liu, H. Tang, W. Deng, Y. Kuang, *Mater. Chem. Phys.* 85 (2004) 396.
45. Z. He, J. Chen, D. Liu, H. Zhou, Y. Kuang, *Diamond & Related Materials* 13 (2004) 1764.
46. D.J. Guo, H.L. Li, J. *Electroanal. Chem.* 573 (2004) 197.

47. Z. Liu, X. Lin, J.Y. Lee, W. Zhang, M. Han, L.M. Gan, *Langmuir* 18 (2002) 4054.
48. B. Rajesh, V. Karthik, S. Karthikeyan, K.R. Thampi, J.M. Bonard, B. Viswanathan, *Fuel* 81 (2002) 2177.
49. C. Wang, M. Waje, X. Wang, J.M. Tang, R.C. Haddon, Y. Yan, *Nano Lett.* 4 (2004) 345.
50. C. Du, J. Yeh, N. Pan, *Nanotechnology* 16 (2005) 350.
51. C. Yu, C. Masarapu, J. Rong, B. Wei, H. Jiang, *Adv. Mater.* 21 (2009) 4793.
52. M. Endo, Y.J. Kim, K.C. Park in "*Carbons for Electrochemical Energy Storage and Conversion Systems*" ed. F. Béguin, E. Frackowiak, CRC Press, Boca Raton, Fl, USA, 2010, pp. 469–507.

Chapter 10

Manganese Oxide/Carbon Nanotube Nanocomposites for Electrochemical Energy Storage Applications

Sang-Bok Ma and Kwang-Bum Kim
Department of Materials Science and Engineering, Yonsei University, 134 Shinchon-dong, Seodaemun-gu, Seoul 120-749, Korea
kbkim@yonsei.ac.kr

Nanolayer manganese dioxide was successfully deposited on carbon nanotube by a spontaneous redox reaction between permanganate ion and carbon surface. The synthesis mechanism of the nanolayer manganese dioxide on carbon nanotubes was proposed via the *in situ* monitoring of the solution chemistry. Nanolayer manganese dioxide showed a high specific capacity and high rate capability. Nanosized lithium manganese oxides were also successfully synthesized and well dispersed on entangled carbon nanotubes' web by microwave-hydrothermal. Nanosized lithium manganese oxide showed excellent high rate capability and good structural reversibility. The remarkable electrochemical properties of the nano-architectured material were attributed to shorter lithium diffusion length, good dispersion, large ionic

Nanotechnology in Advanced Electrochemical Power Sources
Edited by S. R. S. Prabaharan and M. S. Michael

ISBN 978-981-4241-43-4 (Hardcover), 978-981-4303-05-7 (eBook)
www.panstanford.com

supply high electrode/electrolyte interfacial area, and high electrical conductivity by introducing the entangled carbon nanotubes' web. This demonstrated that nano-design of electrode materials, having high electrochemical utilization, high electronic conductivity, and large ion supply, is crucial to enable the electrode materials to have high specific capacity and high rate capability. This synthetic strategy opens a new avenue for the design of nanoelectrode materials on the entangled carbon nanotubes' web in an effective and facile way.

10.1 Introduction

Electrochemical capacitors (ECs) are energy-storage devices with a specific energy and specific power lying somewhere between batteries and conventional dielectric capacitors.[1] The applications of ECs include electric vehicles (EVs), uninterruptible power supplies (UPS), DC power systems, renewable energy, and mobile devices. ECs are classified into two areas according to their charge storage principle.[2] The first is an electrical double layer capacitor (EDLC) using activated carbon (AC) with high surface area and the other is a pseudocapacitor using metal oxides and conducting polymers.

The pseudocapacitance of a pseudocapacitor is the result of a capacitive relationship between the level of charge acceptance and the change in potential from a Faradaic redox reaction between the electrode material and electrolyte.[1] Among the various transition metal oxide materials used in pseudocapacitors, amorphous and hydrated ruthenium oxide has been reported to show a remarkably high specific capacitance ($\sim$720 F/g) compared with other oxides.[3–5] However, its commercial use is limited by its expense. Therefore, considerable effort has been devoted to identifying alternative and inexpensive metal oxide electrode materials with acceptable electrochemical properties. Manganese dioxide (MnO_2) is one of the potential electrode materials for pseudocapacitors in terms of its electrochemical behavior, low cost, and environmental compatibility.[6–11] The pseudocapacitive properties of manganese oxide have been investigated mainly in aqueous solutions.[6–26]

Carbon nanotubes (CNTs) have been studied extensively as electrode materials (100 ∼ 200 F/g) for ECs,[27,28] as additives for improving the electrode performance of metal oxides,[12–16,29] and as deposition substrates for metal oxides in pseudocapacitors.[17,18,30,31] They have good chemical stability, good conductivity, and a large surface area. In addition, CNTs are strongly entangled, providing a network of open mesopores.

Because the pseudocapacitive reaction of MnO_2 is known to be a surface reaction, only the surface or a very thin surface layer of the oxide can participate in the pseudocapacitive reaction. Recent studies of the composite electrodes made from manganese oxide and carbonaceous materials have reported a high specific capacitance and rate capability of manganese oxide, particularly when a small amount of manganese oxide is dispersed uniformly over conductive and porous carbonaceous materials with a high surface area.[12–26] Therefore, in the synthesis of MnO_2/CNT nanocomposites, it is ideal to deposit a very thin layer (∼ nm) of MnO_2 onto CNTs with a control of the coating layer thickness, surface coverage, and phase in order to improve the electrochemical utilization of MnO_2. An increase in the effective interfacial area between MnO_2 and an electrolyte can increase the electrochemical utilization of MnO_2 in the MnO_2/CNT nanocomposites. Greater chemical contact and increased contact area between MnO_2 and CNT can improve the electric conductivity of the electrode due to the high electric conductivity of CNT.

In this chapter, in order to make MnO_2 more attractive for pseudocapacitor electrode materials, MnO_2 was coated spontaneously on CNTs by the simple immersion of the CNTs into a $KMnO_4$ aqueous solution. The synthesis mechanism of the heterogeneous nucleation of MnO_2 on the CNTs was investigated by *in situ* monitoring of the solution chemistry.

Early in the MnO_2 introduction as an electrode material for ECs, most of the research is based on aqueous electrolytes. MnO_2 exhibit good electrochemical reversibility and high rate capability in aqueous electrolyte systems. However, ECs based on aqueous electrolytes suffer from a low specific energy (Wh/kg) because of a small potential window of approximately 1.0 V. Because there is a growing demand for ECs with high specific energy for applications such as EVs, UPS, DC power systems, and mobile devices in recent years, the specific energy of ECs has become a very important research subject. Recently, a few studies on ECs based on organic electrolytes have been reported in literatures.[32,33] Organic electrolytes have an advantage of a wide operating potential window, enabling ECs to deliver a higher specific energy. This chapter showed the excellent electrochemical properties of the MnO_2/CNT nanocomposites in an organic electrolyte for pseudocapacitor applications.

Lithium metal oxides such as $LiCoO_2$ and $LiMn_2O_4$ have also been considered as good candidates for positive electrode materials of hybrid ECs (HECs), which are composed of a battery-type electrode (faradaic reaction) and a capacitor-type electrode (nonfaradic reaction). HECs require both high rate capability and high capacity for positive electrode materials.

Nanosized lithium metal oxide is expected to improve the high rate capability of lithium metal oxides because of an increase in the effective interfacial area between a nanosized lithium metal oxide and an electrolyte, and its shorter lithium diffusion length during charge/discharge.[34] As nanosized materials tend to be agglomerated, good dispersion as well as improving conductivity of an electrode composed of nanosized oxides is very important to achieve high rate capability.[35]

In this chapter, CNTs were used to disperse nanosized lithium metal oxide and connect the nanoparticles along one-dimensional conduction path due to their chemical stability, good conductivity, and large surface area. This chapter reports on the synthesis of lithium manganese oxide (LMO) nanoparticles dispersed along CNTs (LMO/CNT nanocomposite) using a microwave-assisted hydrothermal treatment of the MnO_2 coated CNTs in an aqueous LiOH solution. The LMO/CNT nanocomposite showed an excellent high rate capability and good structural reversibility for energy storage applications. The charge storage mechanism and the structural reversibility of the LMO were examined by *in situ* and *ex situ* synchrotron technique.

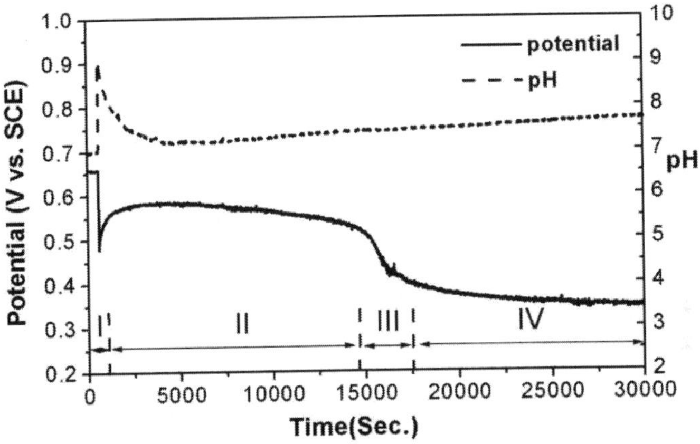

Figure 10.1 Electrode potential (E) and pH *vs.* time curves measured during the reduction of MnO_4 ions in the 200 ml aqueous solution of 0.1 M $KMnO_4$ containing 1.0 g CNT at 70°C.

10.2 Synthesis and Characterization of MnO_2/CNT Nanocomposites

10.2.1 *In situ Monitoring of Reduction Potential and pH*

Figure 10.1 shows the change in the E and pH of the solution with time during the synthesis of MnO_2/CNT nanocomposites in the 200 ml aqueous solution of 0.1 M $KMnO_4$ containing 1.0 g CNT at 70°C. Considering the changes of E and pH during the synthesis, Fig. 10.1 can be divided into four stages.

When CNTs were added to the solution, a rapid drop in the electrode potential and an increase in pH were observed in the first stage (stage I). CNTs can be envisaged as roll-up graphene layer sheet, regarding their molecular structure.[36] The basic sites, which consist of delocalized π electrons, on the surface of the CNTs are responsible for the rise in pH in the first stage, as they can act as Lewis bases capable of complexing protons to its π structure.[37]

$$C_\pi(\text{CNT}) + 2H_2O \rightarrow C_\pi H_3O^+ + OH^- \qquad (10.1)$$

where $C\pi$ is the idealized graphized surface on CNT having maximum π electrons.

The E of the solution for the reduction of MnO_4^- ions to MnO_2 is a function of pH, as shown in the following reactions.[38] According to Eq. 10.2, the MnO_4^- ion needs both protons and electrons for its reduction to MnO_2,

$$MnO_4^- + 4H^+ + 3e^- \rightarrow MnO_2 + 2H_2O \qquad (10.2)$$

$$E = E^0 + \frac{RT}{nF} \ln \frac{a_{MnO_4^-} \cdot a_{H^+}^4}{a_{MnO_2} \cdot a_{H_2O}^2} \qquad (10.3)$$

where E is the electrode potential, E^0 is the standard potential, R is the gas constant, T is the absolute temperature, F is the Faraday constant, n are moles of electron participating in a redox reaction, and a_x is the activity of x species. The sharp decrease in E can be attributed to the sharp rise in pH of the $KMnO_4$ solution caused by the CNT addition.

Right after the rapid drop in the E and the increase in pH in the first stage, protons, adsorbed at basic sites on the CNT surface during the first stage, are expected to be desorbed as MnO_4^- ions were absorbed, thereby producing a heterogeneous MnO_2 film on the CNTs. Protons desorbed from the surface of the CNTs contributed to the slight decrease in pH and the slight increase in E seen during the initial second stage. It should be noted that the pH stayed relatively constant after the initial decrease in the second stage. This is possibly related to the reduction reaction of MnO_4^- ions, as this reaction consumes protons from the CNT surface.

The second stage, which is characterized by a potential plateau with respect to time, represents the reduction of MnO_4^- ions to MnO_2 by the CNTs. The potential plateau in the second stage corresponds to the reduction potential of MnO_4^- ions in the aqueous solution containing the CNTs. At the end of the plateau, that is, when all the MnO_4^- ions are reduced, E shifts downward significantly during the third stage (stage III) and is indicative of the complete reduction of MnO_4^- ions.

In the fourth stage (stage IV), a slight decrease in E and small increase in pH is observed. The measured E corresponds to the equilibrium in an aqueous solution after the reaction for given electrochemical cells. No further reaction is followed.

10.2.2 *UV-VIS Analysis*

Figure 10.2 shows the changes in the concentration of MnO_4^- ions after the addition of 1.0 g CNT into 0.01 M $KMnO_4$ 200 ml solution at 50°C. After the addition of CNT into $KMnO_4$ solution, a small amount of the solution containing MnO_4^- ions, MnO_2, and CNT was sampled at regular intervals and cooled rapidly to 4°C to prevent any further reaction. UV-VIS spectroscopy was used to measure absorption spectrum for the MnO_4^- ions in order to investigate the change in the concentration of MnO_4^- ions in the sample solutions. For the UV-VIS spectroscopy measurements, it should be noted that the reduction reaction of MnO_4^- ions to MnO_2 by the CNTs was performed at the low $KMnO_4$ concentration of 0.01 M and the low reaction temperature of 50°C to quench the reduction reaction effectively in the sampled solution during the preparation step for the UV-VIS spectroscopy.

Following the calibration of UV-VIS spectra with standard $KMnO_4$ solutions, the concentration of MnO_4^- ions in the sampled solutions was calculated and plotted against the reaction time in Fig. 10.2. The change in E of the solution is presented in Fig. 10.2 for ease of comparison. The concentration of MnO_4^- ions decreased monotonously with the reaction time. It should be noted that the time for the complete removal of MnO_4^- ions from the solution coincides exactly with the time for the negative shift of E in the third stage at around 150 min in Fig. 10.2. This confirms that the E plateau in the second stage in Fig. 10.2 represents the reduction of MnO_4^- ions to MnO_2 by the CNTs.

10.2.3 *Effect of pH*

Figure 10.3 shows the changes of E and pH in 200 ml solutions of 0.1 M $KMnO_4$ at different initial pHs containing 1.0 g CNT at 70°C. Before the CNTs were added to the solution, the solution pH was adjusted by 0.01 M HCl and the corresponding initial solution E was 0.66, 0.96, and 1.12 for pH 7, 2.5, and 1, respectively, as shown in the inset of Fig. 10.3(a) and (b). The aqueous $KMnO_4$ solutions with the different pHs have different E values because E is a function of the solution pH according to Eq. 10.3.

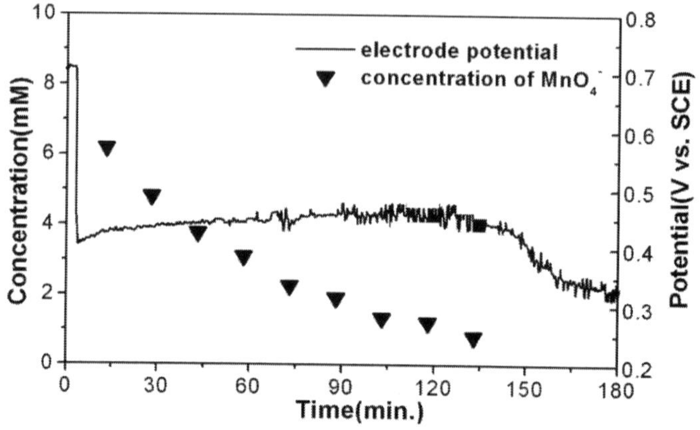

Figure 10.2 Concentration of MnO_4 ions and electrode potential (E) *vs.* time during the synthesis of MnO_2/CNT nanocomposite in the 200 ml aqueous solution of 0.01 M $KMnO_4$ containing 1.0 g CNT at 50°C.

After the CNTs were added to the solution, as shown in the inset of Fig. 10.3(a) and (b), pH increased in all the solutions due to the adsorption of protons on the surface of the CNTs, which caused the rapid drop in E. For the solutions of initial pH 2.5 and pH 7, the solution pH instantly jumped to pH 9 and pH 8 immediately after the addition of CNT, respectively, and then gradually decreased to pH around 7 over time. For the solutions of the initial pH 1, however, their pH increased gradually to pH 7 over time without an imminent sharp jump in pH after the addition of CNT because of the very high proton concentration in the initial solution.

As expected from the pH change in Fig. 10.3(b) Eq. 10.3, the change in the solution E of the initial pH 2.5, 7, and 1 over time can be divided into four stages during the synthesis of MnO_2/CNT nanocomposites. For the solution of the initial pH 1, however, the solution pH increased without a peak and gradually approached pH around 7 in the later stage, although correspondingly, E sharply decreased immediately after the addition of CNT without a dip in the first stage and approached the reduction potential in the second stage. Different initial pHs of the solutions did not affect the overall change in the solution E and pH with time during the reduction of MnO_4^- ions to MnO_2 by CNT.

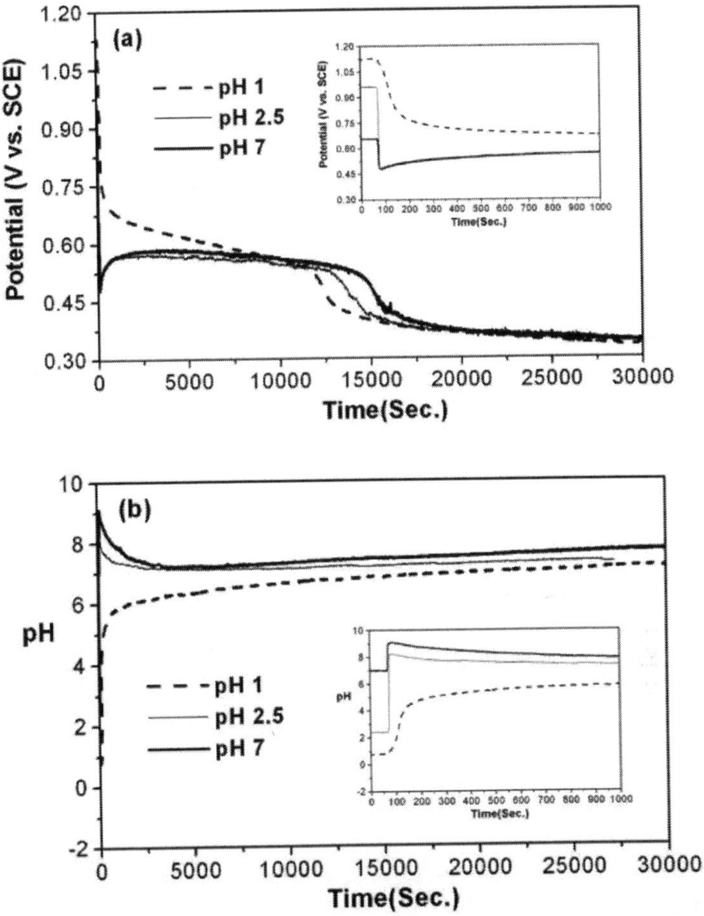

Figure 10.3 (a) E *vs.* time curves and (b) pH *vs.* time curves measured during the reduction of MnO_4 ions by 1.0 g CNT in the 200 ml aqueous solution of 0.1 M $KMnO_4$ at 70°C under different pH conditions of the initial solution.

It should be pointed out that the reduction reaction time between the moment of CNT addition and the time for the negative shift of E in the third stage, that is, the time required for the complete removal of MnO_4^- ions from the solutions, decreased with decreasing pH. This indicates that the reduction reaction rate increases with decreasing solution pH. As more protons in the solution become available for the given concentration of MnO_4^- ions with decreasing solution pH, an increase in the reaction rate for the reduction of MnO_4^- ions to MnO_2 on CNT is expected with a decrease in pH of the initial solution.

10.2.4 *CNT as a Reducing Agent*

Choi *et al.* reported the spontaneous reduction of metal ions to metallic form on the sidewalls of the CNTs in aqueous solutions of noble metal ions.[39] They reported that Au and Pt could be deposited spontaneously on the sidewall of the CNTs by a direct redox reaction between the CNTs and metal ions via the immersion of the CNTs in $HAuCl_4$ (Au^{3+}) and Na_2PtCl_4 (Pt^{2+}) solutions, respectively. They explained that the principle for the spontaneous reduction of the metal ions on the sidewalls of the CNTs is the difference in the reduction potential between the CNTs and the metal ions. Similarly, the spontaneous formation of MnO_2 from MnO_4^- ions on the CNT sidewall may also be explained by the difference in the reduction potential between the CNT and MnO_4^- ions. The work function of the CNTs was determined to be nearly 5 eV. The Fermi level of the CNTs is approximately +0.5 V above the potential of the standard hydrogen electrode (SHE), which is well above the reduction potential of MnO_4^- ions, which is +1.692 V (*vs.* SHE). The relative potential levels may explain the spontaneous electron transfer from the CNTs to the MnO_4^- ions.

MnO_2 is known to be thermodynamically the most stable form of manganese species in an aqueous solution containing MnO_4^- ions. MnO_4^- ions, therefore, tend to oxidize water with the concurrent evolution of oxygen and spontaneous precipitation of MnO_2.[40] Although spontaneous, the reduction reaction of MnO_4^- ions by water is very slow kinetically. Two identical solutions of 0.01 M $KMnO_4$, one with CNTs and the other without CNTs, were prepared and allowed to sit at room temperature without further treatment and observed for the formation of MnO_2. A few days later, the solution with the CNTs became clearer from the original deep purple, which indicates that MnO_4^- ions were reduced to MnO_2 by the CNTs. However, no visible change in color was observed in the solution without the CNTs. It is clear that the CNTs dispersed in an aqueous solution containing MnO_4^- ions act as a reducing agent for MnO_4^- ions and promote the reduction process even at room temperature.

Previous approaches to the synthesis of metal and metal oxide nanoparticles on the CNTs have included various methods such as physical evaporation, electrodeposition, solid state reactions, polyol process, use of interlinkers, spontaneous reduction by CNTs, and other chemical routes. In our approach, MnO_2 was deposited on the CNTs via simple immersion of the CNTs into an aqueous $KMnO_4$ solution without any addition of chemicals such as reducing agent, surfactant, or alcohol that could reduce MnO_4^- ions to MnO_2. In the present study, the CNTs acted as a reducing agent as well as a substrate for the spontaneous deposition of MnO_2 in an aqueous $KMnO_4$ solution. Hence, this process is the simplest method for synthesizing MnO_2 on CNTs. It also provides a useful approach for coating other potential metal oxide materials on CNTs using the relative potential difference.

10.2.5 *Morphology of MnO_2 on CNT*

Figure 10.4 shows SEM images of the CNTs before and after MnO_2 deposition in the 200 ml aqueous solution of 0.1 M $KMnO_4$ with different initial pHs containing 1.0 g CNT at 70°C. Each CNT particle is around 20 ~ 30 nm in size, as shown in Fig. 10.4(a).

After MnO_2 deposition, the CNT particles became a little thicker with size increased to approximately 30 ~ 50 nm. This indicates that the CNTs acted as a substrate for the heterogeneous precipitation of a very thin layer of MnO_2, as shown in Fig. 10.4b–d. The roughness of the MnO_2 deposit on the CNTs increased with decreasing pH of the initial solution. MnO_2 deposited on the CNTs formed in the solution with a pH of 1 [Fig. 10.4(d)] yielded nano-bar MnO_2 while it forms a thin film in a solution having a solution pH of 7 [Fig. 10.4(b)]. This suggests that the reduction rate of MnO_4^- ions to MnO_2 by CNT might affect the morphology of the MnO_2 deposit on the CNTs. As shown in Fig. 10.3(a), the time required for the complete removal of MnO_4^- ions from the solutions decreased with decreasing pH, that is, the reduction reaction rate increased with decreasing solution pH. Therefore, it was observed that the increased reduction rate of

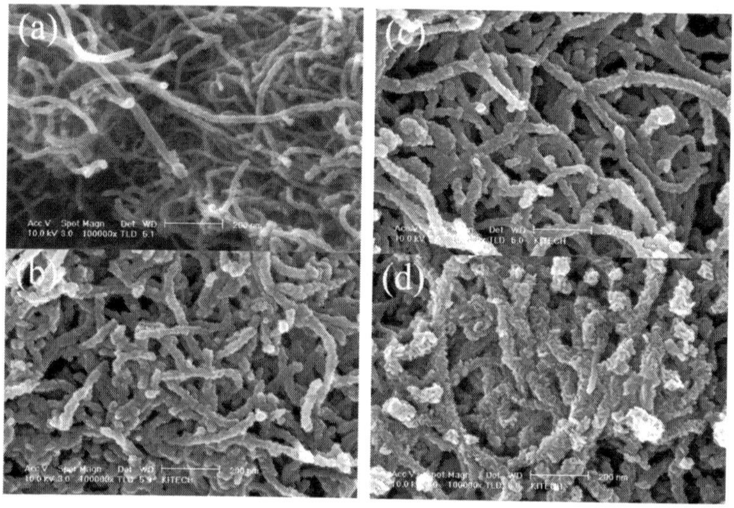

Figure 10.4 SEM images of (a) pristine CNT and MnO_2/CNT nanocomposite prepared in the 200 ml aqueous solution of 0.1 M $KMnO_4$ containing 1.0 g CNT at 70°C under (b) pH 7; (c) pH 2.5; (d) pH 1 of the initial solution.

MnO_4^- ions to MnO_2 by CNT with decreasing initial pH of the solution roughened the morphology of MnO_2.

10.2.6 *Phase and Structure of MnO_2 on CNT*

Figure 10.5 shows the XRD patterns for the CNTs and for MnO_2 on the CNTs prepared in the 200 ml aqueous solution of 0.1 M $KMnO_4$ containing 1.0 g CNT at 70°C with different initial pHs adjusted by 0.01 M HCl, as well as the XRD peaks of the reference oxide, birnessite-type MnO_2 (JCPDS 42-1317). As shown in Fig. 10.5(b), there are three broad peaks at 2 θ of around 12°, 37°, and 66°, in addition to those of the CNTs. These three peaks can be indexed to birnessite-type MnO_2 including an amorphous phase.[41,42] This is consistent with the report by R. M. McKenzie[43] on the birnessite-type MnO_2 synthesized by the dropwise addition of hydrochloric acid to a boiling solution of $KMnO_4$. However, it is difficult to analyze the structure of MnO_2 in detail using XRPD patterns due to their broad and weak peaks.

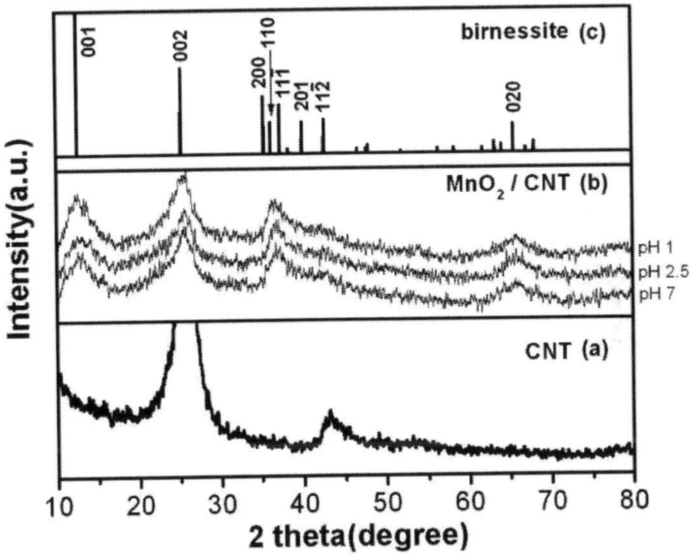

Figure 10.5 X-ray diffraction (XRD) patterns of (a) CNT and (b) MnO_2/CNT nanocomposite prepared in the 200 ml aqueous solution of 0.1 M $KMnO_4$ containing 1.0 g CNT at 70°C, and (c) XRD peaks of birnessite-type MnO_2 (JCPDS No. 42-1317).

Figure 10.6 shows the Raman spectra of MnO_2/CNT nanocomposites synthesized by the reduction of MnO_4^- ions to MnO_2 in the 200 ml aqueous solution of 0.1 M $KMnO_4$ containing 1.0 g CNT at 70°C corresponding to the different initial solution pHs of 7, 2.5, and 1.

Three major features for MnO_2 can be recognized at 500, 575, and 640 cm^{-1}. The two high wave number bands are dominant in all of the spectra, although the bands in the low-frequency region appear with a significantly weaker intensity. The Raman band at 640 cm^{-1} can be recognized as the symmetric stretching vibration (Mn–O) of the MnO_6 groups. The band located at 575 cm^{-1} is usually attributed to the (Mn–O) stretching vibration in the basal plane of MnO_6 sheet.[44,45] Figure 10.6 shows that all the reaction products have the Raman spectra of the birnessite-type MnO_2. Comparison of the Raman spectra of MnO_2 synthesized under various pH conditions in this study indicates that the pH of the initial

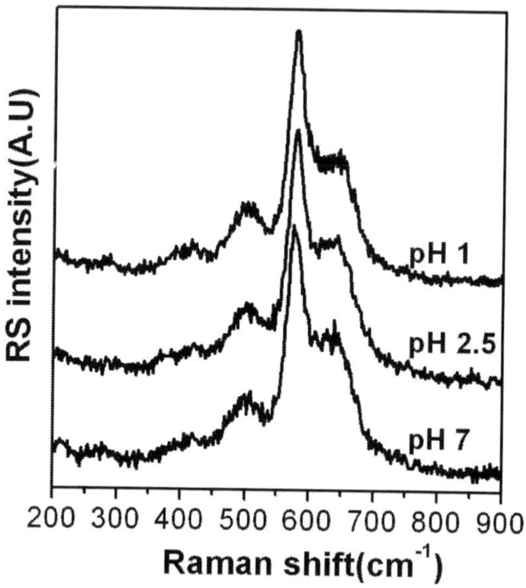

Figure 10.6 Raman spectra of MnO_2/CNT nanocomposite synthesized by the reduction of MnO_4 ions to MnO_2 in the 200 ml aqueous solution of 0.1 M $KMnO_4$ containing 1.0 g CNT at 70°C under different pH conditions of the initial solution.

solution had no influence on the phase and the short range ordering of MnO_2.

10.3 Electrochemical Properties of MnO_2/CNT Nanocomposites

10.3.1 *Electrochemical Properties of MnO_2 Coated onto CNT*

Figure 10.7 shows the first discharge curves of the MnO_2/CNT nanocomposite prepared in the solution at pH 7. This electrode had AB as the conducting agent. The charge and discharge cycles were performed at current densities of 1, 5, and 10 A/g, wherein the weight was based on the total electrode weight, including the electroactive material, conducting agent, and binder. As shown in Fig. 10.7, the first discharge curves of the birnessite-type MnO_2/CNT

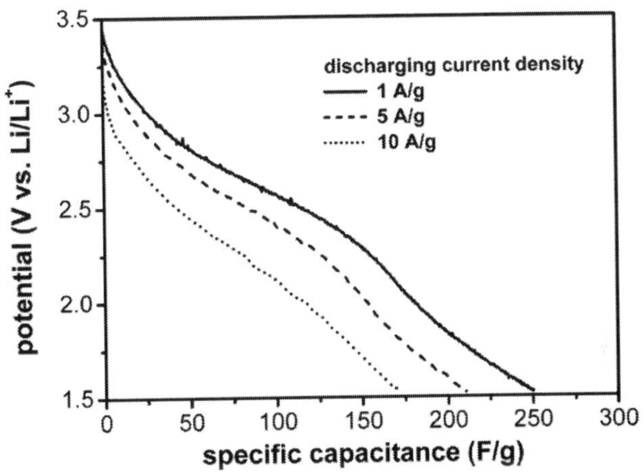

Figure 10.7 Initial discharge profiles of the MnO_2/CNT nanocomposite at different current densities, using AB as the conducting agent in 1 M $LiClO_4$ in PC.

composite showed no obvious plateau, indicating a typical capacitive behavior. The specific capacitance of the electrode was calculated from the discharge profile using the following equation:

$$C = I\Delta t/V \tag{10.4}$$

where C (F/g) is the specific capacitance of the electrode, Δt (s) is the discharge time, V (V) is the potential range from the end of charge to the end of discharge, and I (A/g) is the applied current density based on the total electrode material.

The initial specific capacitance of the total electrode material, including the electroactive material, conducting agent, and binder, was 250 F/g at a current density of 1 A/g, which is equivalent to 139 mAh/g. The specific capacitance per unit weight of the MnO_2/CNT composite was 370 F/g, which is equivalent to 207 mAh/g. Kawaoka *et al.*[26] reported the synthesis of MnO_2/Ketjen Black (KB) with a specific capacity of 165 mAh/g. The specific capacitance of the MnO_2 in the MnO_2/CNT composite was 580 F/g. This can be converted to 320 mAh/g for MnO_2 because the electrode contains 43 wt.% of MnO_2. It should be noted that the specific capacitance of 580 F/g was measured at a high discharge rate of 1 A/g in 1 M $LiClO_4$ in PC. The theoretical specific capacity can be calculated from the following reaction:

$$MnO_2 + xLi^+ + xe^- \rightarrow Li_xMnO_2 \quad (10.5)$$

where x is the number of moles of Li ions and electrons participating in the reaction.

When 1 mole of Li ions and electrons participate in the reaction, the Mn(IV) in MnO_2 is reduced to Mn(III), and the theoretical specific capacity of MnO_2 is 308 mAh/g. Xu *et al.*[46] reported a specific capacity of 436 mAh/g for amorphous MnO_2 between 1.5 and 3.5 V (*vs.* Li/Li^+) in 1 M $LiClO_4$ in PC at a low current density of 20 $\mu A/cm^2$, which corresponds to 1.63 moles of Li^+ per mole of MnO_2. The specific capacitances of 211 and 171 F/g were achieved even at higher discharge current densities of 5 and 10 A/g, respectively, which indicates the very good high rate capability of the MnO_2/CNT nanocomposite. The electrochemical properties of MnO_2 on the CNTs were attributed to the nanometer-scale coating of MnO_2 on the CNTs, the high interfacial area between the MnO_2 and electrolytic solution, and the interconnected pores in a three-dimensional entangled structure of CNTs.

10.3.2 *Effect of Initial Solution pH*

Figure 10.8 shows the change in the specific capacitance of the MnO_2/CNT nanocomposites prepared in $KMnO_4$ solutions at different initial pHs with initial cycling. As shown in Fig. 10.8(a), the initial specific capacitance of the MnO_2/CNT nanocomposite was 250 F/g at 1 A/g for the solution at pH 7 and decreased slightly as the initial pH of the $KMnO_4$ solution was changed from 7 to 1. The specific capacitance of all three MnO_2/CNT nanocomposites approached 150 F/g at the tenth cycle regardless of the initial pH of the solution used for their preparation. However, at a high current density of 10 A/g, each of the MnO_2/CNT nanocomposites showed an initial specific capacitance of approximately 171 F/g, and an identical decrease in specific capacitance during cycling regardless of the initial pH, as shown in Fig. 10.8(b).

During discharge of the MnO_2/CNT nanocomposite electrode, the Mn(IV) in MnO_2 was reduced as lithium ions became intercalated into MnO_2. The diffusion length of lithium ions in MnO_2 can be

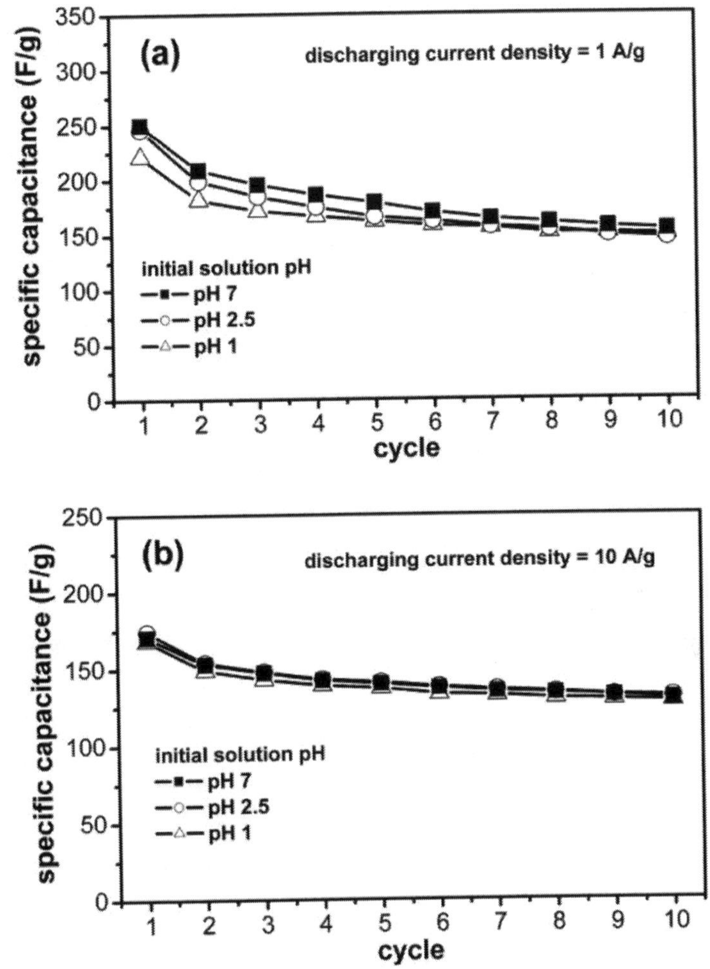

Figure 10.8 Change in the specific capacitance of the MnO_2/CNT nanocomposite during cycling at (a) 1 A/g and (b) 10 A/g, using AB as a conducting agent.

determined using the following equation:[25]

$$L = (D\,\Delta t)^{1/2} \tag{10.6}$$

where L (cm) is the lithium ion diffusion length, D (cm^2/s) is the solid-state diffusion coefficient of lithium in $Li_xMn_2O_4$, and Δt (s) is the discharging time.

The calculated lithium diffusion length was approximately 30 nm considering the solid-state diffusion coefficient[47] of lithium in $Li_xMn_2O_4$ of 2.8×10^{-13} cm^2/s and the discharging time of 30 s at 10 A/g. The SEM images, which were shown in Fig. 10.4, show that the thickness of MnO_2 on the CNTs was less than the lithium diffusion length of 30 nm, considering the surface roughness of MnO_2 on the CNT prepared in solutions at different pHs. This suggests that all the MnO_2 coated onto the CNTs in the three different MnO_2/CNT nanocomposites can participate in the electrochemical reaction, which leads to an identical discharge behavior, as shown in Fig. 10.8. In addition, the surface roughness of MnO_2 on the CNTs was not large enough to change the electrochemical performance of the MnO_2/CNT nanocomposite.

10.3.3 *Effect of CNT as a Conducting Agent*

When AB was used as the conducting agent for the MnO_2/CNT nanocomposite electrode, the initial specific capacitance was approximately 250 F/g at a current density 1 A/g, as shown in Fig. 10.9.

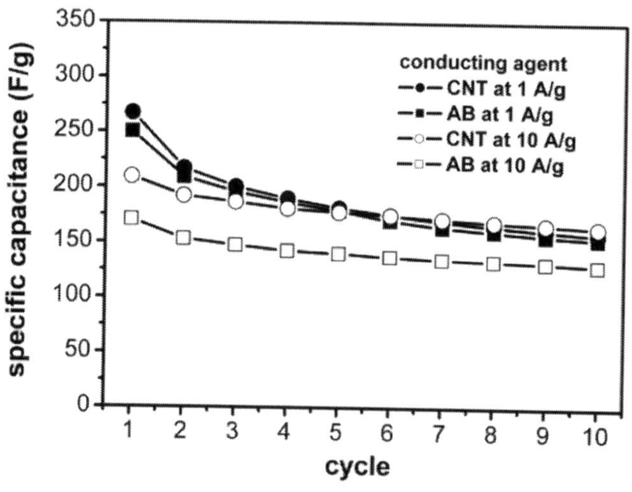

Figure 10.9 Change in the specific capacitance of the MnO_2/CNT nanocomposite during cycling using either AB or CNT as a conducting agent at 1 and 10 A/g.

However, at 10 A/g, the specific capacitance decreased due to an increase in the ohmic drop with increasing current density. In order to improve the high rate capability of the MnO_2/CNT nanocomposite electrode, the AB was replaced with CNT as the conducting agent in the preparation of the MnO_2/CNT nanocomposite electrode.

In Fig. 10.9, the MnO_2/CNT nanocomposite electrodes with CNT and AB as the conducting agent, respectively, showed a similar electrochemical behavior at 1 A/g. However, the initial specific capacitance of the MnO_2/CNT nanocomposite electrode was 210 F/g at 10 A/g for the CNTs, while it was 170 F/g for AB. Furthermore, the MnO_2/CNT nanocomposite electrode with CNT showed a higher specific capacitance than the MnO_2/CNT nanocomposite electrode with AB during cycling, which indicates that the CNTs acted more effectively in forming a conduction network in the composite electrode. This is because CNTs, as a conducting agent in the composite, have a long conduction path on the μm scale and provide many contact points with the MnO_2/CNT nanocomposites. This can further enhance the mesoporosity formed by the MnO_2/CNT nanocomposite and allow the easy access of ions.[48]

10.3.4 *Effect of Heat Treatment of MnO_2/CNT Nanocomposite*

The specific capacitance of MnO_2/CNT nanocomposite decreased from 250 F/g to 160 F/g during cycling at 1 A/g irrespective of the conducting agent used (CNT or AB), as shown in Fig. 10.9. The degradation of MnO_2 results from changes in the composition and crystal structure of MnO_2.[49] In this study, the MnO_2/CNT nanocomposite was heat-treated at 200°C for 24 h in a vacuum to improve the cyclability of the MnO_2/CNT nanocomposite during the charge/discharge cycles. Figure 10.10 shows the discharge curves of the as-prepared and heat-treated MnO_2/CNT nanocomposite electrodes with the AB conducting agent during cycling at 1 A/g. The discharge capacitance of the as-prepared MnO_2/CNT nanocomposite decreased from 209 to 154 F/g after 10 cycles, indicating the rapid degradation of MnO_2, as shown in Fig. 10.10(a). However, there was a slight change in the shape of the discharge curves with the heat-treated MnO_2/CNT nanocomposite during cycling,

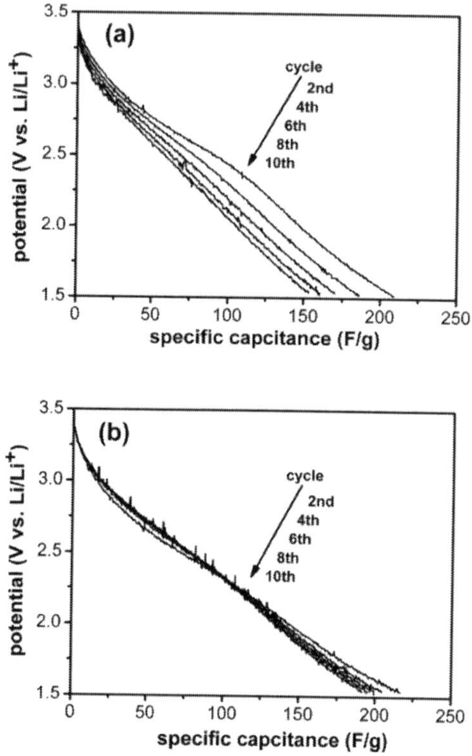

Figure 10.10 Discharge profiles of (a) as-prepared and (b) heat-treated MnO_2/CNT nanocomposite during cycling.

as shown in Fig. 10.10(b). The discharge capacitance of the heat-treated MnO_2/CNT nanocomposite decreased slightly from 217 F/g to 191 F/g over 10 cycles. The improved cyclability of the heat-treated MnO_2/CNT nanocomposite over the as-prepared MnO_2/CNT nanocomposite indicates its structural and chemical reversibility during repeated charge and discharge.

Leroux *et al.*[50] reported the electrochemical behavior of various layered manganese oxides. They reported that interlayer water played a major role in the electrochemical behavior of manganese oxides. In their study, dehydrated manganese oxides at T greater than 200°C showed better retention over 10 cycles and a larger Li^+ diffusion coefficient because of the decreased level of interlayer water. It is believed that, in this study, the improvement in cyclability by heat treating the MnO_2/CNT nanocomposite was mainly due to the dehydration of MnO_2 by heat treatment.[50,51]

In order to determine the effect of heat treatment on the structural stability of the MnO_2/CNT nanocomposite, a series of normalized *in situ* Mn K-edge XANES spectra were measured during the charge/discharge cycles at a low current density of 44 mA/g. Figure 10.11 shows the *in situ* Mn K-edge XANES spectra recorded after the as-prepared and heat-treated electrodes were charged to 4.0 V (*vs.* Li/Li^+) and discharged to 1.5 V (*vs.* Li/Li^+) during the cycles. In all XANES spectra, there was a weak absorption peak (pre-edge) at nearly 6542 eV and a strong absorption peak at nearly 6560 eV. The weak pre-edge absorption was attributed to the electric dipole forbidden transition of a 1s electron to an unoccupied 3d orbital, which is partially allowed because of the electric quadrupole coupling and/or 3d-4p orbital mixing arising from the noncentrosymmetric environment of the slightly distorted MnO_6 octahedral framework.

The main absorption edge features near 6560 eV were assigned to the purely dipole-allowed 1s $\rightarrow$ 4p transition.[52] Figure 10.11 shows that charging the electrode to 4.0 V (*vs.* Li/Li^+) produced a shift in the absorption edge toward a higher energy, indicating an increase in the average oxidation state of the manganese ions in the electrode during charging. When the electrode was discharged to 1.5 V (*vs.* Li/Li^+) for reduction, the absorption feature of the electrode shifted back to a lower energy, which indicates a decrease in the average oxidation state of the manganese ions in the electrode during discharging. The *in situ* XANES results of the MnO_2/CNT nanocomposites showed that the electrochemical reaction of the MnO_2/CNT nanocomposite electrode during charge/discharge at a constant current was not due to double-layer charging by the CNTs, but due to the Faradaic redox reaction between MnO_2 and Li^+ accompanied by a concomitant change in the oxidation state of Mn ions. This confirms that the reaction shown in Eq. 10.5 represents the electrochemical reaction of the MnO_2/CNT nanocomposite.

In order to estimate the oxidation state of Mn in the MnO_2/CNT nanocomposite at the charged and discharged states, the XANES

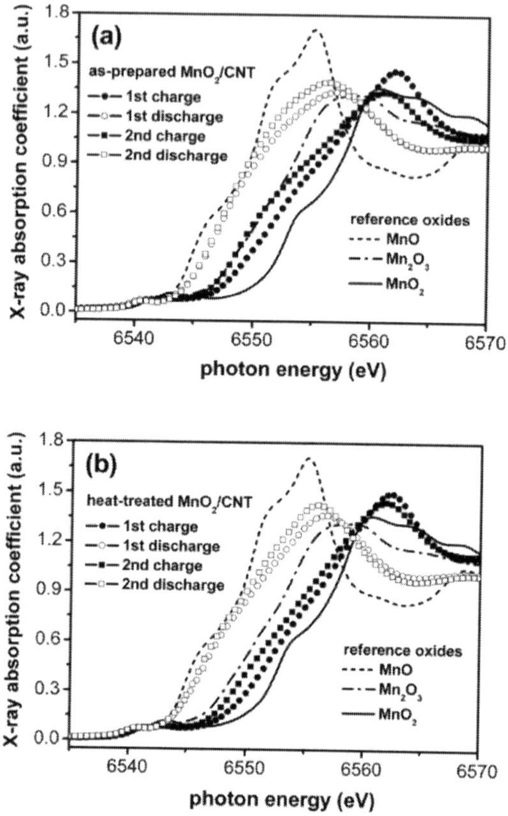

Figure 10.11 *In situ* Mn K-edge XANES spectra of (a) as-prepared and (b) heat-treated MnO_2/CNT nanocomposite during cycling.

spectra for the MnO_2/CNT nanocomposite were compared with several reference oxides such as MnO (Aldrich), Mn_2O_3 (Aldrich), and MnO_2 (Aldrich). The XANES spectra for the charged MnO_2/CNT electrode were located between the XANES spectra for Mn_2O_3 (dash dot line) and MnO_2 (solid line), while the XANES spectra for the discharged MnO_2/CNT electrode was located between MnO (short dash) and Mn_2O_3 (dash dot line). Therefore, the oxidation state of Mn in the charged MnO_2/CNT nanocomposites was between 3+ and 4+, and that for the discharged MnO_2/CNT nanocomposites was between 2+ and 3+. It is well known that manganese oxides exhibit various phases with different structures and oxidation states. Manganese oxides may have different XANES features even for the same oxidation state of Mn. Accordingly, it is difficult to extract the oxidation state of Mn from the XANES spectra for MnO_2/CNT nanocomposites in Fig. 10.11. However, the XANES spectra provide the qualitative and quasi-quantitative information by which the change in oxidation state can be estimated.

As shown in Fig. 10.11(a), the XANES features for the second charged states (■) of the as-prepared MnO_2/CNT nanocomposite did not shift back to the first charged state (•). This suggests that the change in the oxidation state of manganese ions as a result of the insertion/extraction of lithium ions is irreversible during charge/discharge. It is possible that not all the lithium ions inserted into the MnO_2 structure during the first discharging step had been extracted in the second charging step.

However, the XANES feature at the second charged states of the heat-treated MnO_2/CNT nanocomposite shifted back to the first charged state, as shown in Fig. 10.11(b). The heat-treated MnO_2/CNT nanocomposite showed a higher absorption edge than that for the as-prepared MnO_2/CNT nanocomposite, which corresponds to a higher specific capacity of heat-treated MnO_2/CNT nanocomposite during the second discharge. This indicates that the electrochemical redox reaction of the heat-treated MnO_2/CNT nanocomposite through the insertion/extraction of lithium ions is more reversible, which leads to better cyclability than the as-prepared one.

10.4 Nanosized Lithium Manganese Oxide Dispersed on Carbon Nanotubes

10.4.1 *Synthesis of LMO Dispersed on CNT*

Figure 10.12 shows TEM images of the MnO_2/CNT and LMO/CNT nanocomposites. Nanolayer MnO_2 with 10 nm $\sim$ 20 nm thickness was uniformly and entirely coated onto CNTs by a spontaneous direct redox reaction between CNTs and permanganate ions in an aqueous solution, as shown in Fig. 10.12(a). The LMO/CNT

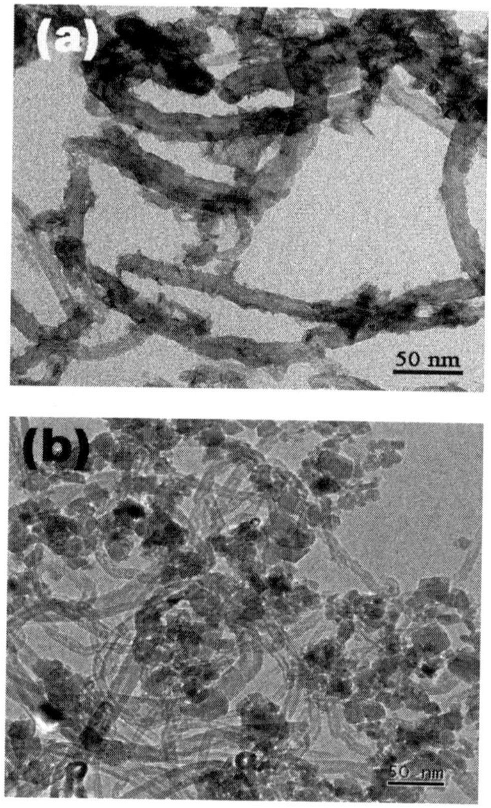

Figure 10.12 TEM images of (a) MnO_2/CNT nanocomposite and (b) LMO/CNT nanocomposite synthesized by a microwave-assisted hydrothermal reaction using MnO_2/CNT nanocomposite and aqueous LiOH solution (c) nanocrystallized LMO nanoparticles.

nanocomposite (64 wt% of LMO) was synthesized through a microwave-assisted hydrothermal reaction of MnO_2/CNT nanocomposite in aqueous LiOH solution. In Fig. 10.12(b), LMO nanoparticles of 20 nm $\sim$ 30 nm in diameter are seen to be formed along CNTs in the entangled CNT network without agglomeration. The formation of LMO nanoparticles from the MnO_2 on CNTs by a microwave-assisted hydrothermal reaction was attributed to the following reactions:

$$MnO_2 + 2H_2O \rightarrow Mn^{4+} + 4OH^- \quad (10.7)$$

$$8Mn^{4+} + 4Li^+ + 36OH^- \rightarrow 4LiMn_2O_4 + 18H_2O + O_2 \quad (10.8)$$

During the microwave-assisted hydrothermal synthesis of LMO/CNT nanocomposite, MnO_2 on CNTs was dissolved as aqueous Mn^{4+} ions[53,54] and then reprecipitated as nanocrystalline LMO nanoparticles preferentially on the external surface of CNTs as shown Fig. 10.12(b). LMO nanoparticles were well dispersed along and in contact with CNTs.

Microwave-assisted hydrothermal synthesis is a simple and fast method to synthesize the LMO nanoparticles on CNTs.[55] Conductive electrons of CNT can be accelerated by absorbing microwaves.[56] Accelerated electrons increase the vibration of the carbon lattice, resulting in a higher temperature on the surface of CNTs than the aqueous media. Because of the selectively heated CNTs, LMO particles tend to precipitate heterogeneously on the CNT surfaces in a short time of microwave irradiation. LMO nanoparticles could be prepared in less than 30 min by microwave-assisted hydrothermal process, while Jiang *et al.*[54] reported the synthesis of $LiMn_2O_4$ by a hydrothermal treatment of MnO_2 for as long as 1~7 days.

It should be noted that the LMO nanoparticles in the nanocomposite were formed on the surface of CNTs and well dispersed along CNTs, which would act as a 1-D conducting agent, as shown in Fig. 10.12(b), because the LMO nanoparticles were precipitated from Mn^{4+} ions as soon as MnO_2 coated onto CNT dissolved as Mn^{4+} ions under microwave irradiation. Considering the thickness of MnO_2 thin layer on CNTs in Fig. 10.12(a) and the size of LMO nanoparticles in the nanocomposite of Fig. 10.12(b), some CNTs were not covered with LMO in the LMO/CNT nanocomposites. When nanosized metal oxide covers the entire surface of CNTs as in the case of Fig. 10.12(a), the metal oxide/CNT nanocomposite will not be able to utilize high conductivity of CNTs. Some exposed CNTs are more likely to provide conduction paths in contact with LMO-coated CNTs and contribute to the electrical conductivity of LMO/CNT nanocomposite electrode. LMO/CNT nanocomposite composed of the CNTs in intimate contact with LMO nanoparticles and exposed CNTs is expected to exhibit the improved high rate capability.

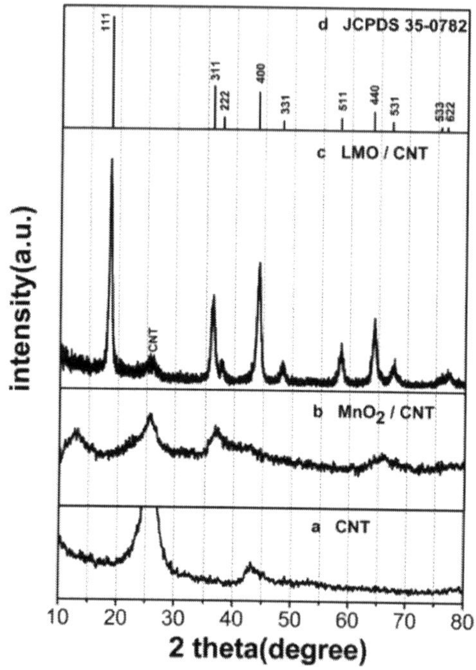

Figure 10.13 XRD patterns of (a) CNT; (b) MnO_2/CNT nanocomposite; (c) LMO/CNT nanocomposite; and (d) XRD peaks of spinel $LiMn_2O_4$ (JCPDS No. 35-0782).

10.4.2 *Phase and Structure of LMO Dispersed on CNT*

Figure 10.13 shows the XRD patterns for the CNTs, MnO_2/CNT, and LMO/CNT nanocomposites as well as the XRD peaks of the reference oxide spinel $LiMn_2O_4$ (JCPDS 35-0782). MnO_2/CNT nanocomposite [Fig. 10.8(b)] had three broad peaks at around 12°, 37°, and 66°, indicating the presence of birnessite-type MnO_2.[57] LMO nanoparticles in the LMO/CNT nanocomposite show much shaper XRD peaks, indicating a higher crystallinity than that of the MnO_2/CNT nanocomposite. The main XRD peaks can be indexed to spinel $LiMn_2O_4$, by considering the standard data of JCPDS No. 35-0782. This confirmed the successful synthesis of the LMO/CNT nanocomposite from the MnO_2/CNT nanocomposite by the microwave-assisted hydrothermal reaction.

The unit cell parameter and volume of spinel LMO nanoparticles calculated by the XRD data using Unitcell software are 8.166 Å and 544.5 $Å^3$, while the standard data are 8.247 Å and 560.9 $Å^3$ of JCPDS No. 35-0782, respectively. The Li/Mn ratio of the spinel LMO nanoparticle was 0.74, as determined by inductively coupled plasma (ICP) spectrometry. This was considered to be the main reason for the smaller unit cell volume of spinel LMO nanoparticles in this study than that from the standard data of JCPDS No. 35-0782. The structure of spinel $LiMn_2O_4$ has the space group Fd3m, consisting of close-packed oxygen in which lithium and manganese ions occupy tetrahedral 8a and octahedral 16d sites, respectively.[58–60] When excess lithium ions occupy some octahedral 16d sites, $Li_xMn_2O_4$ became lithium-rich spinels and the unit cell parameter a_0 decreases with an increase in lithium content of spinel $Li_{1+x}Mn_{2-x}O_4$ due to the smaller ionic radii of Mn^{4+} (0.53 Å) than Mn^{3+} (0.65 Å).[60,61]

10.4.3 *Electrochemical Properties of LMO Dispersed on CNT*

Figure 10.14 shows the discharge curves of the LMO/CNT nanocomposite. The charge and discharge were performed between 3.5 and 4.5 V (*vs.* Li/Li^+) in 1 M $LiClO_4$ in propylene carbonate (PC) at 1.6 and 16 C-rate, respectively. The discharge curves of the LMO/CNT nanocomposite show typical discharge behaviors of spinel $LiMn_2O_4$. The initial specific capacity based on LMO weight was 99.4 mAh/g at 1.6 C-rate. A specific capacity of 99.1 mAh/g was achieved even at 16 C-rate, which indicated the excellent high rate capability of the LMO/CNT nanocomposite. Although the ohmic drop increased slightly with C-rate, LMO/CNT nanocomposite electrode had a low equivalent series resistance. Lithium-rich LMO is known to be good at high rate capability. In this study, we could further improve its high rate capability of Li-rich LMO through the synthesis of LMO nanoparticle/CNT nanocomposite with the following features: (1) the shorter diffusion length of the nanosized LMO; (2) the well-dispersed LMO nanoparticles on CNTs; (3) the large interfacial area between the LMO and electrolytic solution; (4) the high electrolyte accessibility due to the interconnected pores of the entangled CNTs; and (5) the high electrical conductivity of CNT.

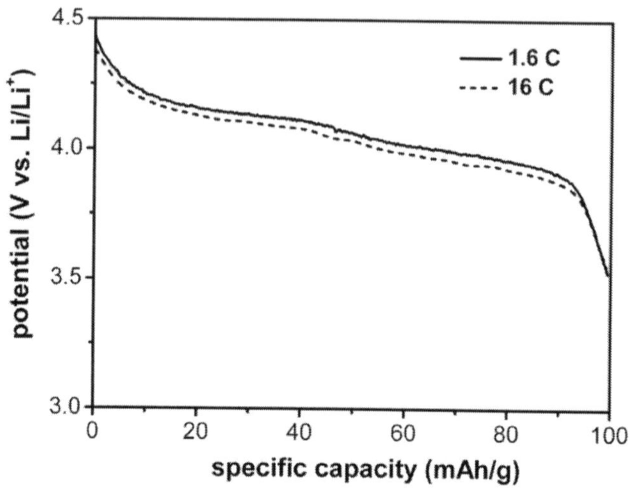

Figure 10.14 Discharge profiles of the LMO/CNT nanocomposite at different C-rate between 3.5 and 4.5 V (*vs.* Li/Li^+) in 1 M $LiClO_4$ in PC.

Cyclic voltammetry (CV) was used to determine the capacity fading of LMO/CNT nanocomposite electrode. The electrode potential was scanned at a high scan rate of 10 mV/s (approximately equivalent to 36 C-rate) between 3.5 V and 4.5 V (*vs.* Li/Li^+) in the both anodic and cathodic directions. Figure 10.15 shows the cyclability of LMO/CNT nanocomposite electrode. The specific capacity was 107 mAh/g based on LMO weight at the first cycle, and this was well maintained to 104 mAh/g at the 50th cycle to give a capacity retention of 97%. In the previous reports, lithium-rich spinel compounds with a smaller cell volume usually show good rechargeability because of its higher oxidation state of manganese ion, which is capable of reducing Jahn–Teller effect.[62,63] In addition, well-dispersed LMO nanoparticles in LMO/CNT nanocomposite have better accommodation of the strain of Li-ion insertion/extraction even at the high potential scan rate.

10.4.4 *Structural Stability of LMO Dispersed on CNT*

In order to investigate the structural stability of LMO/CNT nanocomposite in a short range order during electrochemical cycling, a series

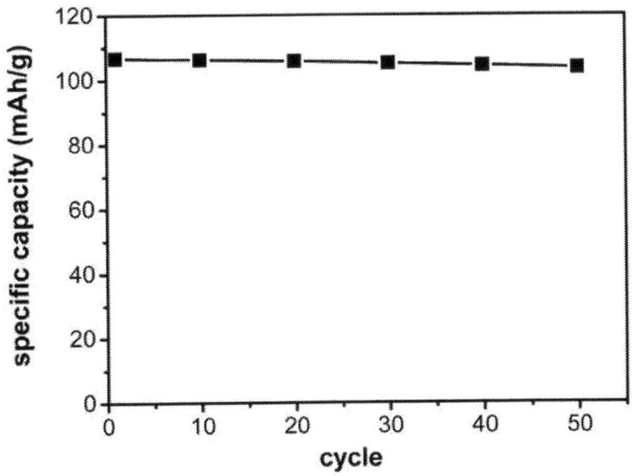

Figure 10.15 Change in the specific capacity of LMO on CNT during cycling at a scan rate of 10 mV/s between 3.5 and 4.5 V (*vs.* Li/Li^+) in 1 M $LiClO_4$ in PC.

of normalized *ex situ* Mn K-edge X-ray absorption spectra (XAS) was measured. The LMO/CNT nanocomposite electrode was cycled at a scan rate of 10 mV/s between 3.5 and 4.5 V (*vs.* Li/Li^+) in both anodic and cathodic directions and the scan stopped at 1st, 10th, and 30th discharged states (3.5 V).

Figure 10.16(a) shows the *ex situ* Mn K-edge X-ray absorption near edge structure (XANES) spectra of the as-prepared and the cycled LMO/CNT nanocomposite electrodes. The XANES spectra of all three samples after cycles shift back to lower energies, even beyond the initial position of the pristine sample, indicating that the manganese ions are reduced back to the initial state, and further.

It is interesting to note that manganese ions are further reduced than the initial oxidation state during early cycling and the edge structure does not show any substantial change between 10 and 30 cycles. This suggests that the local structures of LMO/CNT nanocomposite undergo structural changes in a short range order during initial cycles and are stabilized after the several initial cycles. Figure 10.16(b) shows Fourier transform (FT) magnitudes of the Mn K-edge extended X-ray absorption fine structure (EXAFS) spectra

(a)

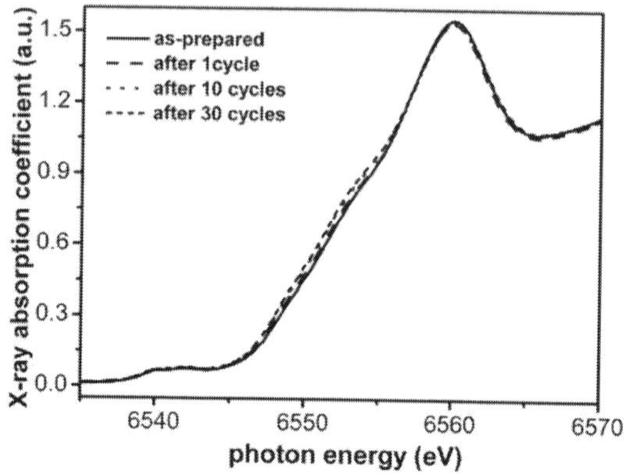

(b)

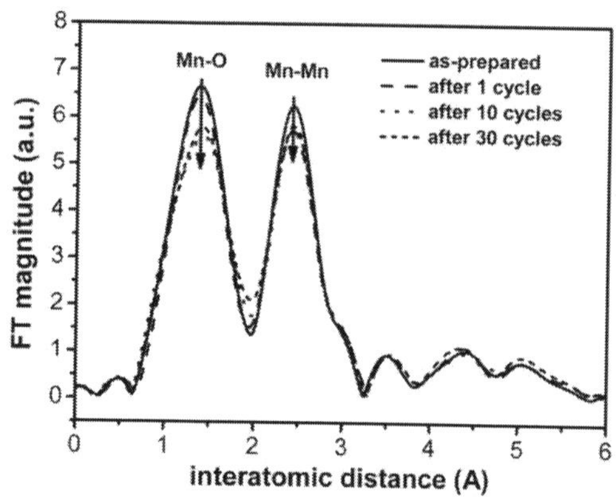

Figure 10.16 *Ex situ* Mn K-edge (a) XANES and (b) EXAFS spectra of LMO/CNT nanocomposite during cycling.

after cycles. The first peak at nearly 1.4 Å is due to Mn-O correlations. The second peak at nearly 2.4 Å is due to Mn–Mn interaction. The most significant change during cycling is observed in the first coordination shell around the Mn atoms. The intensity of the first peak becomes smaller during initial cycles, indicating the more local distortion possibly due to increasing Jahn–Teller ions (Mn^{3+}).[62] In contrast, Fourier transform magnitudes of the EXAFS spectra after 10 and 30 cycles are almost identical. This is consistent with the above XANES result showing more reduction of Mn ions after initial cycles and no further oxidation state change between 10 and 30

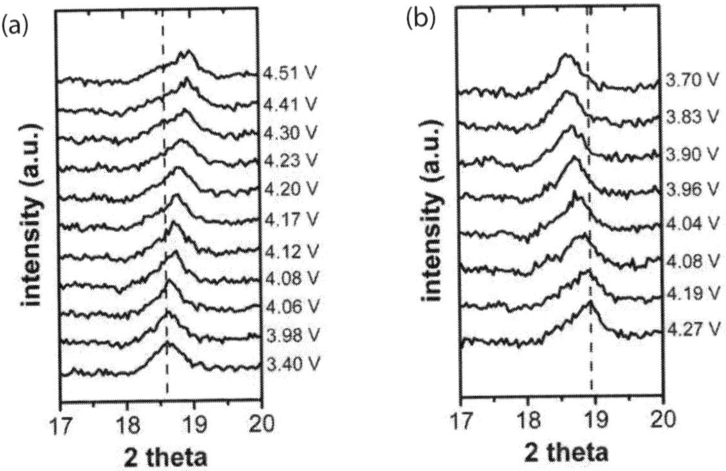

Figure 10.17 *In situ* XRD patterns of LMO/CNT nanocomposite during (a) charge and (b) discharge at 1C-rate between 3.5 and 4.5 V (*vs.* Li/Li^+).

cycles. Consequently, *ex situ* Mn K-edge XAS results indicate that LMO/CNT nanocomposites have a good structural stability in a short range order during electrochemical cycling.

In order to determine the structural change of the LMO nanoparticles in a long-range order during electrochemical cycling, the *in situ* XRD patterns were recorded during charging/discharging.[63,64] Figure 10.17 shows the change in the (111) peak position of the LMO nanoparticles at a high rate charge/discharge of 1 C between 3.5 V and 4.5 V (*vs.* Li/Li^+). When the *in situ* XRD cell was fabricated, the initial open circuit potential (OCP) was 3.40 V (*vs.* Li/Li^+), where 2θ of the (111) peak position was 18.61°. As shown in Fig. 10.17(a), 2θ of the (111) peak position during charging was gradually shifted toward a higher angle and approached 18.94° at 4.5 V (*vs.* Li/Li^+), indicating a decrease in the unit volume of LMO by the lithium extraction. On the contrary, 2θ of the (111) peak position during the discharging was continuously shifted toward a lower angle of 18.61° at 3.70 V (*vs.* Li/Li^+), indicating that the unit volume of LMO increased with the lithium insertion [Fig. 10.17(b)]. The 2θ of the (111) peak position of the discharged LMO nanoparticles was almost identical to that of the as-prepared LMO nanoparticles, indicating that the LMO nanoparticles synthesized by the microwave-assisted hydrothermal reaction have good structural reversibility in a long-range order.

10.5 Summary

In this chapter, we reported on the synthesis and the electrochemical behaviors of nano-architectured manganese oxide for electrochemical energy storage applications. Nanolayer manganese dioxide was successfully deposited on CNT by a spontaneous redox reaction between permanganate ion and carbon surface. The synthesis mechanism of the nano-architectured manganese dioxide on CNTs was proposed via the *in situ* monitoring of the solution chemistry. Nano-architectured manganese dioxide showed a high specific capacity and high rate capability. This demonstrated that a design of electrode materials, having high electrochemical utilization, high electronic conductivity, and large ion supply, is crucial to enable the electrode materials to have high specific capacity and high rate capability. The charge storage mechanism and local/electronic structure change of nano-architectured manganese dioxide on CNT was investigated by *in situ* X-ray absorption spectroscopy during charge/discharge.

Nano-architectured LMOs were successfully synthesized and well dispersed on entangled CNTs' web by microwave-hydrothermal process for electrochemical energy storage applications. Nano-architectured LMO showed excellent high rate capability and good structural reversibility. The remarkable electrochemical properties of the nano-architectured material were attributed to shorter lithium diffusion length, good dispersion, large ionic supply high electrode/electrolyte interfacial area, and high electrical conductivity by introducing the entangled CNTs' web. The structural stability of nano-architectured LMO on CNT was investigated by *ex situ* X-ray absorption spectroscopy and *in situ* X-ray diffraction during charge/discharge. This synthetic strategy opens a new avenue for the synthesis of nano-sized lithium metal oxides on the entangled CNTs' web in an effective and facile way.

Acknowledgments

This work was supported by Korea Science and Engineering Foundation (KOSEF) through the National Research Lab. Program funded by the Ministry of Education, Science and Technology (No. R0A-2007-000-10042-0).

References

1. B. E. Conway, *Electrochemical Supercapacitors; Scientific Fundamentals and Technological Applications*, Kluwer Academic/Plenum, New York (1999).
2. B. E. Conway, *J. Electrochem. Soc.*, 138 (1991) 1539.
3. J. P. Zheng, T. R. Jow, *J. Electrochem. Soc.*, 142 (1995) L6.
4. R. Kotz, M. Carlen, *Electrochim. Acta*, 45 (2000) 2483.
5. I. H. Kim, J. H. Kim, Y. H. Lee, K. B. Kim, *J. Electrochem. Soc.*, 152 (2005) A2170.
6. H. Y. Lee, J. B. Goodenough, *J. Solid State Chem.*, 144 (1999) 220.
7. S. C. Pang, M. A. Anderson, T. W. Chapman, *J. Electrochem. Soc.*, 147 (2000) 444.
8. Y. U. Jeong, A. Manthiram, *J. Electrochem. Soc.*, 149 (2002) 1419.
9. M. Toupin, T. Brousse, D. Belanger, *Chem. Mater.*, 14 (2002) 3946.
10. C. C. Hu, T. W. Tsou, *Electrochem. Commun.*, 4 (2002) 105.
11. R. N. Reddy, R. G. Reddy, *J. Power Sources*, 124 (2003) 330.
12. Y. T. Wu, C. C. Hu, *J. Electrochem. Soc.*, 151 (2004) 2060.
13. Y. K. Zhou, B. L. He, F. B. Zhang, H. L. Li, *J. Soild State Electrochem.*, 482 (2004) 482.
14. G. X. Wang, B. L. Zhang, Z. L. Yu, M. Z. Qu, *Solid State Ionics*, 176 (2005) 1169.
15. E. Raymundo-Pinero, V. Khomenko, E. Frackowiak, F. Beguin, *J. Electrochem. Soc.*, 152 (2005) A229.
16. V. Subramanian, H. Zhu, B. Wei, *Electrochem. Commun.*, 8 (2006) 827.
17. C. Y. Lee, H. M. Tsai, H. J. Chuang, S. Y. Li, P. Lin, T. Y. Tseng, *J. Electrochem. Soc.*, 152 (2005) A716.
18. Z. Fan, J. Chen, M. Wang, K. Cui, H. Zhou, Y. Kuang, *Diam. Relat. Mater.*, 15 (2006) 1478.

19. M. Wu, G. A. Snook, G. Z. Chen, D. J. Fray, *Electrochem. Commun.*, 6 (2004) 499.
20. A. E. Fischer, K. A. Pettigrew, D. R. Rolison, R. M. Stroud, J. W. Long, *Nano Lett.*, 7 (2007) 281.
21. S. B. Ma, Y. H. Lee, K. Y. Ahn, C. M. Kim, K. H. Oh, K. B. Kim, *J. Electrochem. Soc.*, 153 (2006) C27.
22. S. B. Ma, K. Y. Ahn, E. S. Lee, K. H. Oh, K. B. Kim, *Carbon*, 45 (2007) 375.
23. H. Kawaoka, M. Hibino, H. Zhou, I. Honma, *J. Power Sources*, 125 (2004) 85.
24. M. Hibino, H. Kawaoka, H. Zhou, I. Honma, *Electrochim. Acta*, 49 (2004) 5209.
25. H. Kawaoka, M. Hibino, H. Zhou, I. Honma, *Solid State Ionics*, 176 (2005) 621.
26. H. Kawaoka, M. Hibino, H. Zhou, I. Honma, *Electrochem. Solid-State Lett.*, 8 (2005) A253.
27. C. Niu, E. K. Sichel, R. Hoch, D. Moy, H. Tennet, *Appl. Phys. Lett.*, 70 (1997) 1480.
28. E. Frackowiak, K. Metenier, V. Bertagna, F. Beguin, *Appl. Phys. Lett.*, 77 (2000) 2421.
29. J. S. Sakamoto, B. Dunn, *J. Electrochem. Soc.*, 149 (2002) A26.
30. K. W. Nam, E. S. Lee, J. H. Kim, Y. H. Lee, K. B. Kim, *J. Electrochem. Soc.*, 152 (2005) A2123.
31. I. H. Kim, J. H. Kim, K. B. Kim, *Electrochem. Solid-State Lett.*, 8 (2005) A369.
32. S. B. Ma, K. W. Nam, W. S. Yoon, X. Q. Yang, K. Y. Ahn, K. H. Oh, K. B. Kim, *J. Power Sources*, 178 (2008) 483.
33. I. H. Kim, J. H. Kim, B. W. Cho, K. B. Kim, *J. Electrochem. Soc.*, 153 (2006) A1451.
34. A. S. Arico, P. Bruce, B. Scrosati, J. M. Tarascon, W. V. Schalkwijk, *Nat. Mater.*, 4 (2005) 366.
35. C. Jiang, M. Ichihara, I. Honma, H. Zhou, *Electrochim. Acta*, 52 (2007) 6470.
36. M. Knupfer, T. Pichler, M. S. Golden, J. Fink, A. Rinzler, R. E. Smalley, *Carbon*, 37 (1999) 733.
37. K. Kinoshita, *Carbon - Electrochemical and Physicochemical Properties*, New York: John Wiley & Sons (1988).
38. M. Pourbaix, *Atlas of Electrochemical Equilibria in Aqueous Solutions*, Oxford: Pergamon Press (1966).

39. H. C. Choi, M. Shim, S. Bangsaruntip, H. Dai, *J. Am. Chem. Soc.*, 124 (2002) 9058.
40. B. Xue, P. Chen, Q. Hong, J. Lin, K. L. Tan, *J. Mater. Chem.*, 11 (2001) 2378.
41. W. X. Chen, J. Y. Lee, Z. Liu, *Mater. Lett.*, 58 (2004) 3166.
42. Q. Fu, C. Lu, J. Liu, *Nano Lett.*, 2 (2002) 329.
43. R. M. Mckenzie, *Mineral Mag.*, 38 (1971) 493.
44. C. Julien, M. Massot, R. Baddour-Hadjean, S. Franger, S. Bach, J. P. Pereira-Ramos, *Solid State Ionics*, 159 (2003) 345.
45. C. M. Julien, M. Massot, C. Poinsignon, *Spectrochim. Acta Part A*, 60 (2004) 689.
46. J. J. Xu, A. J. Kinser, B. B. Owens, W. H. Smyrl, *Electrochem. Solid-State Lett.*, 1 (1998) 1.
47. E. Deiss, D. Haringer, P. Novak, O. Hass, *Electrochim. Acta*, 46 (2001) 4185.
48. C. Portet, P. L. Taberna, P. Simon, E. Flahaut, *J. Power Sources*, 139 (2005) 371.
49. R. Chen, M. S. Whittingham, *J. Electrochem. Soc.*, 144 (1997) L64.
50. F. Leroux, D. Guyomard, Y. Piffard, *Solid State Ionics*, 80 (1995) 307.
51. M. M. Thackeray, *Prog. Solid St. Chem.*, 25 (1997) 1.
52. K. W. Nam, M. G. Kim, K. B. Kim, *J. Phys. Chem. C*, 111 (2007) 749.
53. Q. Feng, H. Kanoh, Y. Miyai, K. Ooi, *Chem. Mater.*, 7 (1995) 1226.
54. C. H. Jiang, S. X. Dou, H. K. Liu, M. Ichihara, H. S. Zhou, *J. Power Sources*, 172 (2007) 410.
55. S. Komarneni, R. Roy, Q. H. Li, *Mat. Res. Bull.*, 27 (1992) 1393.
56. M. Tsuji, M. kubokawa, R. Yano, N. Miyamae, T. Tsuji, M. S. Jun, S. Hong, S. Lim, S. Yoon, I. Mochida, *Langmuir*, 23 (2008) 387.
57. C. Julien, M. Massot, R. Baddour-Hadjean, S. Franger, S. Bach, J. P. Pereira-Ramos, *Solid State Ionics*, 159 (2003) 345.
58. M. H. Rossouw, A. de Kock, L. A. de Picciotto, M. M. Thackeray, W. I. F. David, R. M. Ibberson, *Mat. Res. Bull.*, 25 (1990) 173.
59. A. de Kock, M. H. Rossouw, L. A. de Picciotto, M. M. Thackeray, W. I. F. David, R. M. Ibberson, *Mat. Res. Bull.*, 25 (1990) 657.
60. Y. Xia, M. Yoshio, *J. Electrochem. Soc.*, 144 (1997) 4186.
61. J. Cho, M. M. Thackeray, *J. Electrochem. Soc.*, 146 (1999) 3577.
62. M. M. Thackeray, *J. Electrochem. Soc.*, 142 (1995) 2558.
63. Y. Xia, M. Yoshio, *J. Electrochem. Soc.*, 143 (1996) 825.
64. S. Mukerjee, T. R. Thurston, N. M. Jisrawi, X. Q. Yang, J. McBreen, M. L. Daroux, X. K. Xing, *J. Electrochem. Soc.*, 145 (1998) 466.

Chapter 11

Manganese Oxides for Supercapacitors

Shinichi Komaba, Naoaki Yabuuchi, and Tomoya Tsuchikawa
Department of Applied Chemistry, Tokyo University of Science, Kagurazaka 1-3, Shinjuku, Tokyo 162-8601, Japan
komaba@rs.kagu.tus.ac.jp

Electrochemical capacitors are attractive for application as power sources. Instead of conventional activated carbon electrodes, we applied electrochemically active manganese oxides as electrodes of asymmetric capacitors. This chapter presents the recent contributions of the manganese-based supercapacitors.

11.1 Manganese Oxides for Electrode Materials

In the past century, considerable effort has been devoted to study the manganese oxides as the electrode materials for the power sources. Crystallization of the manganese oxides is highly influenced by many factors, for example, intergrowth of different structural units, cations incorporation to the vacant sites, structural water, and so on, leading to the complexity of the crystal structures. Typically, the manganese oxides constituted from mainly tetravalent manganese ions are simply denoted as "manganese dioxides (MnO_2)". Manganese dioxides are used as a positive electrode

Nanotechnology in Advanced Electrochemical Power Sources
Edited by S. R. S. Prabaharan and M. S. Michael

ISBN 978-981-4241-43-4 (Hardcover), 978-981-4303-05-7 (eBook)
www.panstanford.com

with a zinc negative electrode, that is, carbon–zinc cells (dry cell) and alkaline cells. Electrolytic manganese dioxide (γ-type MnO_2) is typically used in the alkaline cell, which can be described as an intergrowth structure between pyrolusite (β-type MnO_2) and ramsdellite. In the alkaline cell, manganese oxides are reduced with proton insertion into the bulk of particles for the charge compensation. This process can be described as follows:

$$Mn^{4+}O_2 + xH^+ + xe^- \leftrightarrow H_xMn_{(1-x)}{}^{4+}Mn_x{}^{3+}O_2 \qquad (11.1)$$

In theory, one mole of proton can be inserted into MnO_2, which is accompanied by manganese reduction from tetravalent to trivalent state. MnO_2 can also accommodate lithium ions, which are much larger than proton, to the vacant sites (Eq. 11.2),

$$Mn^{4+}O_2 + xLi^+ + xe^- \leftrightarrow Li_xMn_{(1-x)}{}^{4+}Mn_x{}^{3+}O_2 \qquad (11.2)$$

Heat-treated MnO_2 is used as the positive electrode material for a primary lithium cell, and spinel-type $LiMn_2O_4$ can be expected to be used as a positive electrode for rechargeable and large-scale lithium-ion cells.

Since 1995, amorphous hydrated ruthenium dioxide, $RuO_2{\cdot}nH_2O$, has been extensively studied as the electrode material for the redox capacitor.[1–3] $RuO_2{\cdot}nH_2O$ composite electrode (~100 μm thickness) exhibits extremely high specific capacitance (typically >700 F/g) in acidic media (e.g., H_2SO_4 aqueous solution). Although the ruthenium dioxide is an attractive electrode material, the manganese dioxides are also worth studying as the electrode materials for the redox capacitor, that is, material abundance in nature and its environmental benign. The use of manganese dioxides for the supercapacitor has been reported by Goodenough's research group in 1999.[4] Amorphous hydrated manganese dioxides, $MnO_2{\cdot}nH_2O$, were synthesized by simple precipitation reaction between $KMnO_4$ and $Mn(CH_3COO)_2$. $MnO_2{\cdot}nH_2O$ has a large surface area (303 m^2/g by BET measurement), and a capacitor-like electrochemical response in KCl aqueous electrolyte was reported. A specific capacitance of the amorphous hydrated manganese dioxides reached 200 F/g, which corresponds to 66 $\mu F/cm^2$ as normalized capacitance on the basis of the surface area of manganese dioxides. This capacitance is approximately three times larger value than that of expected value based on a classical electrical double-layer charging (typically $\sim$ 20 μF/cm^2 for a plate electrode). It is noted that surface area of the activated carbon ranged from 1000 to 3000 m^2/g,[5] from which the area normalized capacitance is calculated to be 10–15 μF/cm^2. The observed area normalized capacitance for of $MnO_2 \cdot nH_2O$ is anomalously large compared to the conventional electrical double layer charging. From these observations, it is speculated that Faradaic reaction (or charge-transfer reaction) partly contributes to the capacitor-like response of manganese dioxides, that is, "pseudocapacitance". When we assume that one electron transfer per manganese ions with capacitor-like behavior is achieved as shown in the Eqs. 11.1 and 11.2, the specific capacitance would reach more than 1000 F/g. From these results, for the $MnO_2 \cdot nH_2O$, the Faradaic reaction occurs only at the surface (Eq. 11.3) or is limited to the near surface with the slow kinetics on the transfer of large cations, such as sodium and potassium ions,

$$(\mathrm{Mn}^{4+}\mathrm{O}_2)_{\mathrm{surface}} + xC^+ + xe^- \leftrightarrow \mathrm{C}_x^+(\mathrm{Mn}_{1-x}{}^{4+}\mathrm{Mn}_x{}^{3+}\mathrm{O}_2)_{\mathrm{surface}}, \quad (11.3)$$

where C$^+$ denotes alkaline cations, Li$^+$, Na$^+$, K$^+$, and so on, which depends on the electrolyte solution utilized. Analysis on the manganese dioxide thin films by X-ray photoelectron spectroscopy (XPS) sheds light on the reaction mechanism of the manganese dioxides, especially near the surface.[6] In this case, manganese 3s and 2p core level shift was observed for oxidation and reduction state, corresponding to 0.9 and 0.0 V *vs.* Ag/AgCl, respectively. As the energy of the photoelectron for Mn 3s core level is extremely low (<100 eV), the information of the electronic structure for the manganese ions is relatively limited from the very near surface of the oxides. It was revealed that Mn 3s core level shifted, depending on the applied potential. In contrast, less change was observed for the Mn 2p core level, suggesting that manganese ions are predominantly reduced near the surface of the particles. Furthermore, formation of the hydroxyl group (–OH) was found from O 1s core level when the manganese dioxides discharged to 0 V *vs.* Ag/AgCl. Proton partly plays the role of the charge compensation. It is noted that existence of Na ions was also confirmed by XPS, indicating insertion of Na ions near the surface of manganese dioxides. These observations also imply that the electrochemical performance of manganese oxides is

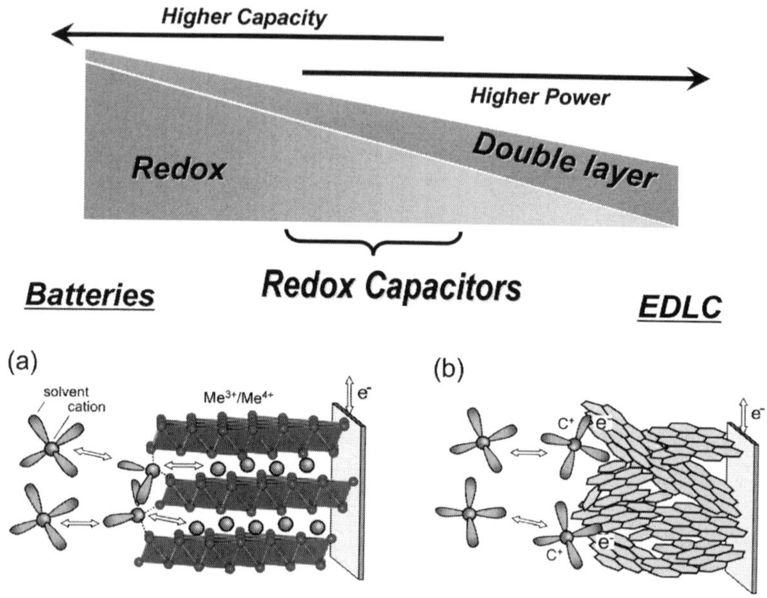

Figure 11.1 A conceptual schematic illustration of the redox capacitor, which effectively utilizes the charge storage mechanisms of both (a) battery and (b) EDLC. For the battery, transition metals are oxidized/reduced accompanied by cation incorporation for the charge compensation. For the EDLC, cations/anions are adsorbed at the negatively/positively charged interface during charge/discharge.

affected by not only its crystal structure but also the morphology and surface structures of the manganese dioxides.

The concept for the "redox capacitors" in comparison to the conventional electric double layer capacitor (EDLC) and batteries is summarized in Fig. 11.1. The redox capacitor effectively utilizes the both electrical double layer charging and Faradaic reaction of the manganese (metal) oxides as described above. There are merits for the redox capacitors, that is, they perform the higher capacity than EDLC and the higher power than batteries.

As for the electrode materials for the supercapacitors, electrical conductivity is one of the important physical properties to achieve higher power density. Ruthenium oxide shows excellent electrical conductivity, for example, $\sim 2 \times 10^4$ S/cm for the polycrystalline RuO_2,[7] and nearly 2 S/cm as the hydrous form.[8] The electrical conductivity of the $RuO_2 \cdot nH_2O$ is comparable to the carbon-based electrode materials. In this regard, the major drawback of manganese dioxides as the electrode materials for the redox capacitor seems to be poor electrical conductivity (5×10^{-6} S/cm for cryptomelane.[9]) The poor electrical conductivity may suggest that the solid-state diffusion length for the electrons and/or cations is limited, resulting in the reduced power and energy density. In fact, manganese dioxides prepared as the thin film electrode (<100 nm) demonstrate clear improvement of the specific capacitance of the manganese electrode, which becomes comparable value (~700 F/g)[10,11] with that of ruthenium dioxides. This is also consistent with the XPS study of the manganese oxides. The improvement in the energy density has been achieved even as a composite electrode by embedding manganese dioxides into the mesoporous carbon[12] and ultraporous carbon[13] frameworks, which allow the three-dimensional path for electron conduction with maximized interface area between electrolyte solution and manganese oxides in the composite structure. The specific capacitance of 600 F/g can be achieved on the basis of the embedded manganese dioxides in the carbon form.[12]

11.2 Synthesis of Manganese Dioxides as Electrode Materials of Redox Capacitor

As discussed briefly in the previous section, the electrode materials for the supercapacitor, size, morphology, surface structures, and crystal structures influence the pseudo-capacitance. For the synthesis of the manganese oxides for the supercapacitor, precipitation, sol-gel, and electrodeposition methods are often utilized. For the case of the precipitation method, $KMnO_4$ and divalent manganese salt, which work as oxidation and reduction agents, respectively, are used as starting materials. Amorphous phase or poorly crystallized cryptomelane (α-MnO_2) is typically obtained by the precipitation method. The α-MnO_2 has larger vacant sites (so-called 2×2 tunnel) in comparison to the β (1×1 tunnel) and γ (1×1 and 1×2 tunnels) forms. The cation insertion into the vacant sites, however, is limited near the surface of the manganese in KCl or Na_2SO_4

electrolyte. The morphology of the manganese oxides seems to be affected by a source of the divalent manganese salts used, relating to the difference in the pH of solution.[4,14] Specific capacitance of the manganese dioxides prepared by the precipitation method reaches 200 F/g for the composite electrode. Organic chemicals, such as ethanol[15] and ethylene glycol,[16] can be used for the reducing agent to synthesize nano-structured manganese dioxides, instead of the divalent manganese salts. Morphology of MnO_2 is modified by the selection of the reducing agent.[15,16] The manganese oxides with high surface area and high porosity can also be prepared by a sol-gel method from $KMnO_4$,[17] and electrochemical deposition method from divalent manganese sources with anodic oxidation.[10,18] Amorphous and hydrous manganese oxides with nanostructured morphology can be obtained by the electrochemical deposition method.[10] It is interesting to note that birnessite-type manganese dioxides (δ-MnO_2: layered structure) synthesized at 600°C (BET surface area 3 m^2/g) showed the specific capacitance of 80 F/g at 5 mV/s in 0.1 mold/m^3 K_2SO_4 solution, which corresponds to more than 2000 $\mu F/cm^2$ of area normalized capacitance on the basis of the surface area of manganese dioxide.[19] This result indicates that K^+ and/or H^+ are inserted into the interstitial space between MnO_2 slabs. The layered structure with two-dimensional diffusion path would be able to effectively accommodate large cations, such as potassium ion, in comparison to the materials having one-dimensional (tunnel type) cation diffusion path.

Recently, our group also reported nano-structured birnessite as the electrode material for the supercapacitor, prepared by the electrochemical conversion from spinel-type Mn_3O_4.[20,21] Crystalline Mn_3O_4 powder was synthesized by simple heat treatment of electrolytic manganese dioxide in air at 1050°C for 72 h and was ground with planetary ball-milling for 12 h. The Mn_3O_4 electrodes were prepared with poly(vinylidene fluoride) binder and carbon (Ketjen Black) on stainless mesh. The cyclic voltammograms of the Mn_3O_4 electrode containing the carbon additive and polymer binder, in the potential range of -0.1 and 0.9, are shown in Fig. 11.2. The Mn_3O_4 electrodes exhibited a pseudo-capacitive behavior in the Na_2SO_4 solution, in spite of the general consensus that Mn_3O_4 is

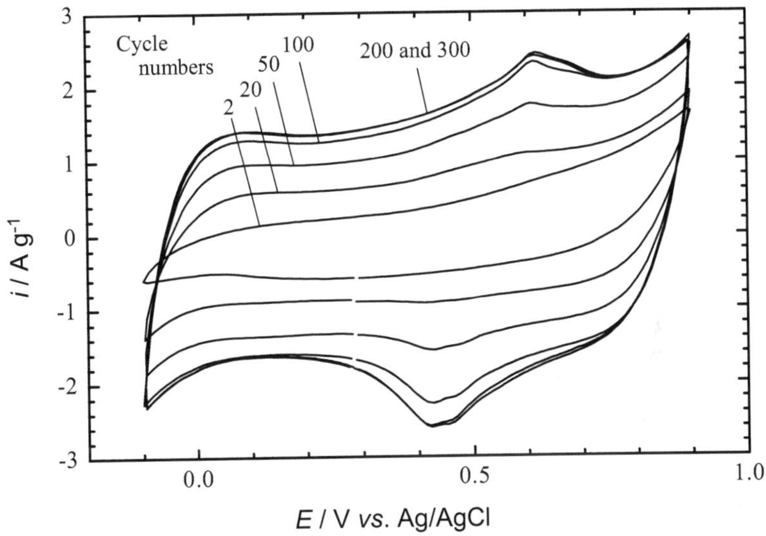

Figure 11.2 Cyclic voltammograms of a Mn_3O_4 electrode in 1 mol dm^{-3} Na_2SO_4 aqueous solution at a rate of 10 mV s^{-1}.

not electrochemically active. In addition, the specific capacitance gradually increased during the 300 cycling test, which is similar phenomenon to Mn_3O_4 thin film.[22] Small oxidation and reduction peaks appeared around 0.7 V and 0.4 V, respectively. This behavior depended on the experimental conditions: potential range, scan rate, electrolyte solution, and particle size.

Figure 11.3 shows the variation in the specific capacitance of Mn_3O_4 calculated from the voltammograms in Fig. 11.2. The capacitance of the Mn_3O_4 electrodes clearly has increased by the successive potential cycling. The Mn_3O_4 demonstrated the maximum value of approximately 190 F/g of oxide after approximately 300 cycles in 1 mol/dm^3 Na_2SO_4 aqueous solution. Electrochemical conversion process depends on the source of the manganese oxides. Mn_2O_3 shows a similar trend for the increase in specific capacitance, which is induced by the electrochemical cycling. To achieve the same capacitance with Mn_3O_4 as the electrode materials, however, much longer electrochemical cycles were needed for Mn_2O_3 (Fig. 11.4).

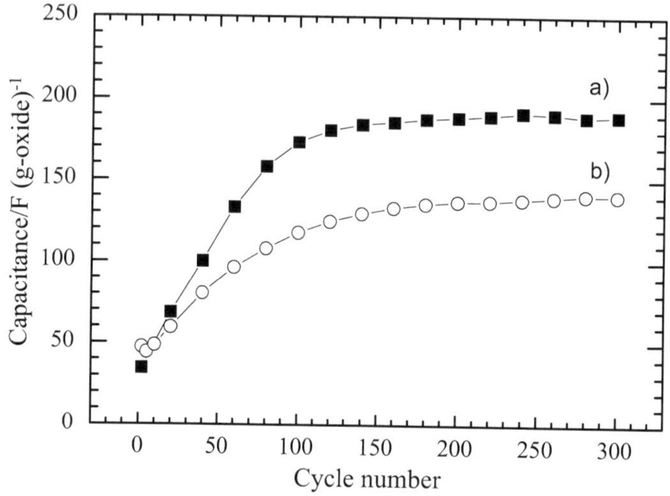

Figure 11.3 The variation of the specific capacitance of Mn_3O_4 electrodes in (a) 1 mol/dm^3 and (b) 0.1 mol/dm^3 Na_2SO_4 aqueous solutions.

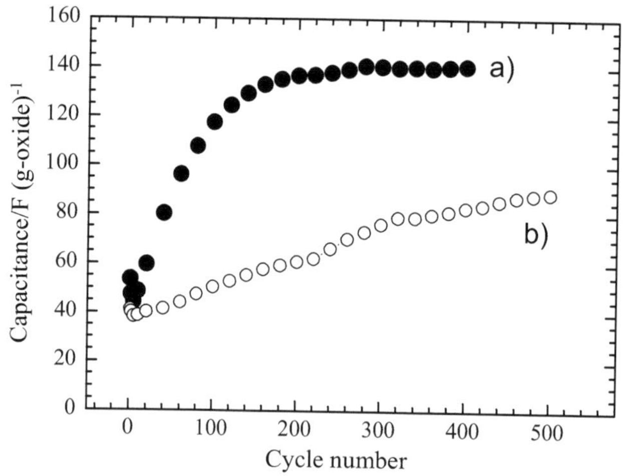

Figure 11.4 Specific capacitance *vs.* cycle number plots of a) Mn_3O_4 and b) Mn_2O_3 electrodes treated by ball milling. The capacitances were evaluated from the voltammagrams at 10 mV s^{-1} in 0.1 mol dm^{-3} Na_2SO_4 aqueous solution.

As shown in Fig. 11.4, the specific capacitance of spinel Mn_3O_4 reached the maximum value of 140 F/g after approximately 200 cycles in 0.1 mol/dm^3 Na_2SO_4 aqueous solution at 10 mV. The specific capacitance of bixbyite Mn_2O_3 electrodes continuously increased during the 500 cycles under the same experimental condition. The Mn_2O_3 electrode shows 80 F/g after the 500 cycles, which is still much lower capacitance than that of the Mn_3O_4 electrode.

Broughton *et al.* described that an Mn thin film electrode was transformed into MnO, Mn_3O_4, Mn_2O_3, and then finally into a low crystallinity MnO_2 product by electrochemical oxidation, resulting in a high specific capacitance,[11,23] and an Mn_3O_4 layer was transformed into Mn_2O_3 and then MnO_2 when applying an anodic potential.[24] Considering these reports, the Mn_3O_4 powder should also be transformed into hydrated manganese (di)oxide by electro-oxidation during the potential cycling, similar to a thin film of Mn_3O_4.[11] It is noted that the electrochemical stimulation in 1 mol/dm^3 Na_2SO_4 results in the capacitance of 190 F/g, which is larger than that in 0.1 mol/dm^3 Na_2SO_4 (140 F/g) as seen in

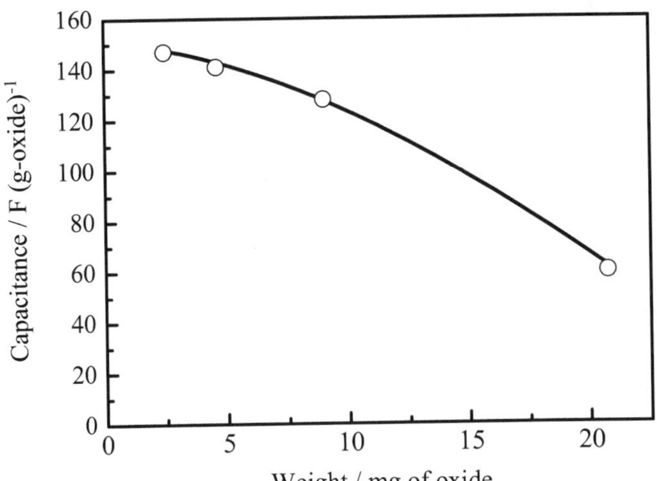

Figure 11.5 Relation between weight of a Mn_3O_4 electrode and maximum capacitance during 500 cycles at a rate of 10 mV s^{-1} in 0.1 mol dm^{-3} Na_2SO_4 aqueous solution.

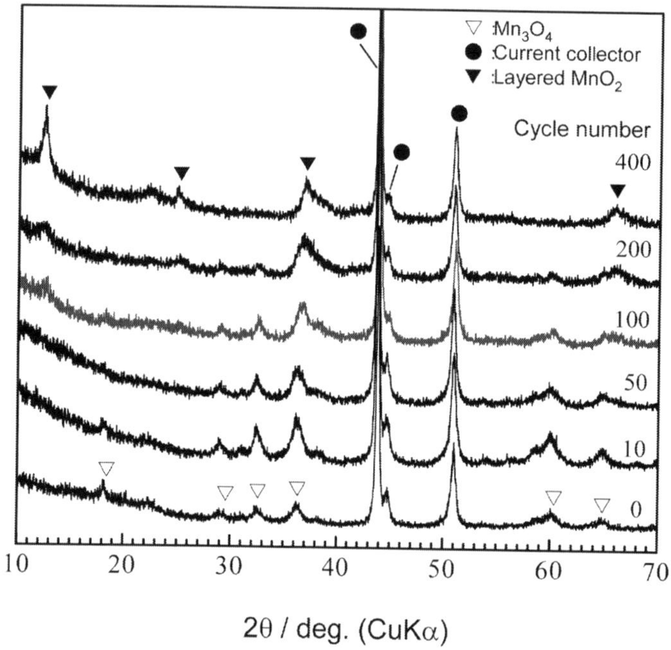

Figure 11.6 XRD patterns of the Mn_3O_4 electrodes for different cycling stages obtained in 1 mol/dm^3 Na_2SO_4 aqueous solution.

Fig. 11.3 with the identical electrochemical cycling conditions. The specific capacitance of Mn_3O_4 electrodes is affected by not only the electrolyte solution but also the thickness of the electrode related to the sample loading condition (weight per in the 0.1 M Na_2SO_4 aqueous unit area) as shown in Fig. 11.5. Specific capacitance of approximately 140 F/g for 3–5 mg loading reduced to approximately 60 F/g for 20 mg loading condition solution.

Structural evolution during the electrochemical stimulation was examined by X-ray diffraction (XRD) method in Fig. 11.6. It was confirmed that hydrated lamellar manganese dioxide, birnessite, was formed by the electrochemical stimulation of Mn_3O_4 in Na_2SO_4 solution. As shown in Fig. 11.6, diffraction peaks appeared after the electrochemical stimulation and all peaks are indexed as birnessite phase[19,20,25] except for those of current collector and conductive agent. As the broad diffraction peaks were observed,

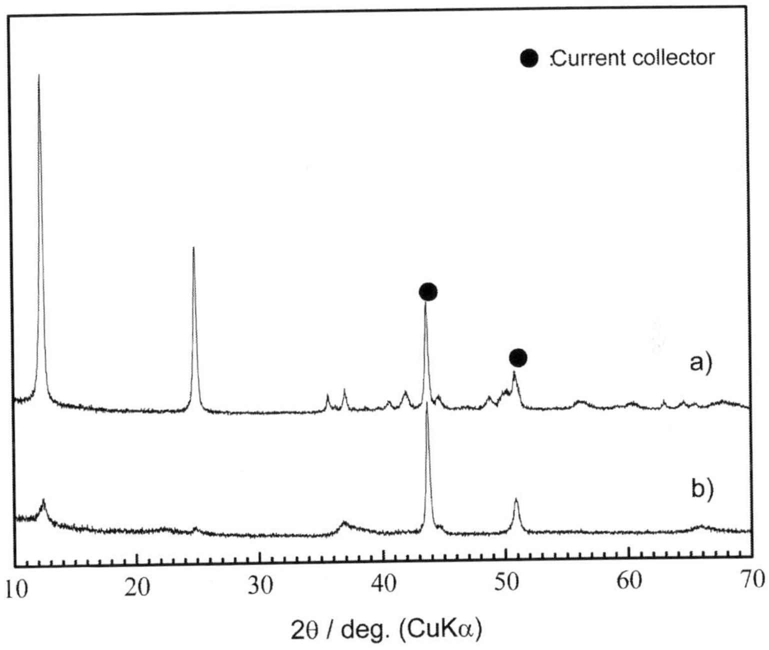

Figure 11.7 XRD patterns of birnessite-type MnO_2 electrodes prepared by (a) chemical oxidation from manganese hydroxide and (b) electrochemical stimulation for the Mn_3O_4 electrode.

it was difficult to distinguish the stacking sequences of birnessite, such as 2H, 3R, and so on.[25] Figure 11.7 compares XRD patterns of birnessite electrodes prepared by the different experimental methods: conventional chemical oxidation process of manganese hydroxide and the electrochemical stimulation for Mn_3O_4 electrode. The intensity of the $00l$ diffraction lines around $2\theta = 12^\circ$ and 25° is relatively lower than that of the birnessite prepared by the chemical oxidation method, indicating that the ***c***-axis orientation perpendicular to the substrate should be less remarkable for the birnessite formed from Mn_3O_4. It is also expected that crystallite size of the birnessite prepared by the electrochemical stimulation is much smaller than that of pristine Mn_3O_4 from the broader peak profiles.

The morphology of the Mn_3O_4 electrode after cycling was observed by electron microscopy as shown in Fig. 11.8. Although

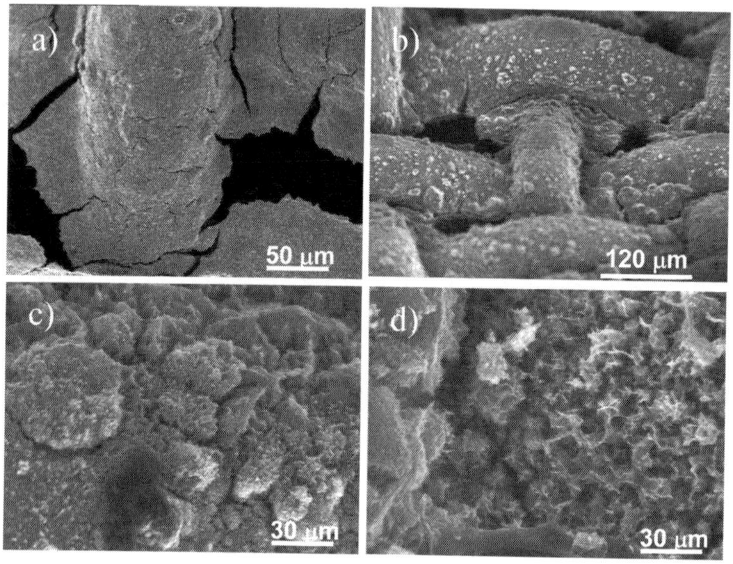

Figure 11.8 SEM images of the Mn_3O_4 electrodes surface: (a) before cycling and after (b) 10, (c) 50, and (d) 200 cycles in 1 mol/dm^3 Na_2SO_4 aqueous solution.

the surface of the pristine electrode that consisted of the Mn_3O_4 particles was flat (Fig. 11.8a), the successive cycling resulted in a much rougher and porous electrode (Fig. 11.8a–d), which could be due to the electrochemical stimulation as mentioned above. This morphology change results in the larger surface area and also causes a remarkable increase in the specific capacitance. The inductively coupled plasma (ICP) spectroscopic analysis of the electrolyte after the electrochemical stimulation revealed that the formation of the birnessite was accompanied with the dissolution of 30% of total manganese in the Mn_3O_4 electrode.

11.3 From Symmetric to Asymmetric Supercapacitor

As the electrolyte solution, concentrated acidic, or alkaline media, for example, H_2SO_4, KOH solution, are often utilized for the carbon and ruthenium dioxide supercapacitor because of the superior ionic conductivity. Manganese dioxides are, however, soluble in the acidic media unlike ruthenium dioxide. Alternatively, 1–2 mol/dm^3 Na_2SO_4 and KCl aqueous solutions are often used as the electrolyte for the manganese dioxide based supercapacitor. It should be noted that such neutral electrolyte solutions with environmental friendly and safety salts offer great advantage compared with the concentrated acid/alkaline electrolytes.

A potential window of the manganese dioxides in mild aqueous electrolyte solution is typically limited in the range of −0.1 V to 0.9 V *vs.* Ag/AgCl because of the decomposition of the water. This fact reduces the energy density of the manganese oxides as a conventional symmetrical capacitor with the maximum terminal voltage of 1 V. The energy (E) accumulated in the capacitor is represented by the following formula,

$$E = \frac{1}{2} C V^2 \tag{11.4}$$

where C and V denote capacitance and voltage of the capacitor, respectively. As the energy density of the capacitor is proportional to the square of terminal voltage, this results in lowering the energy density for the symmetrical supercapacitor consisting of manganese dioxide electrodes, which is due to the limited potential window. The concept of asymmetric (or hybrid) supercpacitor was, therefore, introduced to improve the energy/power density.[26–28] For the asymmetric capacitor, the manganese dioxide electrode is used as the positive electrode with the different negative electrodes, which can be cycled below 0 V *vs.* Ag/AgCl without hydrogen gas generation. Activated carbon electrodes can be used as the negative electrode material, because the electrodes show higher overpotential for hydrogen evolution (∼ −1.0 V *vs.* Ag/AgCl in neutral solution).

Cyclic voltammograms of an asymmetric supercapacitor, consisting of the manganese dioxides as positive electrode and the activated carbon as negative electrode, and their half cells are shown in Fig. 11.9. The asymmetric supercapacitor can be charged up to 2.0–2.2 V (Fig. 11.9b), and its energy density reaches approximately 21–29 Wh/kg,[26–28] which is nearly 10 times larger energy density than that of symmetrical supercapacitor based on the manganese dioxides. The high oxygen and hydrogen overpotentials for the manganese dioxide and activated carbon electrodes, respectively,

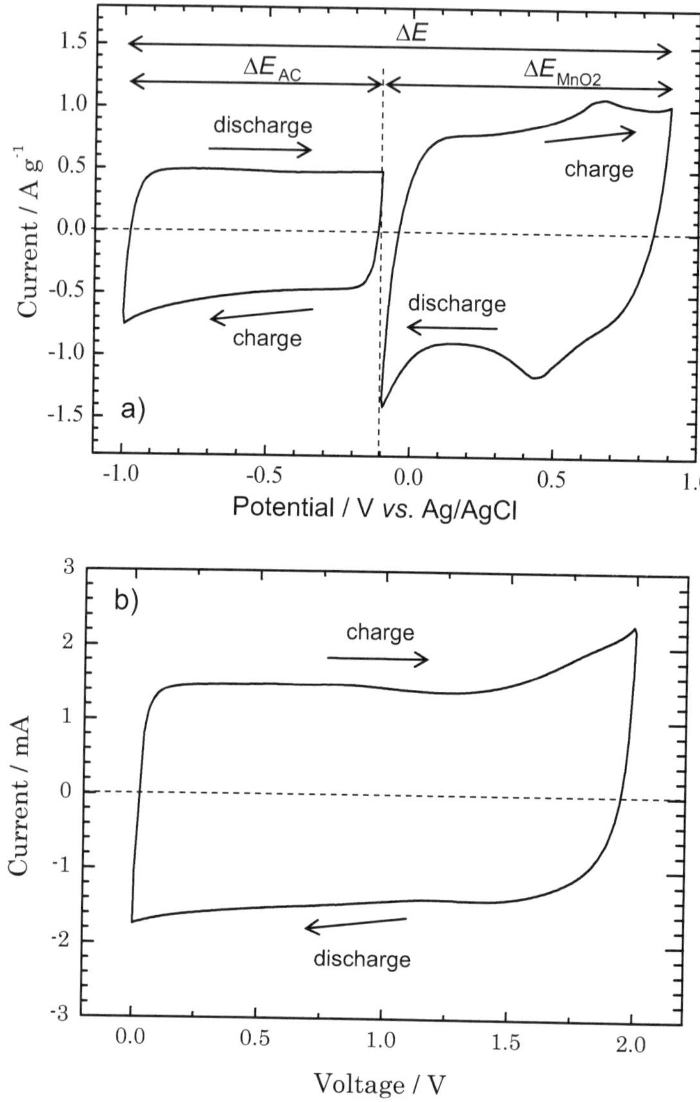

Figure 11.9 Cyclic voltammograms of (a) birnessite electrode (−0.1 V ~ 0.9 V) prepared by the electrochemical stimulation method and activated carbon (−1.0 V ~ −0.1 V) and (b) the asymmetric supercapacitor of the birnessite / 1 mol dm^{-3} Na_2SO_4 / the activated carbon cell.

allow 2.0 V supercapacitor with environmentally benign electrolyte, which contributes the increase in energy density of the asymmetric supercapacitor as expected from Eq. 11.4. Although the specific capacitance of the negative and positive electrodes is lower than $RuO_2 \cdot nH_2O$ electrode, available energy density is comparable with the ruthenium symmetrical supercapacitor (27 Wh/kg)[1] because of its higher terminal voltage as aqueous-based electrolyte system.

11.4 Electrolyte Additives for Improvement of Cyclabiltiy

The asymmetric supercapacitor, consisting of manganese dioxides and activated carbon, is one of the most promising electrochemical energy devices as shown in this chapter. However, there is a drawback to the application of manganese oxides for the capacitors, that is, loss of manganese through partial dissolution of the manganese dioxides, which is considered to be due mainly to the disproportionation of trivalent manganese during electrochemical cycling.[10,17] In general, manganese carbonate and phosphate species are insoluble in water. It is expected that, therefore, the carbonate or phosphate additive into an electrolyte can suppress the dissolution of manganese oxides similar to parkerizing (also called phosphating), which is a method of protecting a steel surface from corrosion and increasing its resistance to wear through the application of an electrochemical phosphate conversion coating. Recently, we demonstrated that such electrolyte additives significantly improve the cyclability of the birnessite-type manganese dioxides as active material for supercapacitor, synthesized by the electrochemical stimulation of Mn_3O_4 powder electrode.[21]

Figure 11.10 shows the galvanostatic oxidation/reduction curves of the birnessite electrode in the additive free and 30 mmol/dm^3 $NaHCO_3$ or Na_2HPO_4 added electrolyte solutions in the potential range of -0.1 V to $+0.9$ V. Among various phosphates and carbonates, $NaHCO_3$ and Na_2HPO_4 were chosen as additives because of the insignificant pH change of the Na_2SO_4 solution and no additional alkali species. When the additive-free (Fig. 11.10a) electrolyte was used, the potential of the birnessite almost linearly varied at constant anodic/cathodic current with the capacitance of

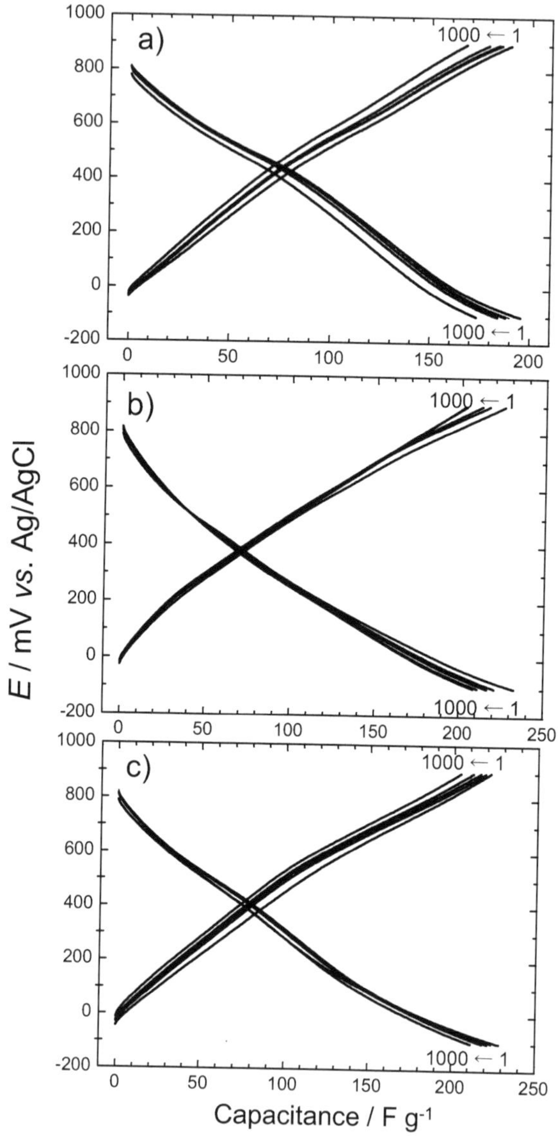

Figure 11.10 Galvanostatic oxidation and reduction curves of he birnessite electrodes synthesized by electrochemical cycling of Mn_3O_4 in (a) additive free, (b) 30 mmol dm^{-3} $NaSCO_3$ added, and (c) 30 mmol dm^{-3} Na_2HPO_4 added electrolytes at a rate of 1 A g^{-1}. Electrolyte used was 1 mol dm^{-3} Na_2SO_4 aqueous solution.

approximately 190 F/g at the initial cycle. The birnessite electrode exhibited supercapacitive behavior, that is, the faradic capacitance and the electric double layer capacitance. After one thousand cycling test, the capacitance decreased and the polarization slightly increased as seen in Fig. 11.10a. Manganese oxide precipitation was visually confirmed in the electrolyte after cycling, indicating that the capacity reduction was due mainly to the loss of active material by the partial dissolution of the birnessite. When the birnessite electrode was tested in the $NaHCO_3$ or Na_2HPO_4 added electrolytes, the capacitance was remarkably increased to approximately 230 F/g at the initial cycle as seen in Fig. 11.10b and 11.10c. There are no additional plateaux in the curves; furthermore, the curves unchanged more than 1000 cycles, indicating no significant degradation of capacity.

The capacitance *vs.* cycle number plots in the electrolytes with and without Na_2HPO_4 additive are shown in Fig. 11.11. The capacitance monotonously decreased in the additive-free electrolyte by the manganese dissolution. On the contrary, the capacitance remained constant after dissolving 30 mmol/dm^3 $NaHCO_3$ and Na_2HPO_4 into the electrolyte solution. Note that no precipitation was visually confirmed after dissolving the above additives. The ionic conductivity of the Na_2SO_4 electrolyte, 87 mS/cm, was slightly increased to 92 mS/cm by adding 30 mmol/dm^3 $NaHCO_3$ or Na_2HPO_4. Higher ionic conductivity is important for high power operation as the capacitor. The pH value of the electrolyte slightly changed because of the additives.

Generally, manganese phosphate and carbonate are insoluble in water. Therefore, if Mn^{2+} ions are dissolved from the electrode surface, its phosphate or carbonate should be promptly deposited on the electrode. In other words, the electrode should be covered with the manganese phosphate or carbonate, so that the dissolution would be successfully suppressed by the additive. Figure 11.12 shows the total amount of manganese released from the birnessite electrode into the electrolyte, which was determined by ICP analyses of the electrolyte solution. Prior to the analyses, the HCl and H_2O_2 solution was added to dissolve all precipitation in the electrolyte. The amount of manganese dissolved in the electrolyte was expressed as the ratios of (dissolved Mn)/(initial Mn in

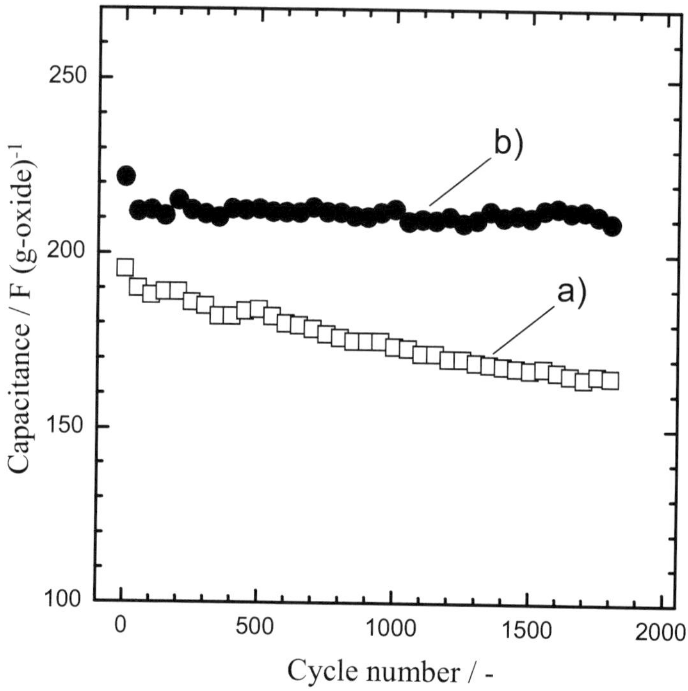

Figure 11.11 Variation in the capacitance of the birnessite electrodes prepared by electrochemical cycling of Mn_3O_4 in (a) additive free or (b) 10 mmol/dm^3 Na_2HPO_4 added electrolytes (1 mol/dm^3 Na_2SO_4).

the birnessite electrode). The dissolved manganese amount was dramatically decreased as the $NaHCO_3$ or Na_2HPO_4 concentration increases, and then the cyclability was enhanced. In both additives, the dissolution of manganese is effectively suppressed when the concentration was 10 mmol/dm^3 or higher in the electrolyte. In the case of the additive-free electrolyte, the amount of Mn dissolved was 4%, while the decrease in capacitance was approximately 15% after 1000 cycles (Fig. 11.11a). Furthermore, the morphology of the birnessite is rough and porous,[20,21] suggesting that only the surface part of the electrode is electroacitve, that is, all the manganese of the electrode was not utilized in electrochemical reaction.

Further investigations are needed to understand the influences of the additives and its origin on the enhanced supercapacitive behavior. There is a general consensus that the manganese dissolution is an essential problem to be solved for the application of electroactive manganese dioxide based electrode to electrochemical capacitors. We believe that there is a general possibility that the new additives realize the higher capacitance and enhanced cycle life of transition metal oxides as an active material for the electrochemical capacitors.

11.5 Summary and Outlook

In this chapter, recent progress and development in the manganese oxides have been briefly reviewed as the electrode materials for the supercapacitors. Manganese dioxides show a variety of the crystal structures, surface structures, and morphology, depending on the synthesis condition of the materials, which resulted in the distinctive electrochemical properties as the electrode materials. Asymmetric supercapacitor consisting of the manganese dioxide and activated carbon electrodes has to be a potentially interesting device, that

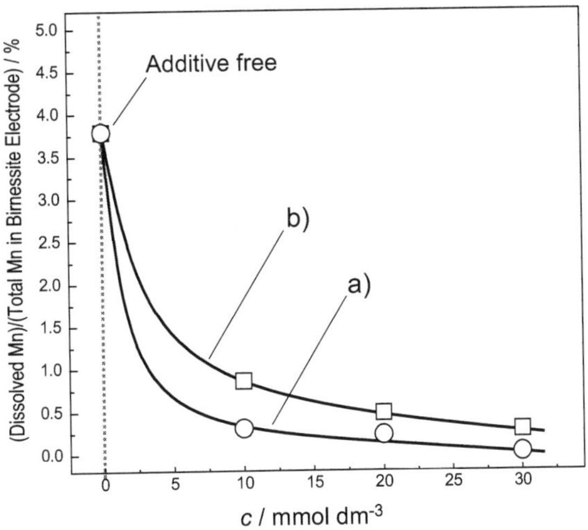

Figure 11.12 Dependence of the amount of dissolved manganese into 1 mol dm^{-3} Na_2SO_4 electrolytes after 1000 cycles on the concentration of (a) $NaHCO_3$ and (b) Na_2HPO_4 additives.

is, simple, low cost, and environmentally friendly. The major drawbacks as the electrode materials are (1) loss of the manganese during electrochemical cycling and (2) poor electrical conductivity. It has been demonstrated that the loss of the manganese oxide from the electrode into the electrolyte solution is drastically suppressed by the phosphate and carbonate additives. It is proposed that the insoluble nature of manganese phosphate and carbonate plays an important role in suppressing the dissolution to the electrolyte. Improvement of the intrinsic electrical conductivity of manganese oxides is the great challenge to further enhancement of energy and power density of the supercapacitors.

References

1. J. P. Zheng, P. J. Cygan, and T. R. Jow, *J. Electrochem. Soc.* **142**, (8), 2699 (1995).
2. J. P. Zheng, and T. R. Jow, *J. Electrochem. Soc.* **142**, (1), L6 (1995).
3. W. Sugimoto, H. Iwata, Y. Yasunaga, Y. Murakami, and Y. Takasu, *Angewandte Chemie-International Edition* **42**, (34), 4092 (2003).
4. H. Y. Lee, and J. B. Goodenough, *J. Solid State Chem.* **144**, (1), 220–223 (1999).
5. W. Xing, S. Z. Qiao, R. G. Ding, F. Li, G. Q. Lu, Z. F. Yan, and H. M. Cheng, *Carbon* **44**, (2), 216–224 (2006).
6. M. Toupin, T. Brousse, and D. Belanger, *Chem. Mater.* **16**, (16), 3184–3190 (2004).
7. O. Abe, and Y. Taketa, *J. Phy D-Applied Phy.*, **24**, (7), 1163–1171 (1991).
8. J. M. Fletcher, W. E. Gardner, B. F. Greenfield, M. J. Holdoway, and M. H. Rand, *J. Chem. Soc. A*, 653–657 (1968).
9. P. K. Sharma, G. J. Moore, F. Zhang, P. Zavalij, and M. S. Whittingham, *Electrochemical and Solid State Letters*, **2**, (10), 494 (1999).
10. S. C. Pang, M. A. Anderson, and T. W. Chapman, *J. Electrochem. Soc.*, **147**, (2), 444 (2000).
11. J. N. Broughton, and M. J. Brett, *Electrochim. Acta* **49**, (25), 4439 (2004).
12. X. P. Dong, W. H. Shen, J. L. Gu, L. M. Xiong, Y. F. Zhu, Z. Li, and J. L. Shi, *J. Phys. Chem., B* **110**, (12), 6015 (2006).
13. A. E. Fischer, K. A. Pettigrew, D. R. Rolison, R. M. Stroud, and J. W. Long, *Nano Letters* **7**, (2), 281 (2007).

14. M. Toupin, T. Brousse, and D. Belanger, *Chem. Mater.*, **14**, (9), 3946 (2002).
15. V. Subramanian, H. W. Zhu, and B. Q. Wei, *Chem. Phy. Lett.*, **453**, (4–6), 242 (2008).
16. P. Ragupathy, D. H. Park, G. Campet, H. N. Vasan, S. J. Hwang, J. H. Choy, and N. Munichandraiah, *J. Phy. Chem. C* **113**, (15), 6303 (2009).
17. R. N. Reddy, and R. G. Reddy, *J. Power Sources* **124**, (1), 330 (2003).
18. S. C. Pang, and M. A. Anderson, *J. Mater. Res.*, **15**, (10), 2096 (2000).
19. T. Brousse, M. Toupin, R. Dugas, L. Athouel, O. Crosnier, and D. Belanger, *J. Electrochem. Soc.* **153**, (12), A2171 (2006).
20. S. Komaba, A. Ogata, and D. Nakagawa, *ITE Letters* **8**, 400 (2007).
21. S. Komaba, A. Ogata, and T. Tsuchikawa, *Electrochem. Commun.*, **10**, (10), 1435 (2008).
22. K. W. Nam, and K. B. Kim, *J. Electrochem. Soc.*, **153**, (1), A81–A88 (2006).
23. B. Djurfors, J. N. Broughton, M. J. Brett, and D. G. Ivey, *J. Electrochem. Soc.* **153**, (1), A64 (2006).
24. B. Messaoudi, S. Joiret, M. Keddam, and H. Takenouti, *Electrochim. Acta* **46**, (16), 2487 (2001).
25. A. Ogata, S. Komaba, R. Baddour-Hadjean, J. P. Pereira-Ramos, and N. Kumagai, *Electrochim. Acta* **53**, (7), 3084 (2008).
26. V. Khomenko, E. Raymundo-Pinero, and F. Beguin, *J. Power Sources* **153**, (1), 183 (2006).
27. T. Brousse, M. Toupin, and D. Belanger, *J. Electrochem. Soc.* **151**, (11.4), A614 (2004).
28. M. S. Hong, S. H. Lee, and S. W. Kim, *Electrochemical and Solid State Letters* **5**, (10), A227 (2002).

Chapter 12

High-Voltage Electrode/Electrolyte Interface in ECs and Hybrid Capacitors

Katsuhiko Naoi and Kazumi Chiba
Institute of Symbiotic Science and Technology, Tokyo University of Agriculture & Technology, Tokyo, Japan
k-naoi@cc.tuat.ac.jp

Electrochemical capacitors (ECs) store electric charges at electrode–electrolyte interface. The performances of ECs depend greatly on many factors such as electrode (material derived, pore distribution, surface functionality, and so on), electrolyte (kind of cation and anion, solvent), and additives for the electrode and electrolyte, interfacial conditions between electrode and electrolytic solution, separator, and so on.[1–7] As for the electrodes, activated carbons (ACs) have been widely used as electrode material because of its high specific surface area (1000–2000 m^2/g), immediate availability, and low cost.[8–12] For electrolytes, nonaqueous electrolytes consisting of acetonitrile (AN) or propylene carbonate (PC) solutions with tetraethylammonium tetrafluoroborate (Et_4NBF_4) are used in commercial ECs because they permit wide operating voltage (2.5–2.7 V).[13]

Nanotechnology in Advanced Electrochemical Power Sources
Edited by S. R. S. Prabaharan and M. S. Michael

ISBN 978-981-4241-43-4 (Hardcover), 978-981-4303-05-7 (eBook)
www.panstanford.com

12.1 Background

12.1.1 *High-Voltage Electrode/Electrolyte Interface*

For the development of ECs, increasing energy density is one of the most crucial matters. As the energy density is a product of capacitance and voltage, increasing capacitance or voltage is necessary to enhance energy density. Increasing voltage is more effective because the energy density increases in proportion to squared voltage. For the improvement of the voltage characteristics of ECs, it is essential to develop higher electrochemical durability at electrode/electrolyte interface.

An effective approach to achieve higher energy is to increase the acceptable working voltage of conventional ECs. Currently, the maximum voltage of ECs is limited to 2.5–2.7 V due to a significant decrease in capacitance and continuous increase in internal resistance at applied voltage over 2.7 V.[5,14] In fact, the float test revealed that the EDLC cells lose its capacitance ($\Delta C < -13$ %) at the initial stage during 0–30 days at 2.5 V, whereas over 2.7 V, for example, at 2.9 V, the capacitance loss became more significant ($\Delta C = -28\%$) during the same mode of life test for the same duration. The undesired faradaic process that leads to a capacitance fade at over a certain cell voltage is the most critical factor that determines the life of ECs.[15] More specifically, the cell voltages between 2.5 and 2.9 V would be the threshold voltage that triggers consecutive and fatal degradation of the EDLC cells[16] as shown in Fig. 12.1.

There have been a few reports concerning such degradation mechanisms. For instance, Hahn *et al.*[17] analyzed gaseous products generated from activated carbon (AC) electrodes (positive or negative electrode) during cyclic voltammetric measurements of EDLC from 0 to 3.5 V in 1 M $TEABF_4$/PC. According to their report, propylene and H_2 were produced for cell voltages ranging from 2.7 to 3.5 V, and CO_2 started to be released from 2.9 V. They speculated that propylene and H_2 were formed by reduction of PC solvents at AC negative electrode when it is polarized at 1.5 V *vs.* Li/Li^+ and CO_2 results from the oxidation of PC at AC positive electrode over 4.3 V *vs.* Li/Li^+. Campana and Koetz *et al.*[18] characterized the surface of a

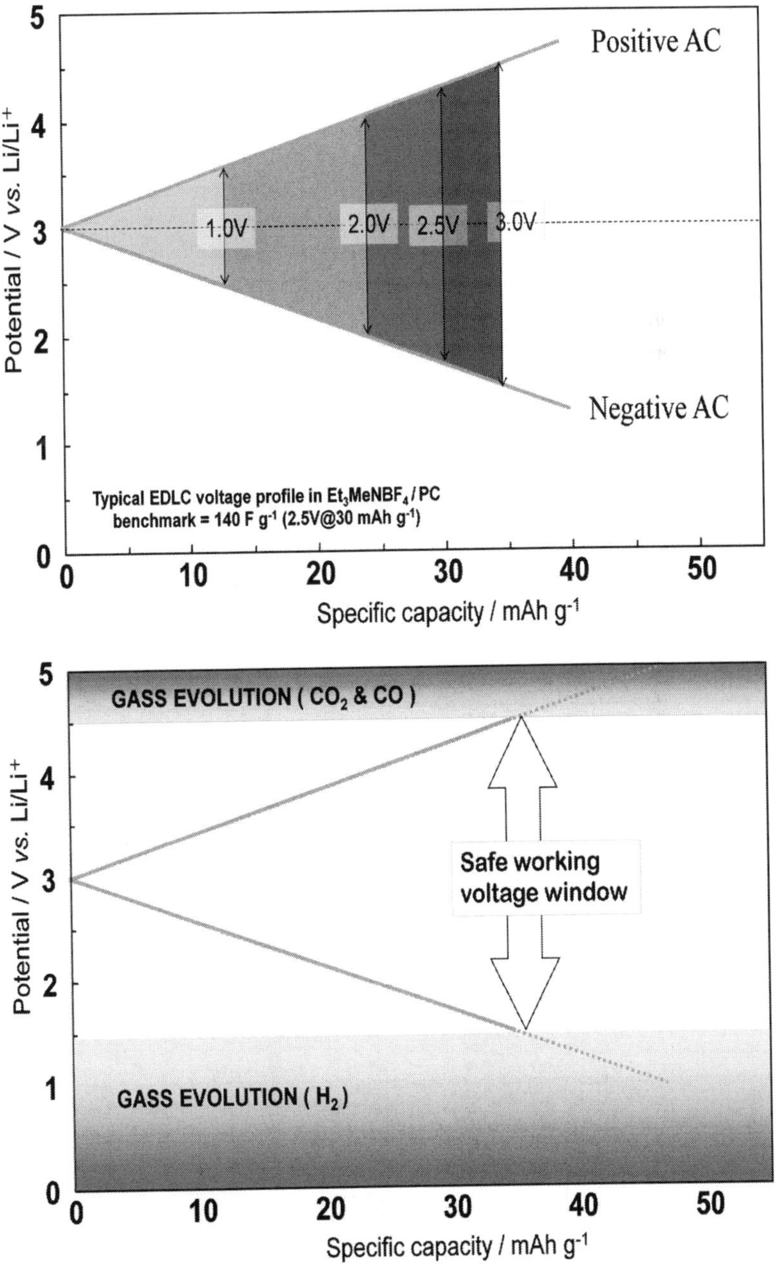

Figure 12.1 Polarization curves and the cell voltage limit for ECs based on AC electrodes.

highly oriented pyrolytic graphite (HOPG) negative electrode in 1 M $TEABF_4$/PC as a model electrode of EDLC by careful *in situ* AFM and *ex situ* SEM observations. They found that TEA^+ insertion, leading to swelling and exfoliation of graphenes, begins at 1.2 V *vs.* Li/Li^+. At a more negative potential of 0.9 V *vs.* Li/Li^+, the SEI of particulate morphology formation starts on the edge planes of the HOPG. From these results, it is indicated that 4.3 V and 1.5 V *vs.* Li/Li^+ are the critical potentials for the positive and the negative electrodes, respectively; beyond these limits, the degradation occurs. Their considerations and the proposed mechanisms are based on the hypothesis of a water-free electrode.

However, one should take residual or unremoved water in the AC electrodes into consideration. Water is certainly released from the positive and the negative AC electrodes into electrolyte after 50 h of float (Fig. 12.2), and other mechanisms are inferred on the same phenomenon. For example, the CO_2 was generated not only by the oxidation of PC but also by its hydrolysis mechanism. Kurzweil *et al.*[19] analyzed the decomposition products in the EDLC

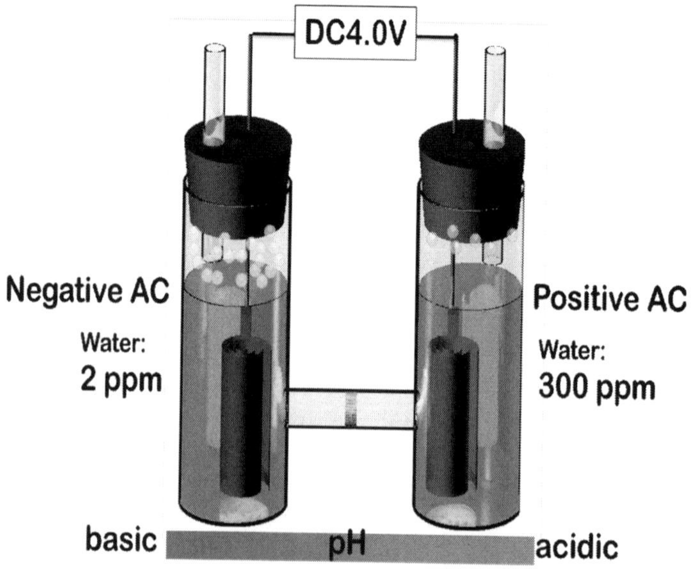

Figure 12.2 Water released from positive and negative electrodes after 50 h of float test.

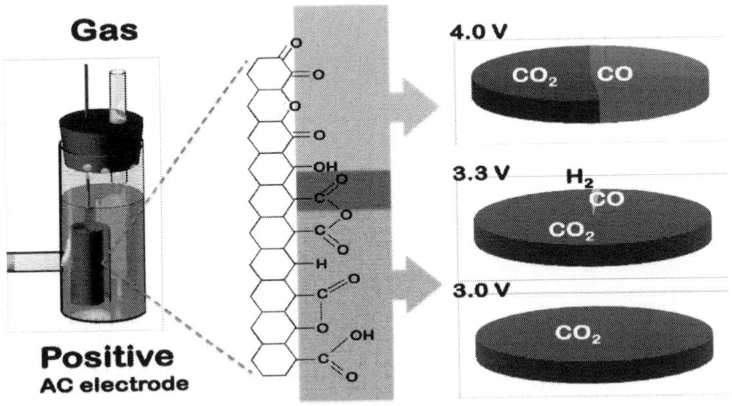

Figure 12.3 Gases evolved from the positive electrode of ECs.

cell after aging at 2.5 V (70°C) or 4.0 V in 1 M TEABF$_4$/AN. They detected some derivatives of acetamides (via hydrolysis of AN), ethylene, and triethylamines formed by Hoffman elimination in the presence of OH^- from residual waters. This is one of the strong evidences that support the hypothesis that AC electrodes contain much unremoved waters that play an important role in degradation processes. However, the behavior of unremoved water in AC electrodes has not yet been clarified as a function of the applied cell voltage in PC-based electrolyte.

Using 1.0 M TEMA (triethylmethylammonium) BF$_4$/PC as an electrolyte, systematic investigations were made on the degradation mechanism of AC electrodes in EDLC cell. Gas and water generation were separately characterized from AC positive and negative electrodes as a function of applied cell voltage of 2.5 up to 4.0 V.

On the positive electrode, the surface functionality determines the kind of gases evolved (Fig. 12.3): easily oxidizable functional groups give CO_2 at 3.0 V, and at higher voltages above 3.3 up to 4.0 V, CO evolution was observed. On the negative electrode, there is another story (Fig. 12.4). At 3.0 V, water is reduced to give H_2 and at the same time generates OH^- that makes an electrolyte alkaline and triggers a lot of other consecutive degradations. One of them is an alkaline-induced hydrolysis of PC, and the other is a Hoffman elimination of TEMA$^+$ both by the species of $OH^-(H_2O)$.

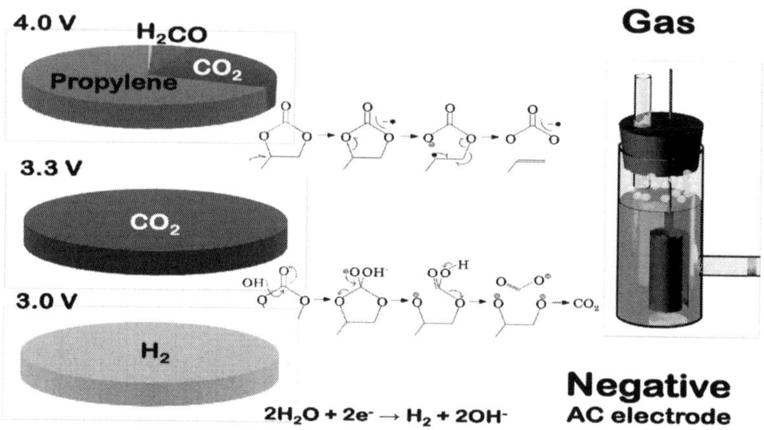

Figure 12.4 Gases evolved from the negative electrode of ECs.

Also, SEI films formed on positive and negative electrodes were observed separately at the same applied voltages (Fig. 12.5). On the positive electrode, exfoliation and the cracks are observed at 3.0–3.3 V as indicated in the cross-sectional SEM micrographs. At much higher voltages, the AC electrode surface begins to be covered by polymeric products (SEI formation). Much thicker SEI layer is accumulated on the negative electrode that increases the internal resistance of the cells tremendously.[7] The reductive decomposition of the PC solvent is particularly serious on the negative electrode when the water molecule remained in the electrolyte.

We could propose a mechanism for the degradation of the AC electrodes by taking into account the water released. Threshold degradation voltage for each positive and negative electrode could be defined on the basis of SEM observation, incremental-voltage-steps EIS analysis, and minute gases/water detection. The following crucial factors are considered at each cell voltage range (Fig. 12.6).

12.1.1.1 Cell voltage range of 2.5–3.0 V (Region I)

No degradation behaviors were observed in both the electrodes. The positive and the negative electrodes are electrochemically stable in the potential range below 4.9 V and over 1.9 V *vs.* Li/Li$^+$, respectively.

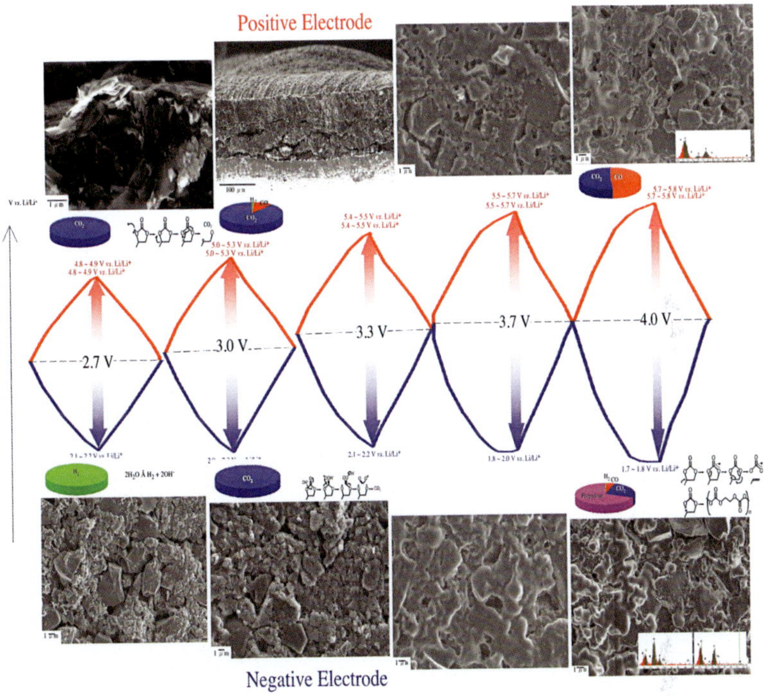

Figure 12.5 Morphological changes of the surface of positive and negative electrodes of ECs.

12.1.1.2 Cell voltage range of 3.0–3.7 V (Region II)

At positive electrode beyond 4.9 V *vs.* Li/Li^+, the surface functional groups are electrochemically oxidized and released as gaseous products of CO_2 and CO. At the same time, the adsorbed waters are blasted out by releasing functional groups from the surface. At negative electrode below 1.9 V *vs.* Li/Li^+, the adsorbed waters are slowly reduced and produce H_2 gas as well as generate equimolar OH^-. The generated OH^- induces Hoffman elimination of TEMA cations and activates hydrolysis of PC producing PG and CO_2 gas. The activation of PC hydrolysis could result in resistive adsorbed layers composed of organic molecules such as PC or PG on the AC surfaces in Region II.

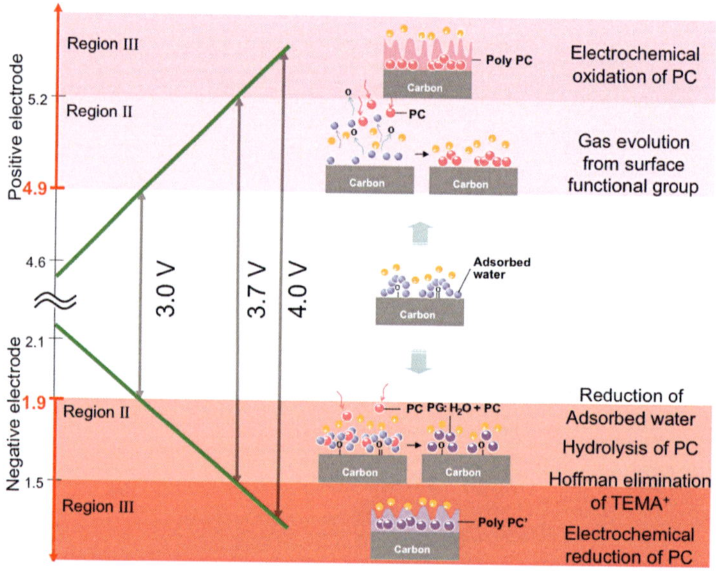

Figure 12.6 Possible SEI models at each potential range of the positive and negative electrodes.

12.1.1.3 Cell voltage range of 3.7–4.0 V (Region III)

PCs are electrochemically oxidized at the positive electrode over 5.2 V *vs.* Li/Li$^+$ and reduced at the negative electrode below 1.5 V *vs.* Li/Li$^+$. The electrochemical oxidation and reduction of PCs result in thick and highly resistive fluorinated poly-PC films on both the electrodes.

12.2 High-Voltage Electrolyte Structure at Interface

The basic structure of electrolytic solution is the electrolyte salts and solvents, and we can see several kinds of electrolytic solutions as follows,

1. Onium salts dissolved in nonaqueous solvents[20,21]
2. Lithium salts dissolved in nonaqueous solvents[22,23]
3. Ionic liquid (the function of electrolyte salt and solvent)[24,25]
4. Sulfuric-acid based aqueous solution[13]

Types i, ii, and iii are currently under practical evaluation widely in high voltage ECs (middle-sized and/or large-sized ECs) in application to HEV and PHEV. Those three types of electrolytic solutions attain two times higher withstanding voltage than that of type iv. Due to such higher voltage characteristics, one could expect higher energy density for ECs, although the ionic conductivity of the electrolytic solutions has poorer (by a factor of 1/2) ionic conductivity than the type iv.[26–30] In this section, the authors will describe mainly the high-voltage characteristics of onium salts (Type i) dissolved in nonaqueous media. The structure of the Type ii salts is almost same as those for Li-ion batteries that are well reported elsewhere. We will render the comments on Type iii salts in other sections.

There remained a few candidates for suitable electrolytic solutions that meet all the following requirements as ECs:

1. High capacitance,
2. High ionic conductivity,
3. High solubility of electrolyte salt,
4. Low viscosity,
5. High thermal and electrochemical stability.

Today, carbonates represented by propylene carbonate (PC) are currently the most widely used solvent in Japan. And, acetonitrile (AN) is a popular solvent in US and in Europe. By using AN instead of PC, one would easily achieve excellent rate performances due to triply enhanced ionic conductivity and much lower viscosity of AN than PC.

A low-viscosity linear carbonate of ethylmethylcarbonate (EMC) or dimethylcarbonate (DMC) can also solve such problems by mixing them with PC. The DMC was found to be a good cosolvent showing high performance caused by a high solubility of spiro-bipyyrolydinium tetrafuloroborate (SBP-BF_4) specifically in DMC. However, the withstanding voltage must be limited below 3.0 V in conventional AN system or such carbonate system because of the poor electrochemical stability of solvents.[30]

Morimoto *et al.* developed a new electrolytic solution having higher withstanding voltage of 3.3 V.[22] Figure 12.7 indicates a new mixture system of sulfones (SLs) and ethylmethylcarbonate (EMC).

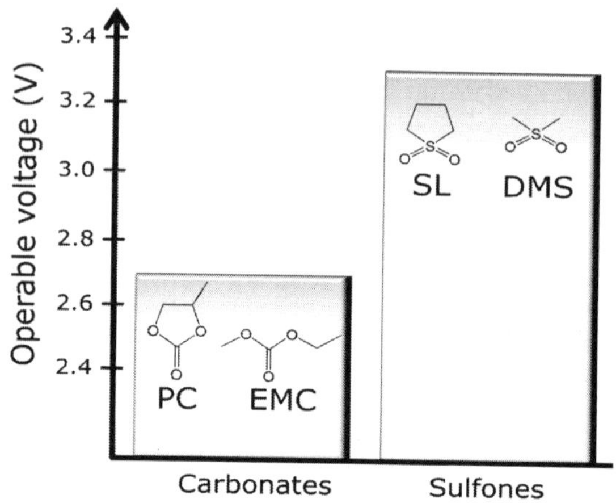

Figure 12.7 Operable voltages for carbonates and sulfones.

The SL, a ring-structured sulfone, is more tolerant to high-voltage operation than any carbonates. However, this new electrolyte system needs to be further improved because the EMC is more volatile than SL and less durable to reflow soldering at elevated temperatures.

Chiba *et al.* investigated new linear-sulfone cosolvents with SL. One of them, dimethylsulfone (DMS), demonstrated an extremely high withstanding voltage.[31] Interestingly, the DMS, the smallest molecule of linear sulfone, has high enough boiling point (>200°C), so its durability to the reflow soldering is assured. They also developed a large-sale manufacturing of the spirobipyyrolydinium tetrafuloroborate (SBP-BF_4) for a salt of conventional carbonates and new SL+DMS mixture solvent system. The SBP-BF_4 shows high ionization due to its rigid and comparatively compact size in ion radius. So, the SBP-BF_4 has large maximum solubility for solvents (>3 M) than other commercialized salts such as tetraethylammonium (TEA), triethylmethylammonium (TEMA), ethylmethylimidazolium (EMI), and diethylmethyl(2-methoxyethyl)ammonium (DEME) systems. The structure of cations and the solubility are indicated in Fig. 12.8 and Table 12.1, respectively.

The high-voltage requirements for ECs enhance the R&D of many new materials for the electrolytic system and interfacial durability.

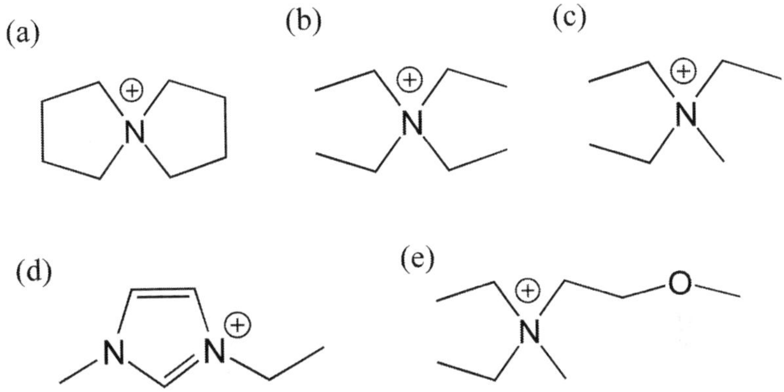

Figure 12.8 Chemical structures of SBP and other commercialized cations: (a) Spirobipyrrolidinium (SBP), (b) Tetraethylammonium (TEA), (c) Triethylmethylammonium (TEMA), (d) Ethylmethylimidazolium (EMI), (e) Diethylmethyl-(2-methoxyethyl)-ammonium (DEME).

Table 12.1 Solubility of each electrolyte salts for various solvents

Solubility (mol dm^{-3} at 303 K)		Traditional	Low viscosity			High ε_f	High stability
		PC	DMC	EMC	DEC	EC	SL
SBP	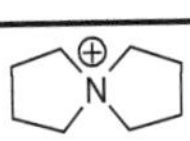	3.6	2.1	<0.	<0.1	3.4	3.7
TEMA	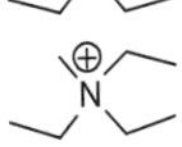	2.2	<0.1	<0.	<0.1	3.3	2.4
TEA	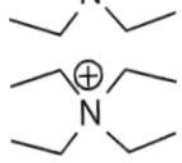	1.1	<0.1	<0.	<0.1	1.9	0.9

12.3 Summary

This chapter discusses various properties of ECs and electrolytic parameters influencing their performance such as cyclability, energy density, and so on. In view of that, various inorganic salts and solvents are considered to improve the voltage window and cycle stability. Authors have performed the comparison studies of two class of solvents between cyclic carbonates (PC, DMC, EMC) and sulfones (SL, DMS) and quaternary ammonium salts have been considered to improve high voltage stability invoking reflow soldering (high temperature) in electronic manufacturing environment. ECs employing these electrolyte systems are expected to exhibit large electrochemical window yet showing a good stability above room temperature operation. Authors considered several critical factors leading to side reaction taking into account AC surface reaction and microscopic changes to the exposed surfaces.

References

1. R. de Levie, *Electrochim. Acta*, 9, 1231(1964).
2. S. Yoon, J. H. Jang, B. H. Ka, and S. M. Oh, *Electrochim. Acta*, 50, 2255(2005).
3. J. H. Jang, A. Kato, K. Machida, and K. Naoi, *J. Electrochem. Soc.*, 153, A321 (2006).
4. S. R. S. Prabaharan, R. Vimala, and Z. Zainal, *J. Power Sources*, 161, 730 (2006).
5. O. Bohlen, J. Kowal, and D. U. Sauer, *J. Power Sources*, 172, 468 (2007).
6. A. Burke, J. Power Sources, 91, 37 (2000).
7. S. Ishimoto, Y. Asakawa, M. Shinya, and K. Naoi, *J. Electrochem. Soc.*, 156(7), A563–A571 (2009).
8. O. Hatozaki, *Advanced Capacitor World Summit 2008 proceedings.*
9. Qu, and H. Shi, *J. Power Sources*, 74, 99 (1998).
10. Y. Honda, T. Haramoto, M. Takeshige, H. Shiozaki, T. Kitamura, and M. Ishikawa, *Electrochemical and Solid-State Letters*, **10**, A106 (2007).
11. K. Kinoshita, *Carbon: Electrochemical and Physicochemical Properties*, Wiley, New York, USA, 187–193.
12. A. Yoshino, T. Tsubata, M. Shimoyamada, H. Satake, Y. Okano, S. Mori, and S. Yata, *J. Electrochem. Soc.*, 151, A2180 (2004).
13. T. A. Centeno, M. Hahn, J. A. Fernandez, R. Kotz, and F. Stoeckli, *Electrochem. Commun.*, 9, 1242 (2007).
14. S. Yoon, J. H. Jang, B. H. Ka, and S. M. Oh, *Electrochim. Acta*, 50, 2255 (2005).
15. J. H. Jang, S. Yoon, B. H. Ka, Y. H. Jung, and S. M. Oh, *J. Electrochem. Soc.*, 152, A1418 (2005).
16. G. G. Amatucci, F. Badway, A. D. Pasquier, and T. Zheng, *J. Electrochem. Soc.*, 148, A930 (2001).
17. M. Hahn, A. Wursig, R. Gallay, P. Novak, and R. Kotz, *Electrochem. Commun.*, **7**, 925 (2005).

18. F. P. Campana, M. Hahn, A. Foelske, P. Ruch, R. Kotz, and H. Siegenthaler, *Electrochem. Commun.*, **8**, 1363 (2006).
19. P. Kurzweil, and M. Chwistek, *J. Power Sources*, **176**, 555 (2008).
20. M. Ue, K. Ida, and S. Mori, *J. Electrochem. Soc.*, 141, 2989 (1994).
21. H. K. Song, J. H. Sung, Y. H. Jung, K. H. Lee, L. H. Dao, M. H. Kim, and H. N. Kim, *J. Electrochem. Soc.*, 151, E102 (2004).
22. T. Morimoto, K. Hiratsuka, Y. Sanada, and K. Kurihara, *J. Power Sources*, 60, 239 (1996).
23. A. Balducci, R. Dugas, P. L. Taberna, P. Simon, D. Plee, M. Mastragostino, and S. Passerini, *J. Power Sources*, 165, 922 (2007).
24. T. Sato, *Electrochemistry*, 72 (9) 711 (2004).
25. H. Ohno, *Ionic liquids: The front and Future of Material Development*, CMC Publishing Co., P234, (2003).
26. A. Chu, P. Braatz, *J. Power Sources*, **112**, 236 (2002).
27. T. Morimoto, *TANSO*, **214**, 202 (2004).
28. A. E. Fischer, K. A. Pettigrew, D. R. Rolison, R. M. Stroud, and J. W. Long, *Nano Letters*, **7**, 281 (2007).
29. M. Min, K. Machida, J. H. Jang, and K. Naoi, *J. Electrochem. Soc.*, **72**, 404 (2004).
30. N. Ando, S. Tazaki, T. Fujii, K. Kojima, T. Matsui, O. Hatozaki, K. Hado, and H. Shibuya, *The 46th battery Symposium in Japan*, 1C12, (2005).
31. K. Chiba, T. Ueda, and H. Yamamoto, *Electrochemistry*, 75(8) 664 (2007).

Index

accumulators 250
ACs *see* activated carbons
activated carbon electrodes 282, 317, 329, 335
activated carbons (ACs) 76, 90, 107, 117, 227, 252, 254, 261, 262, 319, 329–331, 339, 340
adsorption 107, 177, 180, 184, 185, 187, 194, 195, 288
advanced battery 127
air cathode 59, 91, 97, 103–108, 110–112, 118
  carbon-based 95
  porous 105
air electrode 57, 59, 104, 106, 116, 240–242
anode 52, 53, 55, 56, 58, 59, 61, 71, 73, 77, 91, 98–100, 104, 112, 146, 153, 161, 162, 166
  carbon-lithium 99
  lithium metallic 95
  Sn-based 66
anode materials 52, 57, 65, 90, 95, 128, 128–131, 134–136, 139, 140, 147, 153, 156, 161, 163–165, 167, 168, 237
  carbonaceous 128
  composite 137
  dominant 128
  oxide-based 146
aqueous electrolytes 99, 105, 106, 227, 277, 283
asymmetric supercapacitor 255, 269, 328–331, 335
ball-milling 131, 133, 134, 136, 153
  high-energy 76
  planetary 322
  vibrating 154
bands 212, 293
  antibonding 212
  valence 264
batteries 36, 38, 40, 51–53, 57, 58, 60, 61, 86, 89, 91–94, 105, 113, 116, 193, 239, 320
  aprotic 99
  lithium–air 239
  lithium polymer 73
  metal–air 57
  nickel-metal hydride 71
  primary 57, 61, 73
  rechargeable 55, 57, 73, 94, 105, 200
  rechargeable Li-based 65
  secondary 57, 91, 116, 178
  solid-state 100
capacitance 222, 226–229, 231–233, 250–252, 254, 256, 257, 259–261, 263, 282, 283, 295–299, 318, 319, 321–325, 328, 329, 331–334, 340
  electric double-layer 229
  faradic 333
  pseudo 226
capacitance retention 229, 230

capacitors 226, 250–253, 261, 329, 331, 333
asymmetric 317, 329
dielectric 282
electrical double layer 282, 320
hybrid electrochemical 233
super 245
symmetrical 252, 253, 329
capacity 55, 58–60, 71–77, 103–105, 107, 108, 112, 135, 136, 138–143, 153–165, 168, 169, 235–237, 239, 240, 295, 296, 307–309
capacity decay 68, 163, 164, 201
capacity retention 38, 69, 137, 138, 143, 155, 167, 178, 308
carbonates 69, 98, 101, 331, 333, 336, 347, 348
cyclic 349
dimethyl 16
propylene 102, 188, 194, 307, 339, 347
carbon materials 55, 111, 135, 222, 232, 235, 245, 256, 258, 277
carbon matrix 32, 68, 118, 134, 140, 259, 264, 266
carbon nanofibers 128, 134, 135, 227
carbon nanotubes 64, 127–130, 143, 157, 229–232, 250, 252–257, 269–273, 275–278, 281, 283–294, 296, 298, 299, 303–307, 311, 312
amorphous 143
graphitized 276
multi-walled 63, 68, 129, 225, 256, 269
single-walled 103, 129, 229
carbon network 255, 257, 259, 261, 263, 265, 267
carbon substrate 266, 269
catalyst 60, 96–98, 103, 105, 111, 112, 116, 117, 240–242, 245, 256, 271–273, 275, 277
bi-functional 113
nano 117
nanoscale 117
nanosized metal oxide 98
non-precious 118
cathode 52, 53, 56, 57, 59–61, 71, 73, 89, 90, 95, 98–102, 111, 112, 119, 153, 160, 195, 200, 239, 240
high-voltage 153
porous 98, 99
cathode materials 2, 24, 41, 55, 72, 74, 77, 200, 201, 215, 239, 240, 245
air 108, 240
layered 201
polyanionic 24
polyanionic-based 24
cations 26, 40, 100, 104, 253, 321, 339, 348
alkaline 180, 319
cell 9, 10, 16, 23, 40, 55, 95, 97, 102, 104, 106, 107, 112, 118, 120, 121, 153, 160
alkaline 318
battery-fuel 95, 97
carbon-zinc 318
coin type 120
electrochemical 112
hybrid 106
lithium-ion 205
rechargeable lithium 65
two-electrode 262
chemical vapor deposition (CVD) 108, 132, 226, 244
CNTs *see* carbon nanotubes
composites 2, 38, 76, 136, 149, 155, 163, 253, 257–260, 264, 266, 276
graphene-based 107, 234

nitrogen-enriched 258, 259
polymer-ceramic 100
compounds 3, 21, 74, 82, 179, 185, 187, 192, 200, 201, 205, 216–218
inorganic 66
lithiated 180, 201
lithium-rich spinel 308
nanoscale 202
perfluorinated 103
transition metal macrocyclic 111
conductivity 99, 102, 190, 264, 265, 277
electronic–ionic 127
high thermal 223
contact angle 102
coulombic efficiency 59, 61, 67, 131, 136, 138, 148, 162
CV *see* cyclic voltammetry
CVD *see* chemical vapor deposition
cyclic voltammetry (CV) 63, 267, 269, 308
cycling 63, 67, 68, 136, 139, 141, 157, 159, 160, 163, 164, 178, 201, 227, 228, 296–300, 302, 309, 310, 327, 328
charge–discharge 237
cycling performance 77, 131, 136, 140, 148, 151, 156, 161, 163, 166

decomposition 65, 96, 97, 101, 120, 329
catalytic 272
electrocatalytic 97
propylene 272
thermal 161, 224, 226
defects 131, 215, 223, 241, 244
anti-site disorder 217
lateral 133
density functional theory (DFT) 35, 130, 241
deposition 162, 272
chemical 162
chemical vapor 132, 226
electrochemical 136
DFT *see* density functional theory
differential scanning calorimetry (DSC) 82
diffraction peaks 11, 17, 34, 39, 326
direct methanol fuel cells (DMFCs) 111
DMFCs *see* direct methanol fuel cells
DSC *see* differential scanning calorimetry

ECPs *see* electrically conducting polymers
ECs *see* electrochemical capacitors
EDPs *see* electrolyte decomposition products
EIS *see* electrochemical impedance spectroscopy
electrically conducting polymers (ECPs) 251, 252, 276
electric vehicles (EVs) 25, 53, 55, 73, 93, 94, 97, 120, 282, 284
electrochemical capacitors (ECs) 92, 125, 226, 233, 282–284, 317, 335, 339–341, 343–345, 347–350
electrochemical cycling 35, 137, 308, 311, 323, 331, 336
electrochemical impedance spectroscopy 178, 195
electrochemical reactions 5, 10, 19, 37, 61, 69, 75, 104, 178, 240, 244, 298, 301, 334
electrochemical stimulation 325–328

electrode materials 74, 81, 84, 127, 138, 163, 178, 200, 201, 229, 277, 282, 283, 312, 317–323, 335, 336, 339
electrodes 56–58, 63, 64, 163, 164, 177–179, 184–188, 233, 252–254, 260–262, 268, 283-286, 294, 295, 301, 322–329, 333, 334, 339, 340
  as-prepared GNSs 243
  battery-type 284
  capacitor-type 284
  counter 36, 140
  faradaic 228
  gas diffusion 96
  graphene-based 108, 241
  graphene nanosheet 236
  graphite 135
  heat-treated 301
  metal-hydride 270
  metal oxide 95
  nanoarchitectured 168
  oxide-based 185
  porous 328
  porous catalytic 102
  porous iron 63
  powder 331
  synthetic FeS2 73
  thin-film control 68
  electrolyte 56, 57, 61, 63, 64, 98–104, 107, 108, 118, 119, 190, 200, 229, 232, 233, 282–284, 328, 329, 331–336, 339, 342–344
  acidic 262
  additive-free 333, 334
  aprotic 98
  glass-ceramic 100
  glycol 116
  glyme-based 96
  hybrid 99, 105
  mixed 102
  neutral 263
  non-aqueous 58
  polyacrylonitrile-based gel 100
  polymer 158
  solid 158
electrolyte decomposition products (EDPs) 164
electrolytic solutions 179, 182, 183, 252, 259, 260, 263, 319, 321, 323, 326, 328, 331, 333, 336, 339, 346, 347
electron spin resonance (ESR) 199, 200, 202–206, 208, 210, 212, 214, 216, 218, 220, 232
ESR *see* electron spin resonance
etching 131–133, 135
EVs *see* electric vehicles

fast Fourier transform (FFT) 62
FFT *see* fast Fourier transform
fuel cells 53, 91, 97, 107, 111, 116, 125, 250, 271–273, 275, 276

gas diffusion electrode (GDE) 90, 96, 102, 106, 107, 116, 118, 119
GDE *see* gas diffusion electrode
GNS *see* graphene nanosheets
GNS
  decorated 237, 238
  heat-treated 242, 243
  layered 244
graphene nanosheets (GNS) 108, 223–225, 227–229, 235–238, 240–245
graphene sheets 221, 224, 226, 231, 232, 235, 238, 239
  doped 232
  exfoliated 240
  functionalized 241
  layered 227
  multilayer 227

graphite 52, 55, 56, 63, 78, 129, 130, 136, 139, 146, 156, 222, 224, 225, 234–236, 240, 242, 244
  lithiated 55
  lithium-intercalated 222
  multi-layered 224
  pyrolitic 224

highest occupied molecular orbital (HOMO) 264, 265
highly oriented pyrolytic graphite (HOPG) 342
HOMO *see* highest occupied molecular orbital
HOPG *see* highly oriented pyrolytic graphite
hydrothermal method 139, 143, 165, 168
hydrothermal reaction 152, 304, 306, 312
hydrothermal synthesis 28, 31–33, 42, 44, 45, 305

ICP *see* inductively coupled plasma
inductively coupled plasma (ICP) 307, 328

layered double hydroxides (LDHs) 228, 229
lattice gas 3
LDHs *see* layered double hydroxides
Li–air batteries 57–59, 89–96, 95–108, 112–114, 116, 120, 222, 223, 239–243, 245
  aprotic 98
  aqueous 99
  hybrid electrolyte 106
  nonaqueous 94, 106
  rechargeable 96, 101, 102, 105
LIBs *see* Li-ion batteries
Li-ion batteries (LIBs) 52, 54, 55, 71, 127–130, 134, 136–139, 146, 153, 156, 161, 163, 167–169, 237
lithium manganese oxide nanoparticles (LMO nanoparticles) 304, 305, 311, 312
  nanocrystalline 305
  spinel 307
LMO nanoparticles *see* lithium manganese oxide nanoparticles
lowest unoccupied molecular orbital (LUMO) 264, 265
LUMO *see* lowest unoccupied molecular orbital

magnetic field 202, 203, 211
magnetic ions 200, 211, 215, 218
manganese dioxides 282, 317–323, 325, 327–329, 331, 335
  birnessite-type 322, 331
  electroactive 335
  electrolytic 318, 322
  hydrated 318
  lamellar 326
  nano-architectured 312
manganese ions 210, 212, 217, 301, 303, 307–309, 319
manganese oxides 60, 213, 253, 283, 300, 302, 303, 317, 318, 321–323, 329, 331, 335, 336
  active 317
  dehydrated 300
  hydrous 322
  layered 300
  nano-architectured 312

materials 32, 40–42, 52, 53, 55, 65, 78–80, 84, 139, 153, 154, 166, 167, 206, 207, 228, 229, 236, 237, 264, 265, 276, 277
carbonaceous 110
carbon-based 222, 228
electroactive 192, 294, 295
energy storage 245
graphene-based 231, 240, 245
inactive 136, 160
iso-structural 39
lithiated 77
mesoporous 157
nano-architectured 281, 312
nanoelectrode 282
nanometer-sized 195
nanoporous 127, 169
nanotubular 250, 270, 272, 275
platinum-based 111
supercapacitor 252
mechanism 37, 39, 67, 84, 100, 134, 233–235, 244, 267, 342, 344
degradation 178, 340, 343
hydrolysis 342
insertion 195
polaronic conduction 2
storage 226
methanol oxidation 271–277
miscibility gap 13, 15, 18
multi-walled carbon nanotubes (MWNTs) 129, 131–134, 270, 272, 276
MWNTs *see* multi-walled carbon nanotubes

nanocomposites 76, 128, 162, 238, 251, 252, 253, 257, 305
core-shell Si–C 137
graphene/PANI 231
graphene–SnO2 237
metal oxide/CNT 305
tin-based 128
nanotubes 63, 64, 130–132, 142, 152, 157–159, 168, 250–252, 256–259, 269–273, 276
NMR *see* nuclear magnetic resonance
NMR shifts 29, 30
NMR spectra 29
nuclear magnetic resonance (NMR) 29, 82, 134, 200

OCP *see* open-circuit potential
OCV *see* open-circuit voltage
OER *see* oxygen evolution reaction
OER catalysts 97, 113
open circuit potential (OCP) 58, 188, 311
open-circuit voltage (OCV) 9, 17, 18, 57, 58
ORR *see* oxygen reduction reactions
orthophosphate 24
orthorhombic crystal structure 25
oxidation 33, 36, 37, 39, 42, 52, 96, 143, 159, 178, 225, 269, 319, 321
anodic 63, 322
galvanostatic 332
time controlled 42
oxides 56–58, 64–66, 69, 81, 128, 156, 180–182, 185, 194, 213, 282, 283, 319, 320, 322, 323, 325
chromium 156
cobalt 65, 96, 113, 136, 156
graphene 225
hydrous 252
manganese-based 113
metallic 250
molybdenum 156
nanosized 284
nanostructured 128
nickel 156
polypropylene 190

rare earth 113
reduced graphite 225
spinel 55
tin 237
vanadium 24
oxygen evolution reaction (OER) 90, 96, 97, 111, 113, 116, 117, 120
oxygen reduction reactions (ORR) 90, 95, 97, 100, 111, 113, 116

polarization 36, 56, 61, 72, 145, 162–164, 333
polymorphs 25–31, 33, 34, 37, 38
pores 64, 103–105, 107–109, 118, 119, 148, 164, 261, 264, 268
interconnected 296, 307
interstitial 129
intrinsic 129
nanoscale 241
micropore 235
porosity 59, 78, 104, 107, 147, 164, 222, 232, 240, 256, 322
porous carbon cathode 98–100
porous carbon electrode 105, 118
precursors 31, 34, 44, 139, 154, 155, 157–159, 161, 162, 164, 207, 256
organic 32, 256
organometallic tin 140
polymeric 255
process 36, 40, 69, 74, 81, 84, 85, 90, 97, 113, 143, 146, 224–226, 267, 269, 272
anodic 268
bi-phasic intercalation 21
carbonization 255, 265
charge–discharge 21, 142, 143
crystallization 145
deintercalation 153
diffusion-controlled 85
electrochemical 181
electrochemical conversion 323
electrochemical deposition 142
faradaic 340
growth-controlled 85
heat-treatment 137
intercalation 129, 234, 237
lithiation 138
lithiation–delithiation 141
microwave-hydrothermal 312
nonelectrochemical 117
oxidation/reduction 41, 42
polyol 154, 291
solvothermal 207
unzipping 225
pseudocapacitance 226, 228, 229, 231, 233, 255, 282, 319, 321
faradic 229
pseudocapacitors 282, 283

reaction 56, 58, 66, 69, 71, 75, 76, 80–83, 85, 95, 96, 102, 106, 107, 266, 268, 286, 287, 296
anodic 272
catastrophic 95
charge-transfer 319
displacement 69
faradaic 251, 255, 276
half-cell 98
irreversible 69
nonfaradic 284
one-step 101
parasitic 44, 58
pseudocapacitive 283
recombination 269
redissolution 164
reversible 66, 74
solid-state 31, 53, 80, 84, 153, 155
sporadic 43
redox capacitor 318, 320, 321, 323, 325, 327
reduced graphite oxide (RGO) 225

resistance 227, 232, 260, 261, 307, 331
- electronic 262
- internal 340, 344
- parallel 261

resonance signals 202, 205, 215, 217
- anti-site disorder 217, 218
- overlapping 208

RGO *see* reduced graphite oxide

SAED *see* selected area electron diffraction

selected area electron diffraction (SAED) 144

single-walled carbon nanotubes (SWNTs) 103, 129–131, 229

solid solution 5, 6, 12, 13, 15, 16, 18–20, 41, 72

solution 111, 149, 157, 159, 232, 233, 285–292, 294, 296, 298, 322, 326, 331, 333, 339
- acid 147
- acidic 263
- alkali 144
- alkaline 266, 269
- ammonia 159
- aqueous cobalt chloride 158
- aqueous glycine 206
- cyclohexanone 271
- organic 263, 276

spins 202, 203, 209, 211, 213, 217
- electronic 203
- isolated 211

structure 24–26, 28–30, 33–37, 61, 142, 143, 148, 151, 165, 166, 194, 195, 200, 201, 215, 238, 241, 242, 306, 307, 347, 348
- allotropic 222
- band 213
- capillary 269
- composite 321
- core–shell 137, 138
- crystalline 137
- defect-free 244
- edge 309
- electronic 179, 199, 201, 212, 319
- hollow 140, 143
- hybrid 233
- hyperfine 211
- intergrowth 318
- layered 160, 166, 201, 322
- mesoporous 157, 159
- molecular 285
- net-like 162
- organic 180
- orthorhombic 201
- superfine 203

supercapacitors 107, 222, 223, 226–229, 231, 233, 250, 252, 253, 254–256, 258, 317, 318, 320–322, 324, 326, 330–332, 334–336
- graphene-based 227
- manganese-based 317
- oxide-based 231
- ruthenium dioxide 328
- symmetrical 329, 331
- ultrathin 228

SWNTs *see* single-walled carbon nanotubes

synthesis 30, 31, 74, 77, 157, 159, 222, 224, 226, 284, 285, 287–289, 291, 293, 295, 305–307, 312
- high temperature 32
- precipitation 28
- solid-state 31, 34

systems 51–53, 57, 58, 60, 61, 63, 69, 71, 72, 74, 80–83, 94, 95, 97, 203, 205, 253, 255, 256, 347, 348
- aqueous 95
- asymmetric 254
- carbonate 347

catalytic 178
electrolytic 348
hybrid 106, 233
nanoparticle 128
nonaqueous 95, 116
spin 202

TEM *see* transmission electron microscopy
TEM images 141, 161, 162, 165–167, 238, 273, 303, 304
transition metal 24, 26, 37, 40, 43, 52, 56, 65, 156, 179, 181, 205, 212, 218, 228
inactive 141
paramagnetic 29
transition metal oxides 64, 111, 113, 127, 156, 161, 162, 168, 177, 178, 180, 187, 188, 253, 276, 335
transmission electron microscopy (TEM) 62, 137, 164, 273
uninterruptible power supplies (UPS) 282, 284
UPS *see* uninterruptible power supplies

vapor transport 61
voltage 37, 41, 42, 55, 65, 69, 71, 79, 147, 156, 162, 200, 205, 329, 330, 340, 347, 348
DC 134
intercalation cutoff 168
mid-discharge 153
open-circuit 9, 10, 20, 57
polarization 116
squared 340
upper cut-off 39
voltage window 36, 38, 39, 42, 349

X-ray diffraction (XRD) 3, 6, 11, 15–18, 27, 34, 35, 71, 77, 82, 138, 200, 201, 293, 312, 326
XRD *see* X-ray diffraction
XRD patterns 25, 26, 39, 292, 306, 326
XRD peaks 25, 292, 306